Directional Hearing

William A. Yost and George Gourevitch
Editors

Directional Hearing

With 133 Figures

Springer-Verlag
New York Berlin Heidelberg
London Paris Tokyo

William A. Yost
Director
Parmly Hearing Institute
Loyola University of Chicago
Chicago, Illinois 60626
USA

Library of Congress Cataloging-in-Publication Data
Directional hearing.
Includes bibliographies and index.
1. Directional hearing. 2. Auditory perception.
I. Yost, William A. II. Gourevitch, George.
[DNLM: 1. Audiology. 2. Auditory
Perception. 3. Sound Localization. WV 270 D598]
QP469.D57 1987 591.1'825 87-4945
ISBN-13: 978-1-4612-9135-0 e-ISBN-13: 978-1-4612-4738-8
DOI: 10.1007/978-1-4612-4738-8

Softcover reprint of the hardcover 1st edition 1987

Typeset by David E. Seham Associates, Metuchen, New Jersey.

9 8 7 6 5 4 3 2 1

ISBN-13: 978-1-4612-9135-0 Springer-Verlag New York Berlin Heidelberg

Preface

All living creatures interact with their environments. For many animals, directional information on the contents of their surroundings is provided to a greater or lesser extent by the auditory system in conjunction with other sensory systems. The present volume, which evolved from a conference on directional hearing held in 1983 at the meetings of the Acoustical Society of America in Cincinnati, Ohio, introduces the reader to the current facts and theory on how vertebrates, from fish to man, locate sound sources in space.

We invited some of the most creative scientists investigating directional hearing to acquaint the reader with their approaches to this subject. The diversity of topics they have covered characterizes the study of directional hearing and offers an overview not only of the problems confronting their field of inquiry, but also of the strategies that have been developed to resolve them.

In their contributions to this volume, the authors describe the fundamental dimensions of directional hearing, that is, the stimuli that serve sound localization; the anatomical and functional characteristics of the auditory nervous system responsible for it; the various proficiencies achieved by different animals in this task; and, of great importance to man, the practical implications of sound localization. It is, therefore, not surprising to find variation in the style of the chapters from almost "how-to-do-it" manuals to mostly theoretical discourse.

It should be noted that more than half the chapters in this volume are based on human research. This is due, in part, to a historical element in the study of sound localization, namely, it has been examined for a much longer time and in much greater detail in man than in animals; consequently, much of our knowledge of directional hearing rests on the study of humans. In part, it also reflects man's need for applied knowledge in this field to answer medical and environmental questions. A broader understanding of directional hearing—one that is derived not only from knowledge gained from human study, but also from knowledge of the comparative, neurophysiological, and evolutionary aspects of sound localization—must depend on a wide range of animal studies. This book is an attempt to introduce a broad informational base for directional hearing.

The work on this book was made easier by the assistance of the staff of the Parmly Hearing Institute, especially the Institute's administrative assistant, Marilyn Larson.

WAY
GG

Before this book was completed, George Gourevitch, my friend and coeditor, passed away. George died on May 19, 1987. George's scholarly input was crucial in producing this book and in planning the symposium. This book, *Directional Hearing,* is dedicated to George Abraham Gourevitch in recognition of his lifelong commitment to the scholarly study of hearing.

WAY

Contents

Contributors

DAVID A. BERKLEY, Bell Labs, Murray Hill, New Jersey, USA

LESLIE R. BERNSTEIN, Department of Psychology, University of Florida, Gainesville, Florida, USA

JOHN H. CASSEDAY, Duke University Medical Center, Durham, North Carolina, USA

H. STEVEN COLBURN, Department of Biomedical Engineering, Boston University, Boston, Massachusetts, USA

ELLEN COVEY, Duke University Medical Center, Durham, North Carolina, USA

NATHANIEL I. DURLACH, Massachusetts Institute of Technology, Cambridge, Massachusetts, USA

RICHARD R. FAY, Parmly Hearing Institute, Loyola University of Chicago, Chicago, Illinois, USA

ALBERT S. FENG, Department of Physiology and Biophysics, University of Illinois, Urbana, Illinois, USA

GEORGE GOUREVITCH, Hunter College of the City University of New York, New York, New York, USA

ERVIN R. HAFTER, Department of Psychology, University of California, Berkeley, California, USA

DORIS J. KISTLER, Waisman Center, University of Wisconsin, Madison, Wisconsin, USA

George F. Kuhn, Vibrasound Research Corporation, Denver, Colorado, USA

Shigeyuki Kuwada, Department of Anatomy, University of Connecticut Health Center, Farmington, Connecticut, USA

Mark E. Perkins, Department of Psychology, New York University, New York, New York, USA

James A. Simmons, Department of Psychology and Section of Neurobiology, Brown University, Providence, Rhode Island, USA

Constantine Trahiotis, Department of Speech and Hearing Science, University of Illinois, Champaign, Illinois, USA

Frederic L. Wightman, Waisman Center, University of Wisconsin, Madison, Wisconsin, USA

Tom C.T. Yin, Waisman Center, University of Wisconsin, Madison, Wisconsin, USA

William A. Yost, Parmly Hearing Institute, Loyola University of Chicago, Chicago, Illinois, USA

P. M. Zurek, Research Laboratory of Electronics, Massachusetts Institute of Technology, Cambridge, Massachusetts, USA

Part I The Physical and Psychoacoustical Foundations of Directional Hearing

Our knowledge of directional hearing has depended not only on the study of the localization of sounds that originate in real space but also on the localization of sounds that are presented over earphones (lateralization). Even though the perception of sound is not identical under these two conditions, (in the first instance the source of the sound is perceived to be at some position in the environment, whereas in the second instance the source is typically perceived to be at some position inside the head), a wealth of information about directional hearing has been provided by earphone studies.

Since the latter part of the 19th century when Lord Rayleigh made his observations on how humans localize sound sources, it has been known that this ability is strongly dependent on interaural intensity and interaural time differences. Although the emphasis in this section is on the role of these cues in sound localization, other binaural cues are also discussed, some which have been examined only recently. Among them, for example, are spectral cues that arise when complex sounds encounter the body of the listener. Such spectral differences are crucial for the localization of sound in the vertical plane, for the localization of sound achieved with only one ear, and most probably, for a realistic reproduction of the acoustic environment.

Greater understanding of directional hearing has resulted not only because of psychoacoustical studies, but also because of studies of the physical behavior of signals as they impinge on the listener. As a consequence of such work, some localization behavior can be accounted for on the basis of the physical transformation of the signal by the head, pinnae, and trunk of the listener.

The four chapters of this section provide an overview of the fundamental acoustical and psychological phenomena of directional hearing and serve as reference for the more specialized topics of directional hearing discussed in the following sections.

1 Physical Acoustics and Measurements Pertaining to Directional Hearing

GEORGE F. KUHN

The complex geometry of the human torso, head, and external ear encodes direction-dependent information onto the incoming sound field in terms of phase distortions, amplitude distortions, and spectrum distortions. This chapter develops analytical models for some of the well-known localization cues based on these distortions that are the result of reflections and diffractions. Experimental results supporting these models are reported here also.

Sound waves incident onto the ears of a listener are reflected by the head on the side facing the incident wave (commonly known as the head-baffle effect), diffracted to the ear on the shadowed side of the head (commonly known as the head-shadow effect), and transmitted to the eardrum via the pinna cavity system and the ear canal. These reflections and diffractions produce interaural time differences (ITD) and interaural sound pressure level differences (ILD), which are well-known sound localization cues.

The shoulders, via backscattering, and the pinna cavity system, via relative wave motion within the cavity system, produce direction-dependent sound spectra at the ears. These spectra are known to be significant for the sound localization in the vertical plane.

This chapter describes analytical and physical models for the ITDs and the ILDs as a function of frequency and angle of incidence (see Chapters 2 and 3 also). Physical models for the spectral localization cues produced by torso reflections and pinna cavity wave motion are also described. The models are supported by experimental data, collected primarily with an electroacoustic manikin. Some comparisons are also made between the predictions of the models and a listener's ability to localize sounds.

Interaural Time Differences

Analysis of Phase-Derived Low-Frequency ITD

It is well known that sounds incident in the horizontal plane can be localized on the basis of phase-derived interaural time differences (ITD_p).

TABLE 1.1. Equation and terms used to describe the pressure around a sphere.

$$\left[\frac{p_i + p_s}{p_o}\right]_{r=a} = \left(\frac{1}{ka}\right)^2 \sum_{m=0}^{\infty} \frac{i^{m+1}\, p_m(\cos\Theta)\,(2m+1)}{j_m'(ka) - i n_m'(ka)} \qquad (1)$$

where

P_i	=	incident pressure on the spherical surface
p_s	=	scattered pressure
P_o	=	incident free-field pressure
k	=	acoustic wave number
a	=	radius of the sphere
i	=	$\sqrt{-1}$
p_m	=	*m*th order Legendre polynomial
Θ	=	angle relative to the pole facing the incident wave
j_m'	=	derivative of m-th order spherical Bessel function
n_m'	=	derivative of m-th order spherical Neumann function
m	=	summation index
r	=	radial distance from the center of the sphere

These ITD_p are derived from the solution for the acoustic pressure on the surface of a rigid sphere of equivalent radius, *a*, as the average human head. The acoustic pressure on that sphere is given by (Kuhn, 1977) and shown in Equation (1) of Table 1.1. Equation (1) can be simplified to Equation (2) under the assumption that $ka<1$; i.e., for low frequencies:

$$\left[\frac{p_i + p_s}{P_o}\right]_{r=a} \simeq 1 \pm i\left(\frac{3}{2}ka\right)\sin\Theta_{inc} \qquad \begin{matrix} + \text{ for ear toward source} \\ - \text{ for ear away from source} \end{matrix} \qquad (2)$$

It can be shown that at low frequencies, Equation (2) yields

$$ITD_p \simeq \frac{3a}{c}\sin\Theta_{inc} \qquad (3)$$

where *c* is the ambient speed of sound and θ_{inc} is the angle of incidence, as shown in Figure 1-1.

Analysis of High-Frequency ITD_p

Although at high frequencies, sounds incident in the horizontal plane are not generally localized on the basis of ITD_p, an analysis of high-frequency ITD_p leads to an improved understanding of past ITD models and of the group-velocity-derived ITDs (ITD_g) as well as the leading-edge delay of transients ITD (ITD_t).

Since Equation (1) converges increasingly slowly with frequency, a Watson-Sommerfeld transformation (see, for example, Doolittle et al, 1968) converts Equation (1) to a series that converges increasingly rapidly with

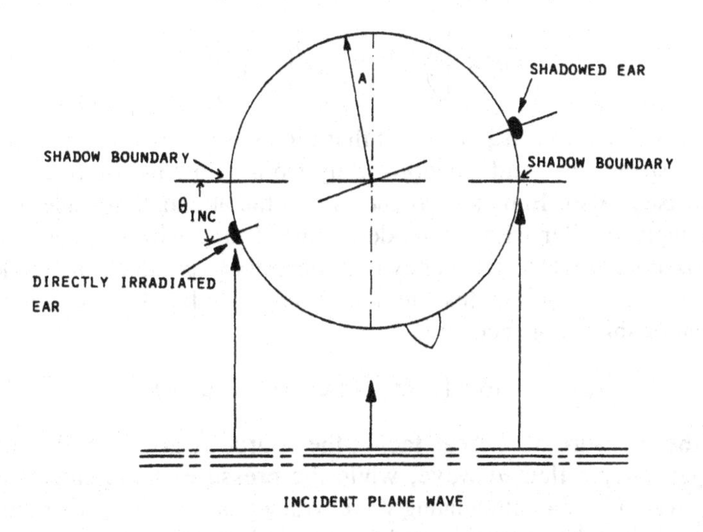

FIGURE 1.1. The geometry of the incident wave field relative to the head.

frequency. Application of that transformation (Kuhn, 1977) allows the pressure in (1) to be formulated as:

$$(P)\Big]_{r=a} \alpha \sum_{m=o}^{\infty} \sum_{\lambda=\pm} \sum_{L=1}^{\infty} \exp\left[-(\Theta_\lambda + 2m\pi)/\Theta_L\right] \times \exp\left[i\omega\left\{a\left(\Theta + 2m\pi\right)\Big/C_L^{ph}\right\} - t\right] \quad (4)$$

where Θ_λ is the angular coordinate on the sphere, m is the number of times the partial wave has circumvented the sphere, Θ_L is the attentuation angle or the angle through which the L-th partial wave must travel to attenuate to (1/e) of its original value, and c_L^{ph} is the phase velocity of the L-th partial wave. Equation (1) describes the pressure on the surface of a sphere as an infinite set, "L," of partial acoustic waves that are emitted at the shadow boundaries of the sphere, traveling clockwise and anti-clockwise, designated by $\lambda = \pm$, around the perimeter of the sphere, any number of times, m.

The phase velocities of these partial waves can be written approximately (Franz, 1954) as

$$\frac{C_L^{ph}}{c} \simeq \left[1 + \frac{6^{-1/3}}{2} (1/ka)^{2/3} q_L\right]^{-1} \quad (5)$$

where $q_L = 1.47, 4.68, 6.95, \ldots$ for $L = 1, 2, 3. \ldots$ Also, the attenuation angle is given approximately by

$$\Theta_L \simeq \left[\frac{1}{2}\sqrt{3}\left[\frac{1}{6}\,ka\right]^{1/3} q_L\right]^{-1} \tag{6}$$

Thus, it is clear from Equation (5) that the phase velocities approach the ambient speed of sound asymptotically from below as the frequency or ka increases. Also, Equation (4) shows that the attenuation angle becomes increasingly smaller with the mode-number L and with frequency or ka. If it is assumed that the frequency is sufficiently large, then the first mode, $L=1$, with $m=o$ dominates the sound field. Under these assumptions, the relationship in (4) becomes

$$(p)_{r=a} \simeq \exp\left[-\Theta^{\pm}/\Theta_1\right] \exp(i\omega\left[a(\Theta^{\pm}/c)-t\right]) \tag{7}$$

Then the pressure at the ear facing the source is given by the incident and specularly reflected wave, while the pressure at the shadowed ear has the form (7) of an attenuating wave with a phase velocity approximately equal to that of the ambient speed of sound which attenuates exponentially with the angle through which it has traversed. This high-frequency solution matches the geometrical acoustics model that was reported by Woodworth (1962) and other authors.

The high-frequency ITD_p is therefore given approximately by

$$ITD_p \simeq \frac{2a}{c}\sin\Theta_{inc} \tag{8}$$

It is clear from the comparison of Equation (3) to (8) that the low-frequency ITD_p is 50% greater than the high-frequency ITD_p.

Analysis of Group Velocity-Derived High-Frequency ITD_g

Gaunaurd (Kuhn and Gaunaurd, 1980) calculated the group velocities of the partial waves that are emitted at the shadow boundary of the sphere and then propagate into the shadow region on the surface of the sphere of radius a. The group velocity of the L-th partial wave is

$$(v_g^L) = 1/[1 + (1/3)\mathrm{Re}\,a_L(ka)^{-2/3} - (1/3)\mathrm{Re}\,\beta_L\,(ka)^{-4/3}] \tag{9}$$

$$\text{where } a_L = 6^{-1/3}\,(0.5 + i\,0.866)\,q_L \tag{10a}$$

$$\text{and } \beta_L = -6^{-1/3}\,(0.5 - i\,0.866)\,[(q_L/180) + (7/20q_L)] \tag{10b}$$

where the q_L's are the zeros of the Airy function, given after Equation (5).

Calculations of these group velocities are shown in Figure 1-2. The group velocity for the first partial wave, $L=1$, is greater than (0.94c) for frequencies greater than approximately 2 kHz for an average-sized adult head.

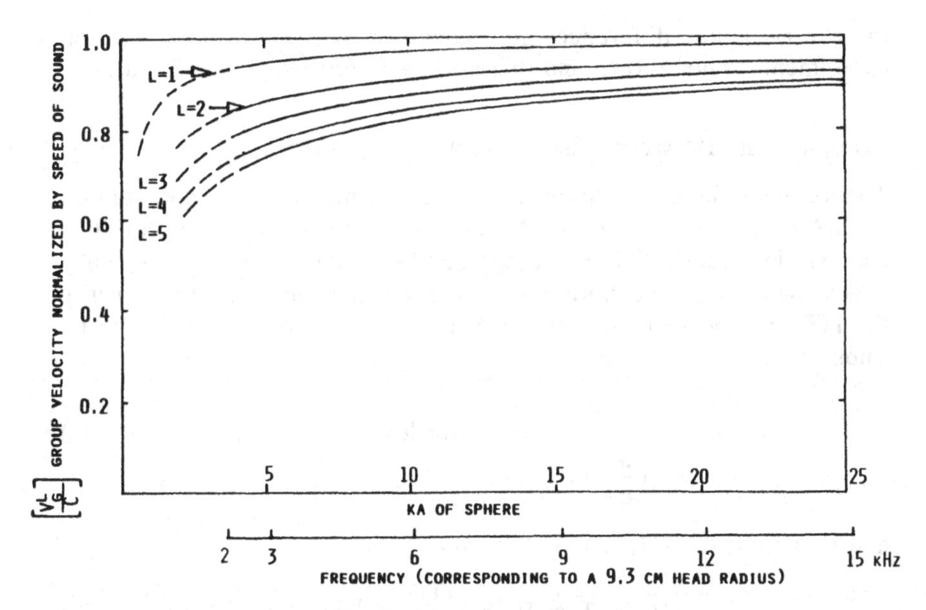

FIGURE 1.2. Group velocities of the L-th partial wave on a rigid sphere.

Further calculations (not shown here) show that the attenuation of these partial waves increases with the mode number, L. Therefore, as a first approximation, it is assumed that the effective group velocity is that of the partial wave L = 1. This group velocity is only a very small amount less than the ambient speed of sound. Therefore, the group delay at high frequencies is approximately equal to

$$ITD_g \simeq \frac{2a}{c} \sin \Theta_{inc} \tag{11}$$

Analysis of the Leading Edge ITD_t

The leading edge of an impulse or transient travels at the speed of sound of the medium. This interaural time delay of the leading edge is therefore given by Equation (12):

$$ITD_t = \frac{2a}{c} \sin\Theta_{inc} \tag{12}$$

It should be understood that the signal immediately after the initial arrival of the leading edge will be amplitude- and phase-distorted, and the precise shape of the signal envelope on the surface of the sphere will depend on the incident signal envelope as well as the angle of incidence. Whether these very minute differences are detectable by humans in localization experiments is not known and therefore need further experimentation.

Comparisons Between Theories and Experiments

Theoretical results are compared with experimental data, collected on KEMAR (Knowles Electronics Manikin for Acoustic Research), for ITD_p and ITD_t in Figure 1-3. The theoretical and experimental data are generally in good agreement. Therefore, it is appropriate and convenient to normalize the ITDs by the variables and to denote these normalized ITDs by Π. Thus, Equation (3) becomes

$$\Pi_L = \frac{ITD_p}{\left(\frac{a}{c}\right)\sin\theta_{inc}} \simeq 3 \text{ (for low frequencies)} \tag{13}$$

and Equations (8), (11), and (12) become

$$\Pi_h \simeq \Pi_g \simeq \Pi_t \simeq \frac{ITD_{p,\,g,\,t}}{\left(\frac{a}{c}\right)\sin\theta_{inc}} \simeq 2, \tag{14}$$

for high frequencies.

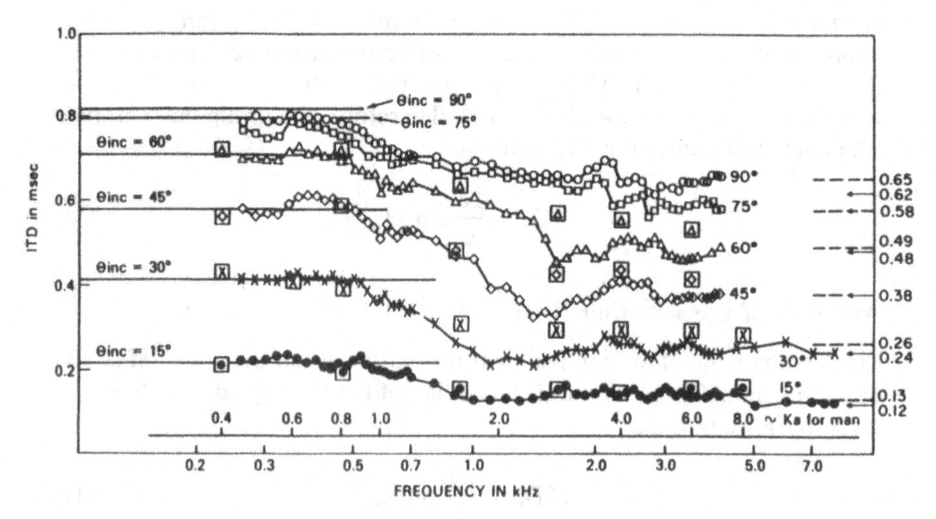

FIGURE 1,3. Comparisons between measured and theoretically predicted interaural time differences. The boxed data points are based on the solutions to equation (1); the solid line is based on equation (2). The arrows on the right ordinate are the time difference measurements made at the leading edge of a tone-burst; the dashed lines on the right ordinate are the high frequency time differences, approximated by equation (8) which also corresponds to high frequency group-delay.

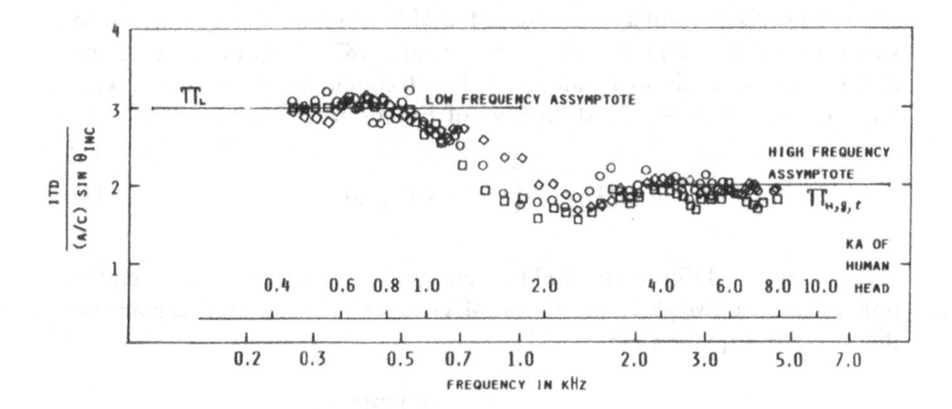

FIGURE 1.4. Normalized interaural time differences measured on KEMAR; circles are for $\Theta_{inc} = 15°$, squares for $\Theta_{inc} = 30°$, diamonds for $\Theta_{inc} = 45°$.

Figures 1-4 and 1-5 show the measured ITD_p and the predicted ITD_p when normalized as prescribed by Equations (13) and (14). It is clear from these results that as a first approximation, the various ITD localization cues reduce to two distinct values, as given by Equations (13) and (14).

Relationships Between the Minimum Audible Angle and Acoustical Theory (See Also Chapters 2 and 3)

Mills (1958) reported that the minimum audible angles (MAA) for humans increase with the angle of incidence (and with frequency above approx-

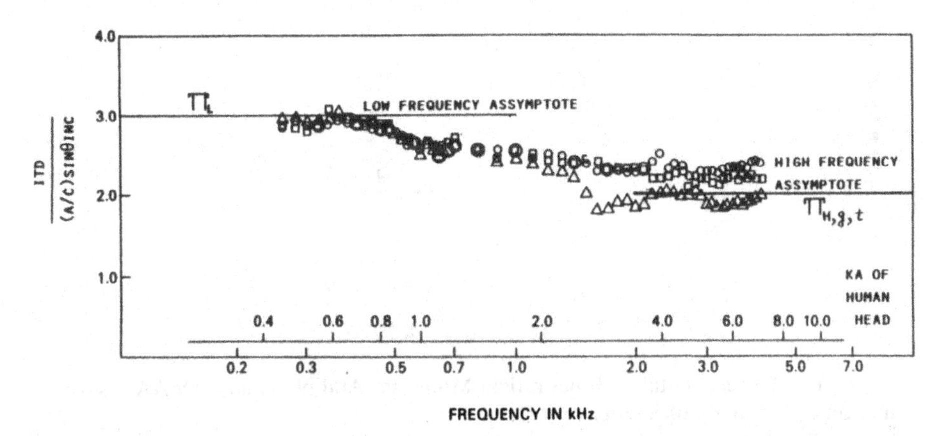

FIGURE 1.5. Normalized interaural time differences measured on KEMAR; triangles are for $\Theta_{inc} = 60$, squares for $\Theta_{inc} = 75°$, circles for $\Theta_{inc} = 90°$

imately 900 Hz). A similar increase of the MAA with the angle of incidence was reported for monkeys by Brown et al (1982). This increase of the MAA with the angle of incidence at low frequencies is consistent with the acoustical theory. The derivative of Equation (3) yields

$$d\,(ITD_p) = \left(\frac{3a}{c}\right)\cos\theta_{inc}\, d\theta_{inc} \tag{15}$$

Equating the differentials $d(ITD_p)$ and $d\theta_{inc}$ to the minimum detectable time difference (MDTD) and to the MAA, respectively, and rearranging the terms in Equation (15), yields

$$MAA = MDTD\left(\frac{c}{3a}\right)\sec\theta_{inc} = \frac{MDPD}{\omega}\left(\frac{c}{3a}\right)\sec\theta_{inc} \tag{16}$$

where MDPD is the minimum detectable phase difference. Thus, the MAA increases as the secant of the angle of incidence. Mills' (1958) and Brown et al's (1982) results are predicted quite well by Equation (16), as shown in Figure 1-6. It appears therefore that the MAA can be predicted for any angle of incidence, given one value of the MDTD or MDPD at one particular angle of incidence for a particular species.

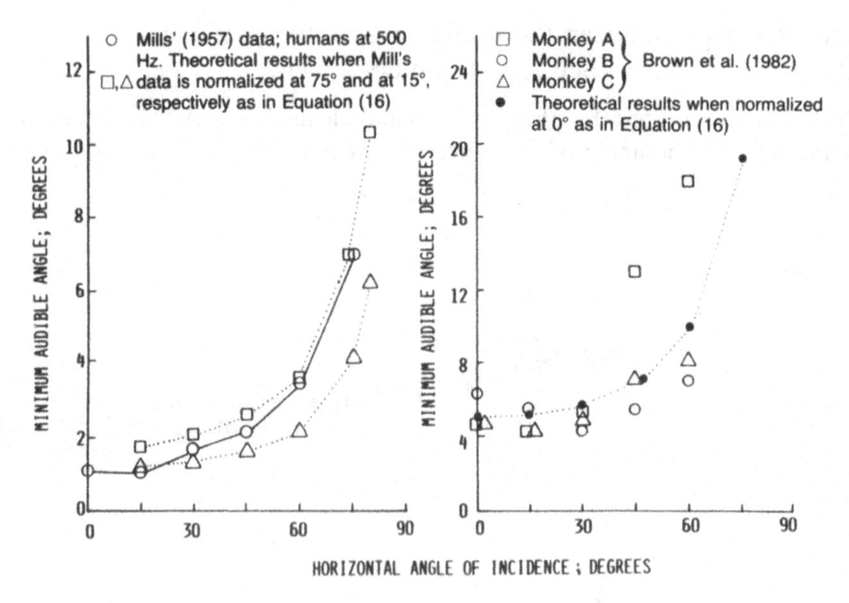

FIGURE 1.6. Experimental and theoretical Minimum Audible Angles (MAA) as a function of the angle of incidence (16).

Interaural Pressure Level Differences

A primary localization cue in the horizontal plane at high frequencies is the interaural level difference (ILD). The ILD is defined as:

$$\text{ILD (db)} = 20 \log \bar{p}_L - 20 \log \bar{p}_R \tag{17}$$

where $\bar{p}_L$ and $\bar{p}_R$ are the pressure magnitudes at the left ear and at the right ear, respectively. The ILD results from three primary mechanisms; one, the backscattering of the sound to the ears from the shoulders or torso; two, the scattering of sound by the head or sphere; three, the directivity produced by the pinna. The effect of the torso on the horizontal directivity is generally less than 2 dB and is therefore not discussed here because of its relatively small effect. The scattering of the sound by the sphere or head is naturally accounted for in Equation (1). Measurements of the pressure at the coupler microphone, when the pinna is replaced by

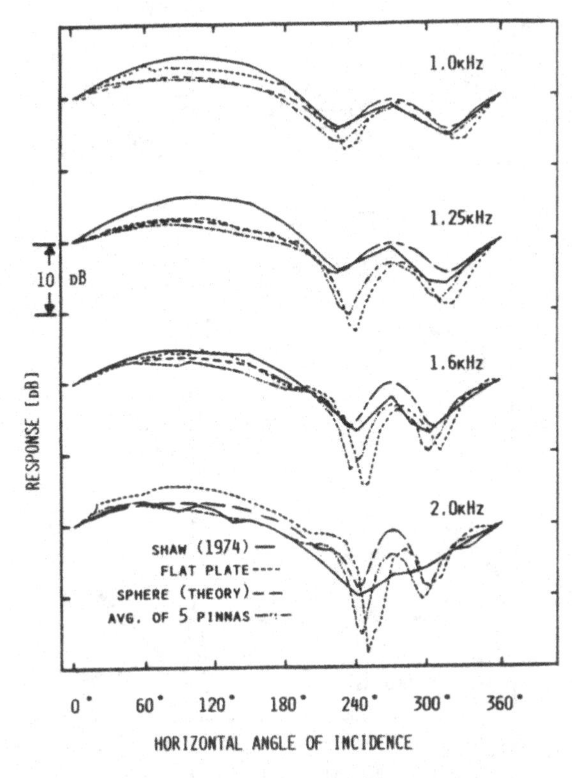

FIGURE 1.7. Comparisons of horizontal directivity of Shaw's composite, theory of a sphere, and coupler measurements in KEMAR with five pinnas and with pinna replaced by a flat plate.

a flat plate, are compared with the theoretical predictions of the pressure in Figures 1-7 and 1-8.

The average response from Shaw's composite (1974) and the average response, measured on KEMAR's coupler microphone, using five different pinnas (Kuhn, 1979) are also shown in Figures 1-7 and 1-8.

It is clear from the results in Figures 1-7 and 1-8 that the horizontal directivity is determined primarily by the head below 2.5 kHz and is therefore predicted reasonably well by the theory of a point-receiver on an equivalent sphere, as the head. The directivities above 2.5 kHz are governed by the pinna and by the head in the illuminated region and in the shadowed region of the head, respectively.

Between 2 kHz and 5 kHz the pinna's size and orientation produce a

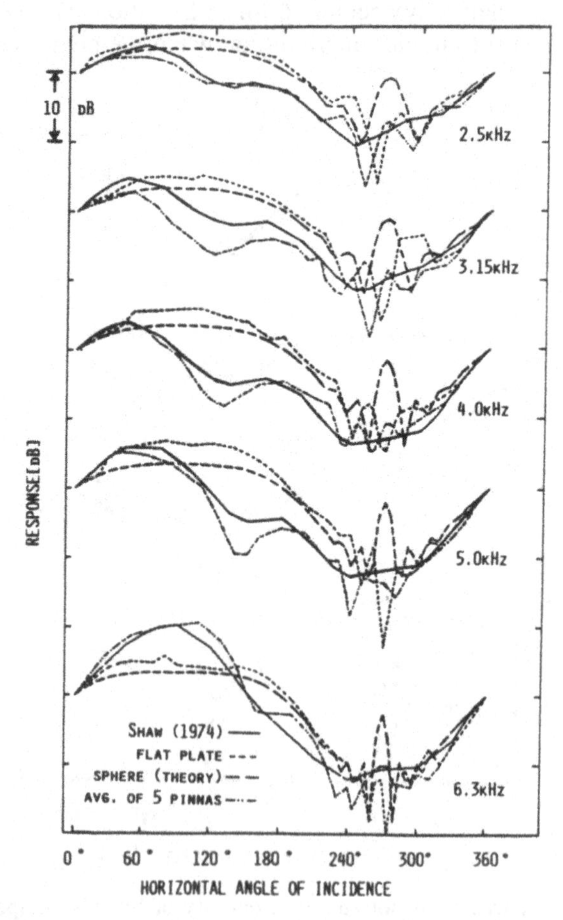

FIGURE 1.8. The same comparisons as shown in Figure 1.7 but for higher frequencies.

directivity pattern that, when combined with that of the head, produces an increased and sharpened directivity at approximately 45° to 60° relative to the median plane. At the same time, the pinna reduces the horizontal directivity by approximately 8 to 15 dB for sounds incident from the rear quadrant between 90° and 180°.

In the shadow region, that is between 180° and 360°, the gross features of the directivity are determined by the head. However, the fine structure of the directivity—that is, the precise location and magnitude of the nulls and peaks of the response—is dependent on the shape of the head and on the geometrical details of the pinna. The directional response of the sphere always has a peak or a "bright spot" at 270° because of the sphere's perfect symmetry. There, the clockwise-moving waves interfere constructively with the anticlockwise-moving waves at the 270° location. Although these peaks and nulls appear also in the measurements made on the head, the precise locations and magnitudes of these nulls and peaks are distributed somewhat differently owing to the lack of symmetry of the head and pinna.

In all cases the horizontal directivity is multivalued, that is, several

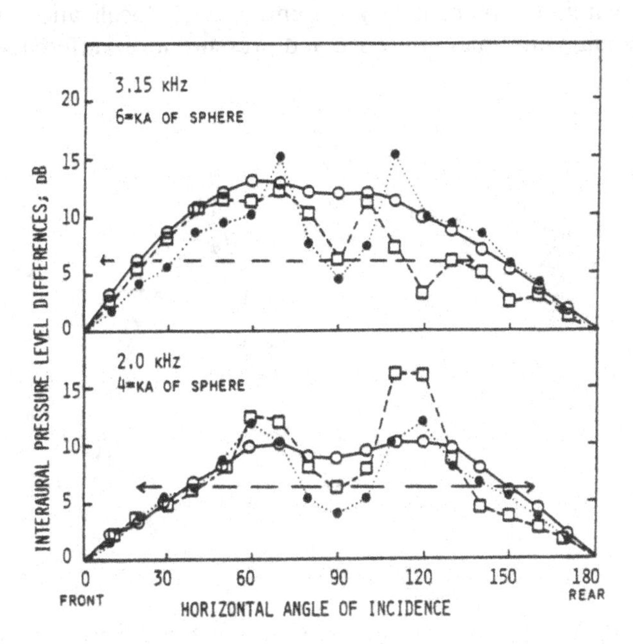

FIGURE 1.9. Interaural pressure level differences (ILD) from Shaw (1974), the open circles; theoretical calculations for a rigid sphere, closed circles; and average measurements for five pinnas using one-third octave bands of white noise, the squares. The dashed horizontal line illustrates an example of identical ILD values for several diffferent angles of incidence.

angles of incidence produce the same ILD. ILDs from Shaw's (1974) composite are compared with the ILDs predicted for a point receiver on a sphere and to the ILDs measured on KEMAR's head and averaged across five pinnas at 2 kHZ and 3.15 kHz in Figure 1-9. For example, at 2 kHz there are as many as four angles of incidence that produce the same ILD localization cue between 45° and 135°. This multiplicity of ILD-based localization cues exists throughout the high-frequency region. Thus, if high-frequency localizations were solely based on ILDs, a large cone of confusion, ranging from approximately 30° to 150°, would be expected. Localization within such a cone of confusion must therefore be resolved through other, additional localization cues such as high-frequency group delays, discussed earlier, or presure-spectrum localization cues of the type discussed below for median vertical plane localization.

As the frequency increases, the intersubject differences in the sound pressure level transformations become larger owing to the large variations in the pinna geometry which play an important role between subjects. This large variability in the pressure spectra for frontal incidence is shown in Figure 1-10. Thus, owing to torso reflections, due to head diffraction, owing to the axial resonances of the ear canal plus concha and to the reflections within the pinna cavity system, spectral localization cues are produced along with time-difference and pressure level-difference cues.

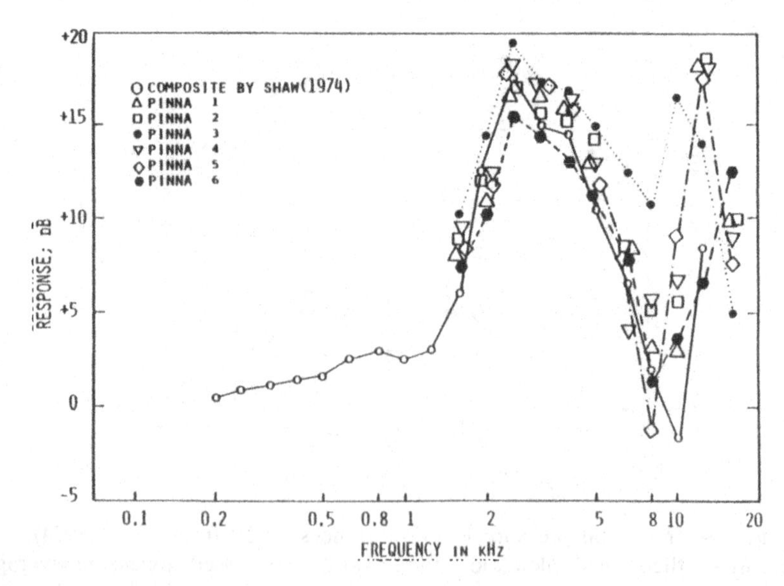

FIGURE 1.10. Sound pressure level transformations to the eardrum (Shaw, 1974) and to the coupler microphone for frontal sound incidence measured on KEMAR using one-third octave bands of white noise.

Directional Effects in the Vertical Median Plane (See Also Chapter 2)

Sound sources in the vertical median plane produce no interaural level or time differences, since the body, head, and pinna are assumed to have perfect symmetry. [Some arguments (Searle et al, 1975; Hebrank, 1976) have been made that pinna disparities produce interaural differences even in the median vertical plane.] Blauert (1969/1970) showed that sounds in the vertical median plane were localized on the basis of the maximum response or "boosted bands" at the ear. Subsequent studies, for example Hebrank (1974) and Butler and Belendiuk (1977), showed that localization corresponds not only to the boosted bands or spectral maxima but also to spectral notches or minima. These spectral minima were shown, via recordings at subjects' ear canal entrance, to rotate from the lower frontal quadrant to the upper frontal quadrant as the frequency increased from approximately 5 to 7 kHz.

To establish the mechanisms that produce these spectral maxima and minima, measurements of the vertical median plane directivity were made with the KEMAR so as to separate the contributions to the ear's directivity by the torso or shoulder from the contributions by the head and the pinna. Figures 1-11 and 1-12 show the vertical, median plane directivities measured between 1.6 and 8.0 kHz on KEMAR and averaged across five different pinnas. Clearly, the vertical median plane directivity is governed by the reflections from the torso at frequencies below 2 kHz whereas above 4 kHz the vertical median plane directivity is governed by the pinna. The maxima in the directivity correspond consistently with the directions of sound localization reported by Blauert (1969/1970 and 1983) as shown in Figures 1-11 and 1-12.

The low-frequency directivity stems entirely from the constructive and destructive interferences of the directly incident sound with the sound reflected specularly by the torso. The high-frequency directivity is produced by the interference pattern produced across the ear canal entrance by the reflection(s) within the pinna cavity system, as is shown below.

The accoustical effects of single or multiple reflections within the pinna cavities on the response of the ear and on sound localization were studied early on by Batteau (1967) and by Shaw and Teranishi (1968). Further studies by Shaw (1975, 1979) showed that higher order modes are produced in the pinna and that pinna-shape distortions have substantial effects on the frequency response of the ear. Time histories of the impulse response of human pinnas in the vertical planes have been reported by Hiranaka and Yamasaki (1983).

Figure 1-13 shows the median-plane vertical directivity of a particular pinna, as measured on the head of KEMAR (no torso). The first null develops for sound incident from the lower frontal quadrant, beginning at

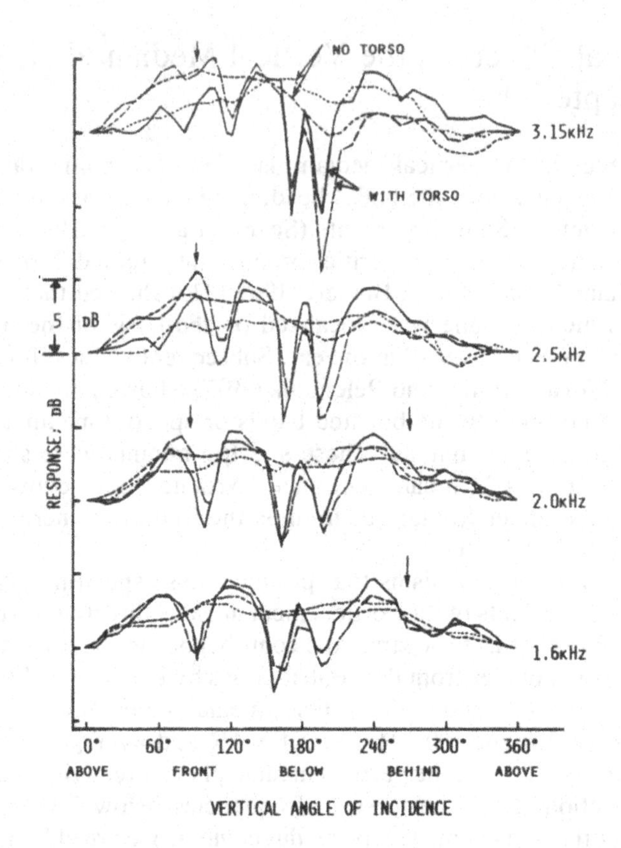

FIGURE 1.11. Effects of pinna and torso on the vertical median plane directivity measured at the coupler microphone (average response of five pinnas): ------- no pinna, no torso; ----- with pinna, no torso; ——— with pinna, with torso;—no pinna, with torso. Blauert's (1969–1970) localization data are indicated by the vertical arrows.

approximately 5 kHz for adult pinnas. As frequency increases, this null rotates to the upper frontal quadrant. However, at the same time as this null rotates, additional nulls are formed producing a complex lobe structure of minima (nulls) and maxima (peaks).

A comparison of Figures 1-12 and 1-13 demonstrates that the difference between the maxima and the minima for individual subjects is much greater than the difference based on averages across subjects. Individual pinna directivities, which produce the averaged results of Figure 1-12, are shown in Figure 1-14. Note that the null that begins at 5 kHz for incidence from below rotates to the upper frontal quadrant as the frequency is increased to 8 kHz, as reported by Butler and Belendiuk (1977). However, there is

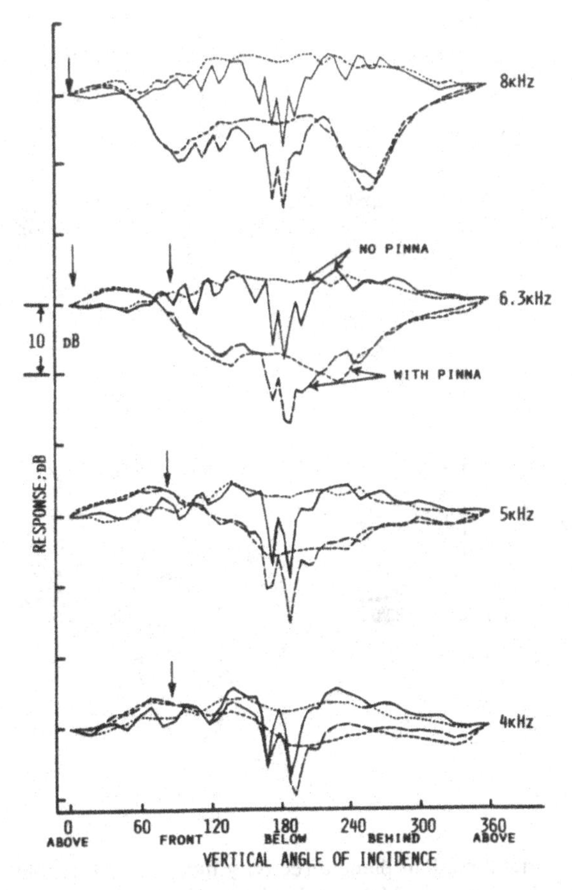

FIGURE 1.12. The same responses as shown in Figure 1.11 but for higher frequencies.

also another null in the lower rear quadrant imaged about the vertical axis, which rotates from below to above, in the rear half plane.

These nulls in the directivity pattern can be explained on the basis of the standing waves produced within the pinna. Figure 1-15 shows the standing wave pattern at 6 kHz produced within a pinna, molded from an adult female, at a directional null, as measured at the coupler microphone. The pressure distribution across the ear canal corresponds to the first higher order mode, the circumferential mode, of a cylindrical duct. The cutoff frequency for this mode in a 7.5-mm diameter ear canal is approximately 26.8 kHz. Therefore, this mode does not propagate to the eardrum and decays exponentially with distance from the ear canal entrance. Only the plane wave component, which is relatively small, propagates to the eardrum and therefore the sharp null in the directivity pattern.

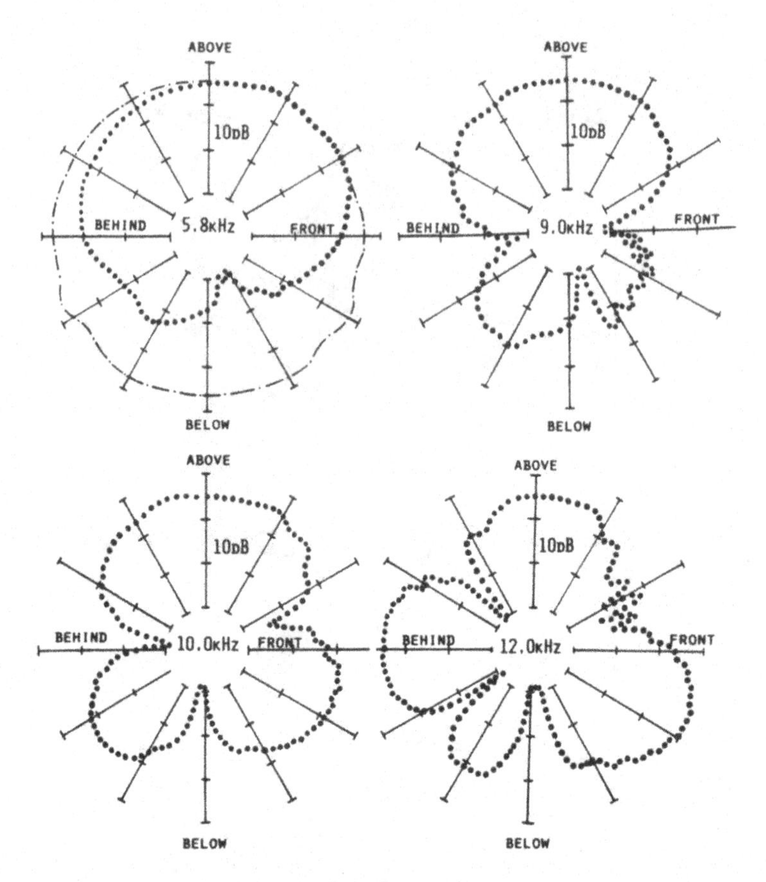

FIGURE 1.13. Vertical median plane directivity measured at coupler microphone of KEMAR (without torso). Male adult pinna; pinna length = 6.78 cm; pinna breadth = 3.50 cm.) —•— shows directivity of the head alone at 5.8 kHz; pinna replaced by a flat plate.

A convenient and mathematically tractable geometry for simulating the pinna's curved surfaces, is the parabola. If the concha wall and helix are combined to form one effective parabola and if the ear canal is placed off the parabola's axis, approximately simulating the pinna, the directivity patterns shown in Figure 1-16 are formed. The precise locations and depths of nulls and their corresponding frequencies can be adjusted systematically by adjusting the parabola's focal length, its height, and the off-axis placement of the ear canal entrance. Calculations (not shown here) of the sound collection and interference pattern produced by a parabolic pinna support this physical description. The calculations also show that the "efficiency" of the pinna is approximately 25% to 60% of an ideal parabolic reflector. Too great an efficiency produces a wave field dominated by the reflected sound that is focused by the parabolic reflector; too small an efficiency

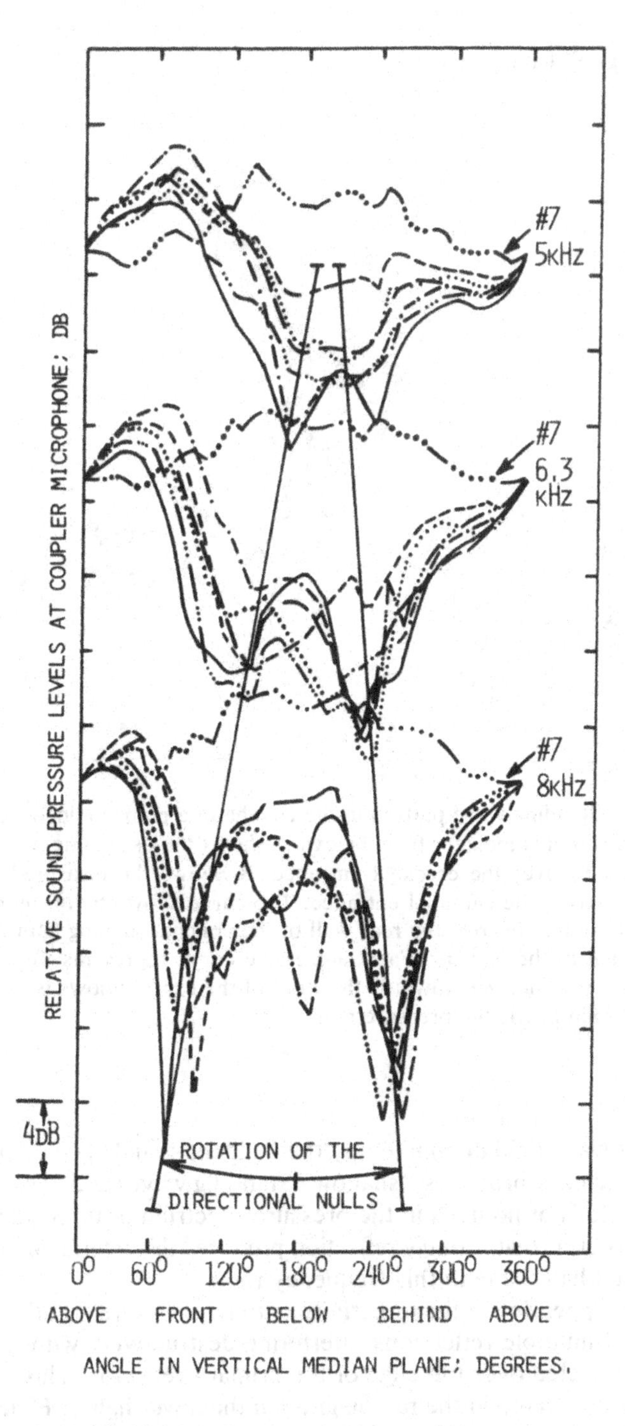

FIGURE 1.14. Vertical, median plane directional response for five pinnas and for small KEMAR pinna at the coupler micorphone of KEMAR's head without torso. Curve #7 id for head alone, where the pinna was replaced by a flat plate. Note the rotation of nulls in front and rear.

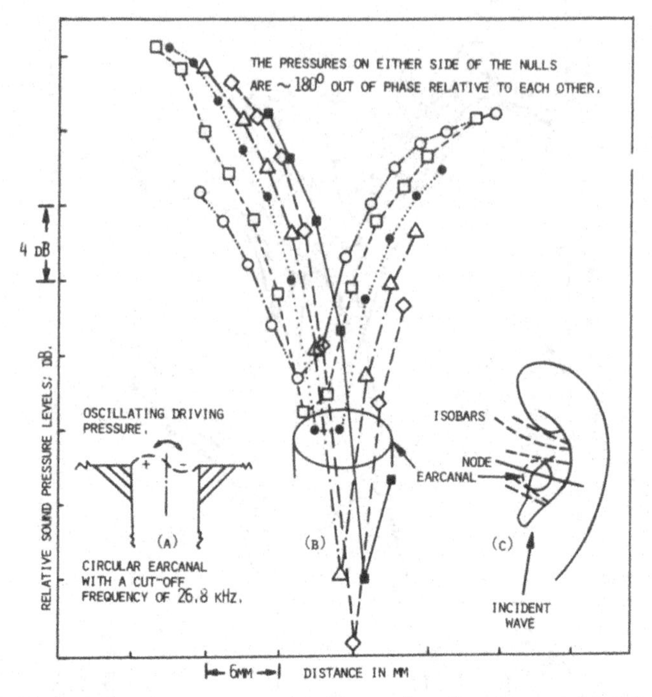

FIGURE 1.15. Standing wave pattern in the concha of a pinna molded from a live subject. The sound is incident from below. **A** shows the oscillating pressure distribution which drives the earcanal entrance. **B** Shows the measured pressure distributions across the earcanal entrance. The curve shown by (■) represents a pressure distribution nearest the rear wall of the concha, running approximately along a normal to the isobars. Each successive curve represents the measured pressures in 2 mm intervals towards the front of the ear. **C** Shows isobars in the concha, the solid line is the pressure node.

produces a wave field dominated by the incident sound. Too great or too small an efficiency produces "shallow" standing waves and consequently "shallow" nulls or no nulls in the pressure spectrum or in the directivity pattern. It is clear that a *single* reflection produces this interference pattern in the frontal half plane in this frequency range.

The nulls appearing in the directivity patterns in the rear half plane are the result of multiple reflections interfering destructively with the wave which is diffracted over the edge of the primary reflector. This wave interaction is illustrated in the ray diagram in the lower half of Figure 1-17. The frontal null, shown in the directivity pattern in the upper half of Figure 1-17, is produced by the primary reflector as described earlier. The directional nulls in the rear half plane are produced by the destructive in-

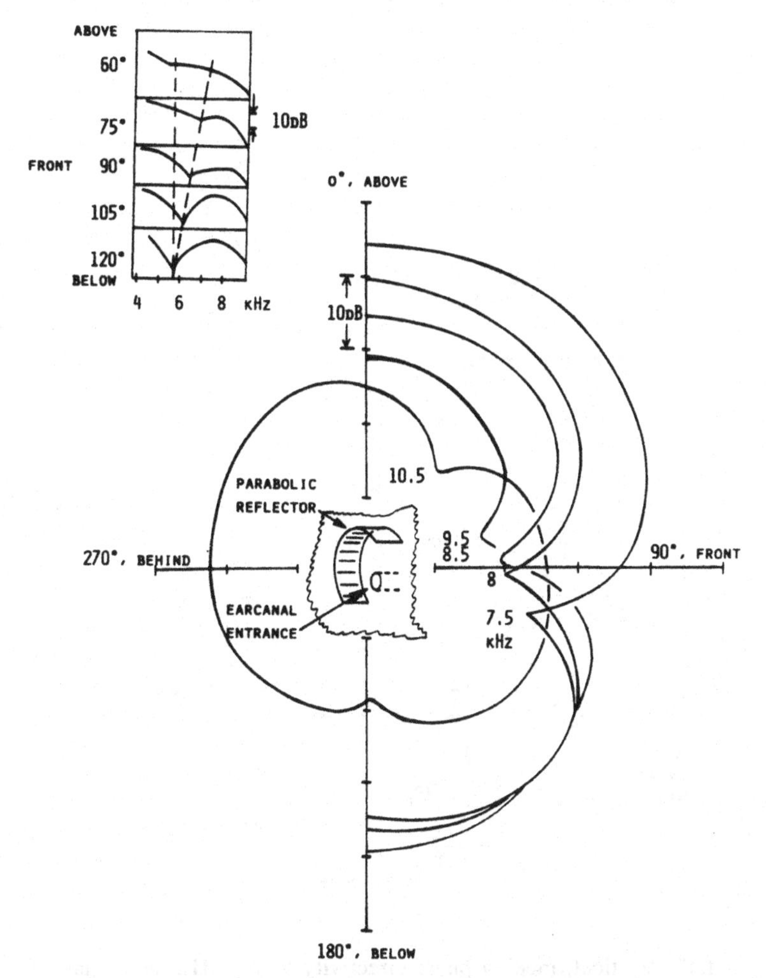

FIGURE 1.16. Rotation of null in the vertical median plane directivity, measured at coupler microphone for a model reflector. The inset shows the rotation of the null in the spectrum at the earcanal entrance (from Butler and Belendiuk, 1977, with permission).

terference of the wave, p_d, diffracted over the edge of the primary reflector with the multiply reflected waves p_{mr}.

Thus, the concha wall and/or helix form the primary reflector to produce the directivity pattern in the frontal half plane via a single reflection up to approximately 12 kHz. In the rear half plane the interaction of a wave *diffracted* by the helix edge with a wave multiply *reflected* by the helix in the fossa region produces the appropriate standing wave pattern across the ear canal entrance forming peaks and nulls in the directivity or pressure spectrum at the eardrum.

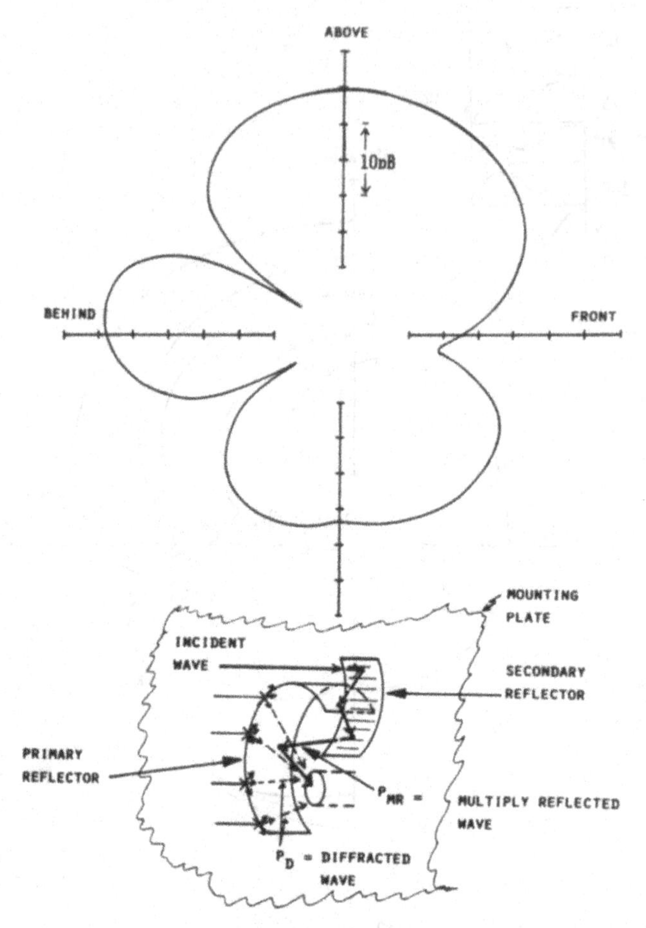

FIGURE 1.17. Vertical, median plane directivity at 7.5 kHz for a dual reflector model.

Summary

Localization cues based on phase-derived ITDs are frequency independent below approximately 600 Hz and are 50% greater than the high-frequency (f⩾2 kHz) ITD_p. The ITD_p rolls off smoothly from 600 to 2,000 Hz, with a small minimum around 1,600 Hz. ITDs are generally well predicted by a theoretical model that treats the human head as a rigid sphere of equivalent perimeter. This theoretical model also predicts the increase in the MAA with the angle of incidence, reported in psycho-acoustic experiments, quite reliably.

Directivities and ILDs for sounds incident in the horizontal plane are relatively unaffected by the pinna for frequencies below ~ 2.5 kHz and are therefore again predicted reliably by the acoustical theory for sound diffraction by a rigid sphere of equivalent perimeter as the head. At frequencies above ~ 2.5 kHz, the pinna sharpens and skews the peak of the horizontal directivity toward the median plane producing a peak between 45° and 60°. In the range between 75° and 240°, the pinna reduces the directional gain. Since the horizontal directivity is multivalued because of the constructive and destructive interferences in the shadow region, the ILDs are also multivalued over a rather large range of angles of incidence. Thus, the ILD localization cues are expected to produce a large cone of confusion, which one might expect to be resolved on the basis of interaural group delays or on the basis of spectral cues. The vertical median plane spectral distribution or directivity at the eardrum is produced by the constructive and destructive interference of the wave incident to the ear, with a wave reflected by the torso at low frequencies or within the pinna cavity system at high frequencies. Below 2 kHz, the directivity is the result of reflections from the torso; above 4 kHz the directivity is primarily the result of one or more reflections within the pinna. The directivity in the frontal, median half plane can be modeled effectively by an ear canal placed judiciously off the axis of a parabolic reflector simulating the concha and/or helix. The directivity in the rear median half-plane results from the interference of the wave that is *diffracted* by the edge of the helix or by the edge of the concha wall with the wave that is *reflected* by the helix-section in the region of the fossa.

Theoretical analyses for parabolic reflectors show that the efficiency of the pinna as a parabolic sound collector lies between 25% and 60% of that for an ideal parabolic reflector. "Poorly shaped" pinnas, because of their inefficiency in focusing the reflected sound, tend to produce shallow interference patterns or a rather smooth spectrum and therefore poor localization cues. Too great an efficiency also produces a shallow interference pattern or smooth spectrum because in this instance the sound field is dominated by the reflected wave, which is focused intensively in the region of the ear canal entrance.

The results and discussions above have corroborated and supported some of the existing psychoacoustical localization models that are based on laboratory studies under static conditions. These mathematical and physical models lay a foundation for systematic, quantitative studies of dynamic localization of complex sounds.

It is easy to show, for example, that source or head rotation in the horizontal plane produces a phase- and amplitude-modulated signal that generates simultaneous time- and amplitude-difference localization cues. Also, the psychoacoustic results for sound localization and the physical measurements of the directivity in the vertical median plane suggest that such localization may be performed in two distinct modes: 1) the detection of the sound at the peak of the directivity where the sensitivity and range are greatest and 2) the tracking of the sound at the nulls of the directivity or pressure spectrum where differential amplitude cues yield the best spatial resolution. Clearly, there is much left to be done in the mathematical-physics as well as in the psychoacoustics, leading to the development of quantitative localization models.

Acknowledgments. Various portions of this work were funded by the National Bureau of Standards, the National Science Foundation, the National Institutes of Health, and by Vibrasound Research Corporation. The assistance of L. Greenspan and G.C. Gaunaurd with the computations is gratefully acknowledged.

References

Batteau, D.W. (1967). The role of the pinna in human localization, Proc. Roy. Soc. (Lond. B) 158, 158–180.

Blauert, J. (1969/1970). Sound localization in the median plane. Acustica 22, 205–213.

Blauert, J. (1983). Spatial Hearing: The Psychophysics of Human Sound Localization. Cambridge, Mass: MIT Press.

Brown, C.H., Schessler, T., Moody, D., Stebbins, W. (1982). Vertical and horizontal localization in primates. J. Acoust. Soc. Am. 92, 1804–1811.

Butler, R.A., Belendiuk, K. (1977). Spectral cues utilized in the localization of sound in the median saggital plane. J. Acoust. Soc. Am. 61, 1264–1269.

Doolittle, R.D., Überall, H., Ugincius, P. (1968). Sound scattering by elastic cylinders. J. Acoust. Soc. Am. 43, 1–14.

Franz, W. (1954). Über die Greenschen Funktionen des Zylinders und der Kugel. Z. Naturforsch. A9, 705–716.

Hebrank, J.H. (1974). The auditory perception of elevation on the median plane. Ph.D. Thesis, Duke University.

Hebrank, J.H. (1976). Pinna disparity processing: A case of mistaken identity? J. Acoust. Soc. Am. 59, 220–221.

Hiranaka, Y., Yamasaki, H. (1983). Envelope representations of pinna impulse responses relating to three-dimensional localization of sound sources. J. Acoust. Soc. Am. 73, 291–296.

Kuhn, G.F. (1977). Model for the interaural time differences in the horizontal plane. J. Acoust. Soc. Am. 62, 157–167.

Kuhn, G.F. (1979). The pressure transformation from a diffuse sound field to the external ear and to the body and head surface. J. Acoust. Soc. Am. 65, 991–1000.

Kuhn, G.F., Gaunaurd, G.C. (1980). Phase- and group-velocities of acoustic waves around a sphere simulating the human head. J. Acoust. Soc. Am. 68(S1), S57(A).

Mills, A.W. (1958). On the minimum audible angle. J. Acoust. Soc. Am. 30, 237–246.

Searle, C.L., Braida, L.A., Cuddy, D.R., Davis, M.F. (1975). Binaural pinna disparity: another auditory localization cue. J. Acoust. Soc. Am. 57, 448–455.

Shaw, E.A.G., Teranishi, R. (1968). Sound pressure generated in an external-ear replica and real human ears by a nearby point source. J. Acoust. Soc. Am 44, 240–249.

Shaw, E.A.G. (1974). Transformation of sound pressure level from the free field to the eardrum in the horizontal plane. J. Acoust. Soc. Am. 56, 1848–1861.
Shaw, E.A.G. (1975). The external ear: New knowledge. In: Earmoulds and Associated Problems. Seventh Danavox Symposium. Stockholm, Sweden: The Almquist and Wiksell Periodical Co.
Shaw, E.A.G. (1979). External ear response and sound localization. In: Localization of Sound: Theory and Applications. R.W. Gatehouse (ed.). Groton, CT: Amphora Press.
Woodworth, R.S., Schlosberg, H. (1962). Experimental Psychology. pp. 349–361. New York: Holt, Rinehard and Winston.

2
A New Approach to the Study of Human Sound Localization

FREDERIC L. WIGHTMAN, DORIS J. KISTLER, AND MARK E. PERKINS

Our interactions with the world around us depend heavily on information supplied by the auditory system. Information about the presence and identity of a sound source is obviously important. However, the location of the sound is often equally important. In everyday life localization seems so automatic and generally so precise that we much more often find ourselves concentrating on "what" rather than "where." Nevertheless, localization is an important auditory function, and the details of how it is accomplished in the auditory system are not well understood.

Systematic study of localization began well over a century ago. Interest in the area has not diminished. In the last decade alone, more than 40 experiments on the subject have been reported in major scientific journals. Despite all this research, our understanding of the auditory processes that mediate localization is incomplete. One possible explanation is that localization experiments are very difficult. Attempts to simplify them have led to experimental results of questionable generality. The major problem arises from the need to bring stimulus variables under experimental control. To accomplish this, localization studies cannot be carried out in any kind of realistic environment. The many reflections and complex interactions of sound waves in a typical room make control of the acoustic stimulus at the ears of a listener nearly impossible. The use of an anechoic room solves some of the problems, but even in this artificial environment, control of the stimulus is a difficult matter.

The aim of early localization studies was simply to determine what features of an acoustic stimulus carried the information about the spatial origin of that stimulus. A simplified geometric analysis of a typical localization task (localization in the horizontal plane) revealed two obvious potential cues (Woodworth, 1938). These are the interaural differences in time of arrival (sound reaches the closer ear as much as 700 μs before the opposite ear) and interaural differences in intensity (the head casts an acoustic "shadow," such that the sound is more intense at the ear closer to the source). A number of acoustical studies (e.g., Feddersen et al, 1957;

see also Chapter 1 of this book) have quantified the dependence of these potential cues on the azimuth of sinusoidal sources. Several psychophysical experiments, conducted with headphones to allow for independent manipulation of the cues, have verified that the interaural difference cues are indeed detectable (Zwislocki and Feldman, 1956; Mills, 1960; see also Chapter 3 of this book). There is considerable indirect evidence that these cues are important for localization. For example, at low frequencies, the interaural time (or phase) difference that is introduced when a stimulus is moved a just-noticeable angle off the midline (Mills, 1958) is about the same as the just-detectable interaural time (phase) difference measured under headphones. The same correspondence holds for interaural intensity differences at high frequencies (see Mills, 1960, for a summary of these points).

From the early localization literature there are two psychophysical studies that laid the groundwork for most of our thinking about the mechanisms underlying localization. The results of these experiments further reaffirmed the primary importance of interaural time and intensity difference cues. The first of these, reported by Stevens and Newman (1936), showed that a listener's ability to localize pure-tone stimuli was much greater at frequencies below 1 kHz and above 4 kHz than at intermediate frequencies. Figure 2-1 (top panel) shows the data from this experiment. Stevens and Newman concluded that in the midfrequency region, listeners could not use either interaural time differences or interaural intensity difference cues, since neither was particularly salient in this region. The second classic experiment was reported by Mills (1958), who studied the minimum audible angle (MAA), or the minimum angular separation, between two pure-tone sound sources that listeners could reliably report. Figure 2-1 (bottom panel) shows the data from this experiment. Two features of these data are important: 1) the loss of localization acuity (as revealed by a higher MAA) in the midfrequency region and 2) the gradual degradation in localization accuracy as the two sources are moved to the listener's side. These results are wholly consistent with those reported by Stevens and Newman and were interpreted as supporting the notion that localization is determined primarily by interaural time and intensity differences.

The simplified geometry assumption, the acoustic measurements, and the results of early psychophysical experiments form the basis of the so-called "duplex theory" of localization, outlined as early as the turn of the century by Lord Rayleigh (Strutt, 1907). This theory holds that localization of low-frequency sounds is dependent on interaural time differences and localization at high frequencies on interaural intensity differences. Unfortunately, this simple theory has serious limitations. One obvious problem is that it cannot explain how localization might be possible without interaural differences. Two conditions in which interaural differ-

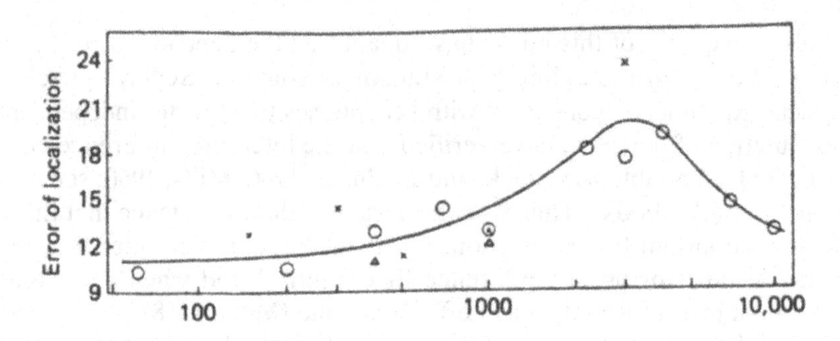

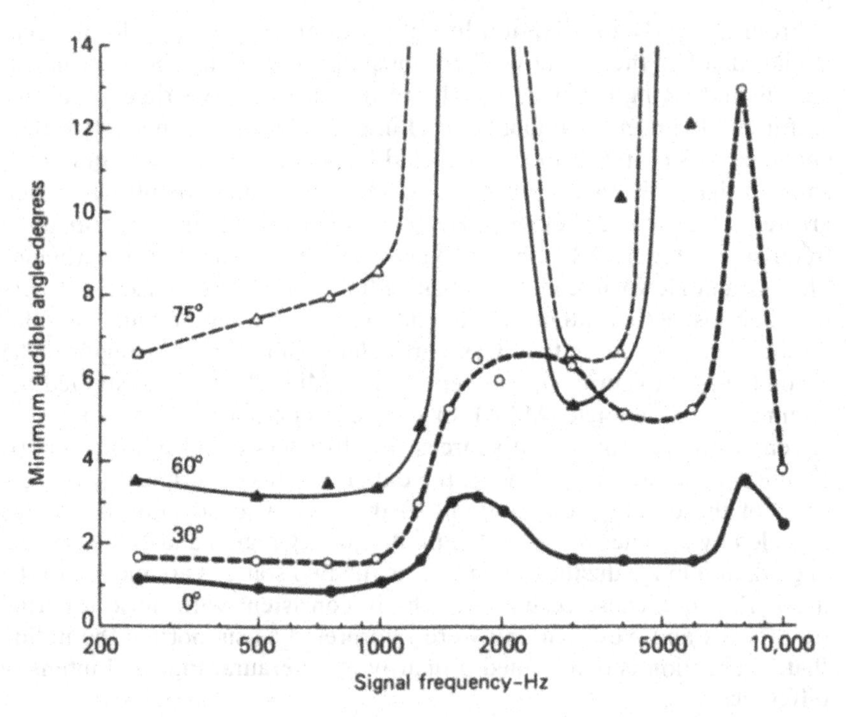

FIGURE 2.1. Average performance of two listeners in free-field localization of pure tones (top panel). The ordinate gives average error in degrees, and the abscissa gives the tone frequency. (From Stevens and Newman, 1936). Also shown (bottom panel; from Mills, 1958) is the minimum audible angle (MAA) as a function of tone frequency (abscissa) and initial sound source position (noted at the left of each curve.)

ences are either absent or minimized are localization in the median (vertical) plane and monaural localization (Hebrank and Wright, 1974a,b). A second problem is that the duplex theory cannot account for those conditions in which sounds seem to emanate from inside the listener's head. This almost always occurs when stimuli are presented over headphones, even though appropriate interaural time and intensity differences are introduced. Clearly, there is more to localization than can be explained on the basis of interaural time and intensity differences.

In the last two decades, the availability of more sophisticated equipment for stimulus control has provoked considerable research on localization cues other than interaural time and intensity differences. Most notable, perhaps, are the studies of the cues provided by a listener's pinnae (Batteau, 1967; Wright, et al, 1974). It has been known for some time that as a result of interactions of a sound with reflections from the convolutions of the pinnae, a direction-dependent filtering is imposed on an incoming stimulus (Blauert, 1983). It is now clear that this spectral shaping is a very important cue for localization (see Butler, 1975, for a review of the research on this issue and Musicant and Butler, 1985). The best demonstration of this is the fact that when the cavities of the pinnae are filled with putty, localization ability is markedly impaired (Gardner and Gardner, 1973). Other recent experiments considered the role of head movements (Thurlow and Runge, 1967), visual cues (Gardner, 1968), a priori knowledge of stimulus properties (Coleman, 1962), and postural variables (Lackner, 1983). The specific contributions of these factors to our perception of auditory space are not well understood, although it is agreed that in certain listening situations they are important.

Although a great deal is now known about localization, there are large gaps in our understanding, and the areas of greatest uncertainty are also the most basic. For example, it is still not entirely clear what characteristics of a sound cause it to be externalized. There is some suggestion that the filtering action of the pinnae is important, but the issue is far from settled. Our inability to address such basic questions almost certainly results from a lack of necessary technology. For example, in spite of the overwhelming experimental advantages of headphone-stimulus presentation, there have been no published reports of actual localization experiments using headphones.

The results of so-called "lateralization" experiments (see Chapter 3), in which stimuli are presented over headphones, are almost never generalizable to actual localization conditions. Lateralization experiments often purport to address issues of localization, but this claim is questionable since the stimuli are rarely externalized. The fact that a subject listening over headphones can discriminate or detect interaural differences may say very little about how discriminations of azimuth and elevation changes are accomplished in free field. Similarly, lateralization paradigms can provide only indirect evidence on the viability of theories of localization such as the duplex theory. Because the stimuli in a lateralization experiment are rarely externalized, and in light of the mounting evidence that cues other than interaural time and intensity differences are important, it is difficult to argue that localization is actually being studied in a lateralization paradigm.

Because of the technical difficulties associated with presenting sounds from actual sources in free field, the scope of most previous localization experiments has been limited. The majority has dealt *only* with localization on the horizontal plane (e.g., azimuth discrimination) or with localization in the median plane (e.g., elevation discrimination). Moreover, while recent research recognizes the complexity of actual localization conditions and the importance of cues such as those provided by the pinnae, there have been only a few attempts to manipulate these cues systematically. Until recently, technological limitations have inhibited such attempts. Schroeder and Atal (1963), and Morimoto and Ando (1982) described a technique using two loudspeakers and digitally generated stimuli whereby the illusion of a sound source at any arbitrary point in space can be created (so long as the position of the listener is known precisely). Bloom (1977) and Watkins (1978) attempted to simulate source elevation changes by altering the spectrum of the source in a manner analogous to pinna filtering. Blauert (1969) and Butler and Planert (1976) made similar attempts to alter the apparent location of a sound by modifying the spectrum. The most significant limitation of these early experiments is that they included no objective, psychophysical tests of the adequacy of the techniques for duplicating the free-field experience.

The primary goal of our research to date has been to develop and test digital synthesis techniques whereby a veridical simulation of free-field listening can be produced over headphones. The availability of such techniques would allow complete control of the acoustic stimulus and thus open the door for systematic study of many unanswered questions about human sound localization. For example, such a technique would permit accurate measurement of the relative importance of interaural time and intensity differences in actual localization conditions. The importance of pinna filtering, prior knowledge of stimulus characteristics, interaural pinna disparity, and individual differences in pinna effects could also be assessed in entirely new ways. To the extent that pinna filtering is important for localization (and we believe it is), a secondary goal is to provide information on how the auditory system extracts information from a spectral contour. This work thus parallels the recent "profile analysis" research reported by Green (1983). The long-term goal of our research is not to redo previous localization experiments with better techniques, but rather to use new methods to answer heretofore unapproachable questions about human sound localization.

New Psychophysical Methods

Our initial efforts have focused on development of both the free-field simulation techniques, and the psychophysical procedures to be used to verify the adequacy of the simulation. The psychophysical studies are described first.

The goal of the psychophysical studies was to develop a measure of overall localization performance that could be used both in free-field and in headphone listening conditions. The same method would then be used to quantify the differences in localization performance between the two conditions. The first procedures we tried were based on multidimensional scaling (MDS) techniques. This choice was motivated by our desire to be able to represent details of the "shape" of auditory space and distortions of that shape which may result from certain stimulus manipulations (e.g., headphone presentation).

The MDS paradigm (Wightman and Kistler, 1980) requires listeners to make judgments about relative distances between sound sources. These apparent distances are scaled according to standard MDS protocol to yield a three-dimensional representation of a listener's "auditory space." In the free-field version of this experiment (Wightman and Kistler, 1980) subjects seated in an anechoic chamber listen to 200-ms white noise bursts presented over digitally equalized loudspeakers. The speakers are arranged in a roughly quarter-spherical array on the subject's right. Initially a triadic comparison paradigm was used to assess apparent intersource distance. On each trial computer-synthesized noise bursts were presented by three loudspeakers, one at a time. Listeners controlled presentation of the sounds (with pushbuttons) until they made a judgment of which two sounds appeared closest together in space and which appeared farthest apart. After both responses were made, a new triad of sounds was made available.

The scaling solutions from the triadic comparisons data matched very closely the actual physical arrangement of the sources, indicating that localization was very precise in these conditions. The "goodness of fit" or correlation between the coordinates of the actual source positions and those derived from the scaling solutions was 0.955 for the poorest of five subjects. This result was encouraging, since it is well known that localization is quite good with wide-band sources.

There arose several problems that led us to abandon the triadic comparisons procedure. The first was that the task is far too time consuming. If all possible triads were to be judged, the 22-speaker array would require more than 18 hours of testing time. By presenting a carefully selected subset of triads, we were able to reduce the time to about six hours per condition and yet maintain stable scaling solutions. This time was also felt to be excessive. In an effort to find an MDS-based procedure that would be faster, we tried a modified "paired-comparison" technique in which listeners simply rate (on a ten-point scale) the apparent distance between two sources presented one at a time. With the paired-comparison procedure, only 30 minutes per run were required (Kistler and Wightman, 1985).

The second problem with the triadic comparisons paradigm was that the scaling solutions were unacceptably resistant to distortion. Goodness of fit was measured as a function of stimulus bandwidth. Given other localization data in the literature, it is to be expected that as the low-pass cutoff of a stimulus is lowered, localization will deteriorate. For many subjects, the scaling solutions derived from triadic comparisons data did not reveal this effect, despite listeners' reports of great difficulty in some conditions. The paired-comparisons technique was much more sensitive. Figure 2-2 compares mean performance of 11 listeners in the triadic-comparisons and paired-comparisons paradigms, and Figure 2-3 shows the performance of the individual listeners in the paired-comparisons conditions.

Another paradigm we developed for quantifying the shape of auditory space requires listeners simply to judge apparent azimuth, elevation, and distance of a sound source. We call this an "absolute judgment" task. In

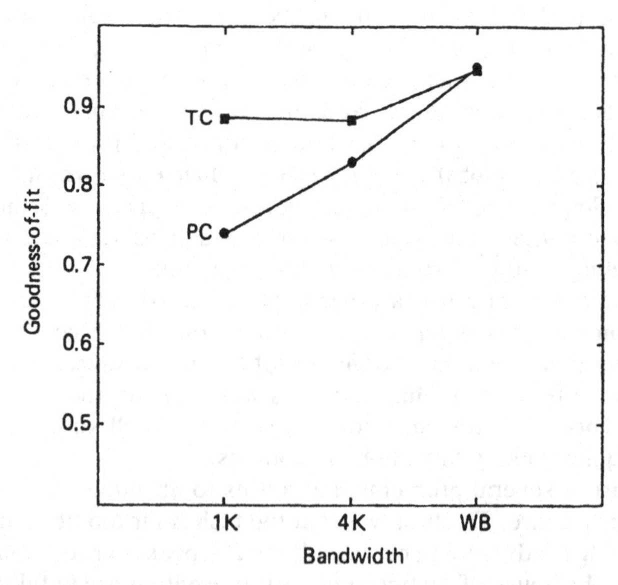

FIGURE 2.2. Average performance of 11 listeners in free-field localization. The solid squares indicate performance in the triadic-comparison paradigm, and the solid circles indicate performance in the paired-comparison paradigm. The ordinate is the "goodness of fit" between the scaling solutions derived from listeners' distance judgments and the actual physical coordinates of the loudspeakers. The abscissa gives the low-pass cutoff of the noise stimulus used in the experiments.

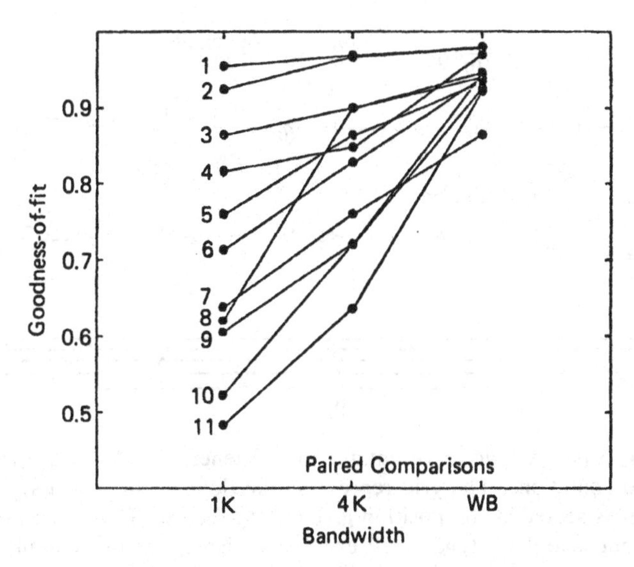

FIGURE 2.3. Individual performance of 11 listeners in the free-field, paired-comparison localization paradigm. Ordinate and abscissa have the same meaning as in Figure 2.2.

this paradigm, the apparent location of a sound source is estimated directly, bypassing the MDS procedures. Although we were initially uncertain about observers' abilities to make such complex judgments, the method has proved successful. After about four hours of training with feedback, three out of four subjects can meet a criterion of at least 0.95 test-retest reliability and 0.95 goodness of fit in the wide-band condition. The method is fast, requiring only about 30 to 40 seconds per trial, or about 15 to 20 minutes per condition.

Figure 2-4 compares the three-dimensional representations of a listener's auditory space obtained from the paired-comparisons paradigm and from the absolute judgment paradigm. Note that distortions of the space are represented differently in the two cases. This difference highlights the need for caution in interpreting the three-dimensional representations obtained from both methods. The exact way in which details of a given listener's auditory spatial experience are represented in the figures is yet to be determined. One obvious problem is that in the scaling solutions, the apparent distance of a sound image is confounded with the degree to which the apparent positions of two sounds are confused. If the locations of two sources are confused, the scaling algorithm places them close to one another in the solution. If the sources that are confused are on opposite sides of the listener (e.g., in front and behind), the scaling solution will represent them both as close to the listener. Moreover, a general uncertainty about source position produces solutions with points clustered about

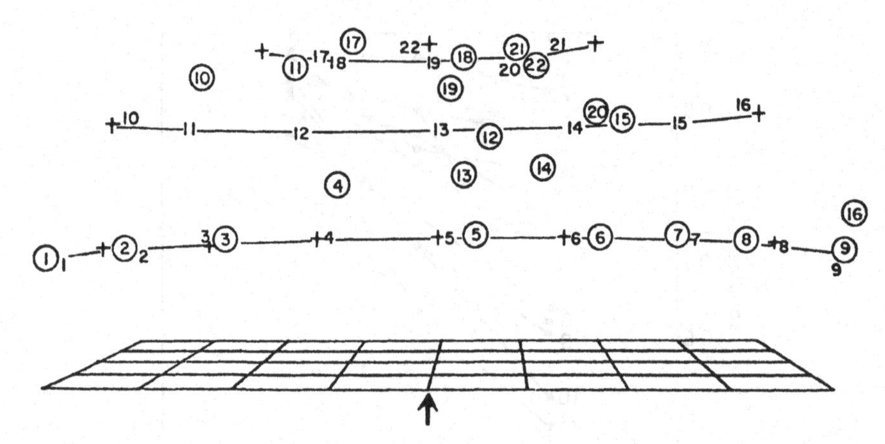

FIGURE 2.4. A perspective representation of a listener's localization performance in free-field conditions. The grid represents the floor of the anechoic chamber. The listener is seated at the position given by the arrow. The observation point is directly opposite the listener's left ear; the loudspeakers are positioned on the observer's right, at positions indicated by the crosses, and are connected by solid lines. The speakers form three arcs around the listener on his right side. One loudspeaker is directly above the listener. The numerals enclosed by circles indicate the perceived positions of the sources as represented by the scaling solution obtained from paired-comparison data. Sources numbered from 1 to 9 are on the lowest arc, 10 to 16 are on the next arc, 17 to 21 are on the top arc, and 22 is directly above the listener. Note that the enclosed numerals are generally close to the appropriate source. The numerals that are not enclosed give the apparent source positions obtained from the absolute-judgment paradigm. These positions give nearly a perfect representation of the actual loudspeaker arrangement.

the origin of the space (the listener's position). Thus the fact that a point is placed near the origin does not necessarily mean that it appears close to the listener. Problems such as this are not evident in the absolute judgment paradigm.

The final issue that was addressed in the preliminary psychophysical studies is listener head movement. The localization literature is quite ambiguous about the importance of head movements in free-field localization. There is some feeling that head movements might help a listener resolve front-back confusions, but there are few data that support this point of view. Because of the fact that head movements would be eliminated in headphone-simulated free-field conditions. we felt it necessary to estimate the influence of head movements in our actual free-field conditions. Thus we tested listeners in the paired-comparisons task with their heads fixed by means of a bitebar. The scaling solutions obtained in this head-fixed condition were indistinguishable from those obtained with no head fixation.

Stimulus Synthesis for Free-Field Simulation

In simplest terms, the goal of the synthesis was to duplicate free-field listening with headphone stimulus delivery. The basic assumption of our approach was that if the acoustic waveforms at a listener's eardrums were the same under headphones as in free field, then the listener's experience would also be the same. Our preliminary trials with synthesized stimuli presented over headphones suggest that at least for the limited range of conditions studied, the assumption is warranted.

Our stimulus synthesis technique can be divided into three phases. The first requires measurement of the acoustic filtering characteristics (transfer function) of a listener's outer ears, with sound sources at various spatial positions. The second involves measurement of the headphone transfer function, with the listener wearing the headphones. In the final phase of of the synthesis, the desired experimental stimuli are digitally filtered. The transfer function of this filter is given by the outer ear transfer function divided by the headphone transfer function. When these filtered stimuli are presented to a listener over headphones, the headphone response cancels, leaving only the free-field outer-ear characteristics imposed on the stimuli. Ideally, the waveform reaching the eardrum is identical to that produced by a free-field stimulus.

The transfer function measurements are made using an adaptation of a technique described by Mehrgardt and Mellert (1977). A burst of pseudorandom noise (flat amplitude, random phase) is presented repetetively to the system under study, and the response of the system to the noise signal is obtained by averaging. The Fourier transform of this response divided by the Fourier transform of the noise signal gives the transfer function of the system. In our case, the noise signal is 40.96 ms in duration and has a flat amplitude spectrum to 40 kHz. It is output periodically via a 14-bit deglitched digital-to-analog converter (D/A) at a rate of 100 kHz. The system response was measured with a miniature electret microphone coupled to a stainless steel probe tube with an inner diameter of 1 mm. The response of the microphone and probe-tube system is shown in Figure 2-5. The amplified microphone output was digitized using a 12-bit A/D at 100 kHz. The responses to 2,000 periods of the signal were averaged, producing a worst-case signal-to-noise ratio of about 40 dB in the range 200 to 20 kHz.

To make the direction-dependent free-field ear canal transfer function measurements, the probe-tube was positioned such that its tip was about halfway down a subject's ear canal and away from the canal walls. The body of the microphone was then taped to the side of the subject's head (Fig. 2-6). The subject was seated in the speaker array, with head held motionless by means of a bitebar. The repeatability of transfer function measurements made in this way was good; within-subjects spectral con-

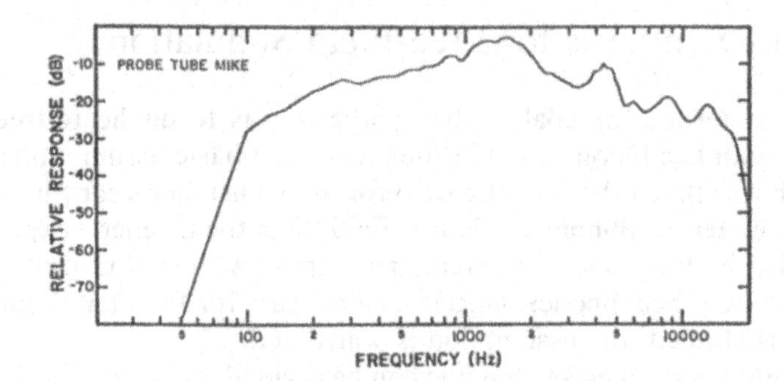

FIGURE 2.5. Acoustic frequency response of the probe tube and microphone used for the outer-ear transfer function measurements. The response is digitally bandpass filtered between 100 Hz and 18 kHz to reduce noise.

fidence limits (95%) were less than 2 dB at frequencies below 10 kHz and no more than 5 dB below 20 kHz. The response of each loudspeaker was measured with a wide-band flat microphone system (Brüel and Kjaer 4133) at a point corresponding to the center of the subject's head. The loudspeaker-response measurements were corrected for microphone characteristics by dividing the measured response by the microphone response (supplied by B&K and verified by electrostatic actuator).

Measurement of the headphone transfer function was made in nearly the same way as the free-field measurements. The probe-tube microphone was mounted in the same way, and the subjects wore headphones (Beyer ET1000, electrostatic). The pseudorandom noise signals were transduced by the headphones. The repeatability of these measurements was not as

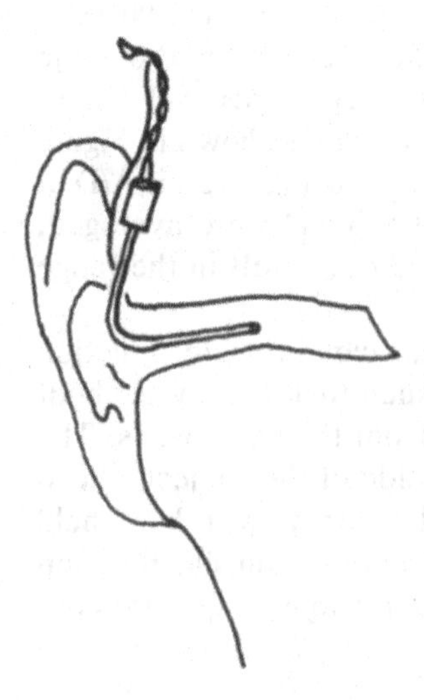

FIGURE 2.6. Diagram of the Probe-tube and microphone assembly in measuring position on a subject, showing placement of the tube in the ear canal.

A

Terms		
$S(J\omega)$	OR S	Loudspeaker transfer function (one for each of 22 loudspeakers)
$P(J\omega)$	OR P	Direction dependent pinna transfer function (one for each of 22 positions)
$A_1(J\omega)$	OR A_1	Auditory meatus transfer function in free field
$M_1(J\omega)$	OR M_1	Condenser microphone transfer function (B & K 4133)

Condition	Stimulus Spectrum	Transformation	Result
1. Measurement of loudspeaker characteristics	1	SM_1	SM_1
2. Correction for microphone characteristics	SM_1	$\frac{1}{M_1}$	S
3. Free-field listening	$\frac{1}{S}$	SPA_1	PA

B

Terms		
$S(J\omega)$	OR S	Loudspeaker transfer function (one for each of 22 loudspeakers)
$P(J\omega)$	OR P	Direction dependent pinna transfer function (one for each of 22 positions)
$A_1(J\omega)$	OR A_1	Auditory meatus transfer function in free field
$M_1(J\omega)$	OR M_1	Condenser microphone transfer function (B & K 4133)
$H(J\omega)$	OR H	Headphone transfer function (Beyers ET-1000)
$A(J\omega)$	OR A_2	Auditory meatus transfer function under headphones
$M_2(J\omega)$	OR M_2	Miniature electret microphone - probe tube transfer function

Condition	Stimulus Spectrum	Transformation	Result
1. Measurement of loudspeaker characteristics	1	SM_1	SM_1
2. Measurement of pinna characteristics	1	SPA_1M_2	SPA_1M_2
3. Measurement of headphone characteristics	1	HA_1M_2	HA_2M_2
4. Headphone listening	$\frac{SPA_1M_2}{HA_2M_2}\frac{1}{S}$	HA_2	PA_1

FIGURE 2.7. Equations used for stimulus synthesis. The top panel shows the steps involved in producing the noise stimuli used for free-field testing. The aim of this procedure is to compensate for individual loudspeaker characteristics. Condition 1 shows the result of the measurement of loudspeaker characteristics (SM) when a white stimulus (1) is transduced. Condition 2 shows the digital removal of the stored microphone characteristic, yielding a pure loudspeaker characteristic (S). In condition 3, a stimulus with a 1/S spectrum is transduced by the loudspeaker and passed through a subject's outer-ear (P) and ear-canal (A). The lower panel shows the steps involved in production of stimuli for simulated free-field listening. The meaning of the various conditions is similar to those in the upper panel. Condition 4 shows how the digitally filtered stimulus, when transduced by the headphones, produces for the subject the same result (PA) as in free-field (top panel).

high as in free field, presumably because placement of the headphones frequently moved the microphone and probe-tube assembly. Since the simulation procedure depends on constant positioning of the measuring microphone in both conditions, the change in microphone position during headphone placement is a problem that will need to be solved in the future.

Stimuli were synthesized by passing digitally computed white gaussian noise through an Finite Impulse Response (FIR) digital filter, implemented by the "fast convolution" technique, which is based on an overlap-and-add Fast Fourier Transform (FFT) algorithm (Stockham, 1966). Transfer functions for 44 different filters were computed, with each filter corresponding to a different source position in free field (the 22 loudspeaker positions in the anechoic chamber) and ear (right or left). The form of the transfer functions and the general flow of the computations are shown in Figure 2-7. For each of the 22 source positions, 15 different 200-ms noise bursts (with 20-ms cosine-squared rise-fall) were computed and stored. Note that the synthesized stimuli are subject specific in that a given subject's outer ear characteristics are superimposed on the stimuli. Thus, for each subject tested, a different set of stimuli was synthesized.

Acoustic Measurement of Outer Ear Characteristics

Position-dependent outer-ear transfer functions have been measured from both ears of eight subjects, for 22 different source positions. Five complete sets of measurements were made on each subject. Although it is not necessary to describe these data in detail, a brief summary is useful.

An example of an uncorrected transfer function is shown in Figure 2-8 (amplitude spectrum only). This function is called "uncorrected," since the responses of the microphone, probe-tube, and loudspeaker have not

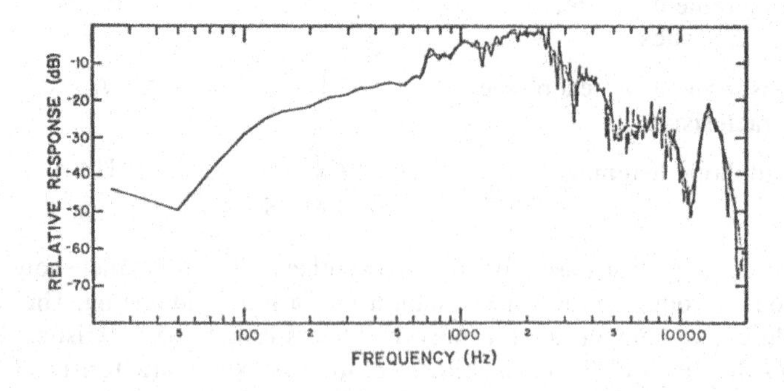

FIGURE 2.8. Example of an uncorrected outer-ear transfer function measured from a subject's ear canal using the probe-tube microphone. The jagged curve shows the influence of echos in the anechoic chamber. The smooth curve running through the middle of the jagged curve shows the effect of our "critical-band" smoothing algorithm.

been removed the function is simply the Fourier transform of the signal recorded from the microphone. The irregularities in this function at high frequencies (i.e., the frequent peaks and valleys) are known to be a result of unwanted reflections in the "anechoic" room from the loudspeakers themselves and from the metal floor frames. Since it is unlikely that such spectral detail is important at high frequencies, the spectra were smoothed with a sliding gaussian filter, the bandwidth of which was set to one half the usual human "critical" bandwidth (Scharf, 1970). The effect of smoothing is also shown in Figure 2-9. (Note: the smoothing was done only for the purpose of analysis of the outer-ear transfer functions and was not included in the stimulus synthesis procedure.)

Although only the amplitude parts of the measured transfer functions are discussed here, our procedure includes measurement of the phase response as well. Extraction of the phase response from the raw transfer functions is complicated by the fact that sizable acoustic delays are included in the measured response. Such pure-time delays are uninteresting components of the measured phase responses, so they were removed. The measured response was decomposed via a Hilbert transform method into a minimum-phase part and an all-pass phase part (Oppenheim and Schafer, 1975, pp 345–353). Pure delay is all-pass, and since the outer ear is mostly minimum-phase, the all-pass part of the measured response in most cases was pure delay. After the delay was removed from the all-pass part of the response (by a simple phase-shift), the minimum-phase and all-pass phase parts were recombined. The continuous phase function was derived from the recombined transfer function using an unwrapping technique adapted from Tribolet (1977). Phase "unwrapping" is necessary because the system phase response extends beyond plus or minus 180 degrees. The usual way of computing phase shift at each frequency (arctangent of imaginary part of transform divided by the real part) only gives the principal value (within plus or minus 180 degrees) of the phase, leaving an ambiguity if the actual phase response extends beyond 180 degrees. Phase unwrapping is a rather complex procedure that resolves, or attempts to resolve, this ambiguity.

A subset of the outer-ear transfer functions obtained from three subjects is shown in Figure 2-9. The purpose of this display is twofold: to highlight

FIGURE 2.9 (pages 40–41). Samples of outer-ear transfer functions measured from three subjects (columns) and eight sound source locations (A–H). The solid curves give the response from the right ear (toward the loudspeakers), and the dashed curves give the response from the left ear. The source positions are noted in terms of azimuth (in degrees, with zero indicating straight ahead), elevation (in degrees, with zero indicating level with the ears), and distance (in meters from the center of the listener's head). The positions are as follows. A (O azimuth, O elevation, 2 distance); B(45,0,2); C(135,0,2); D(180,0,2); E(0,30,1.8); F(90,30,1.8); G(180,30,1.8); H(0,90,1.4).

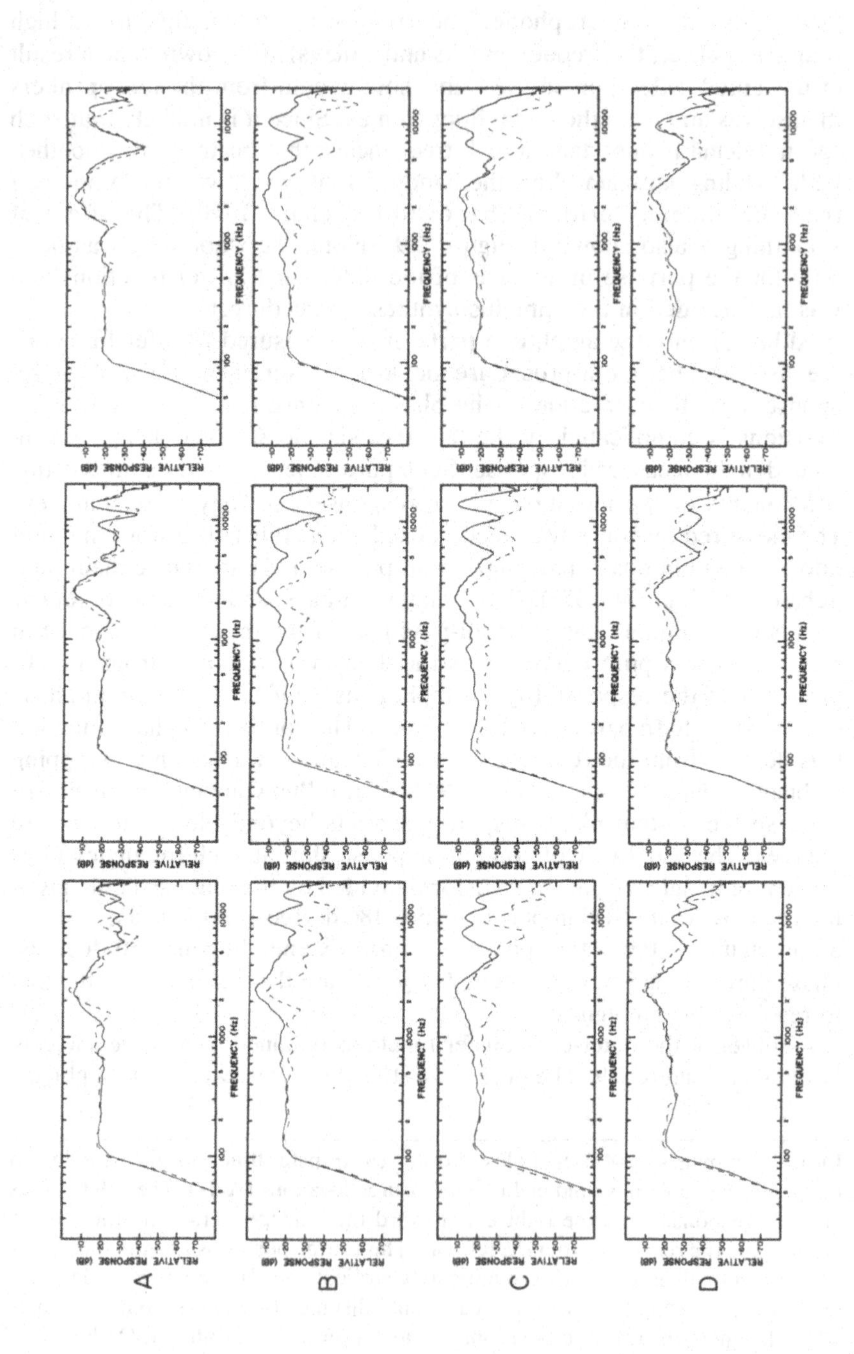
A
B
C
D
RELATIVE RESPONSE (dB)
FREQUENCY (Hz)

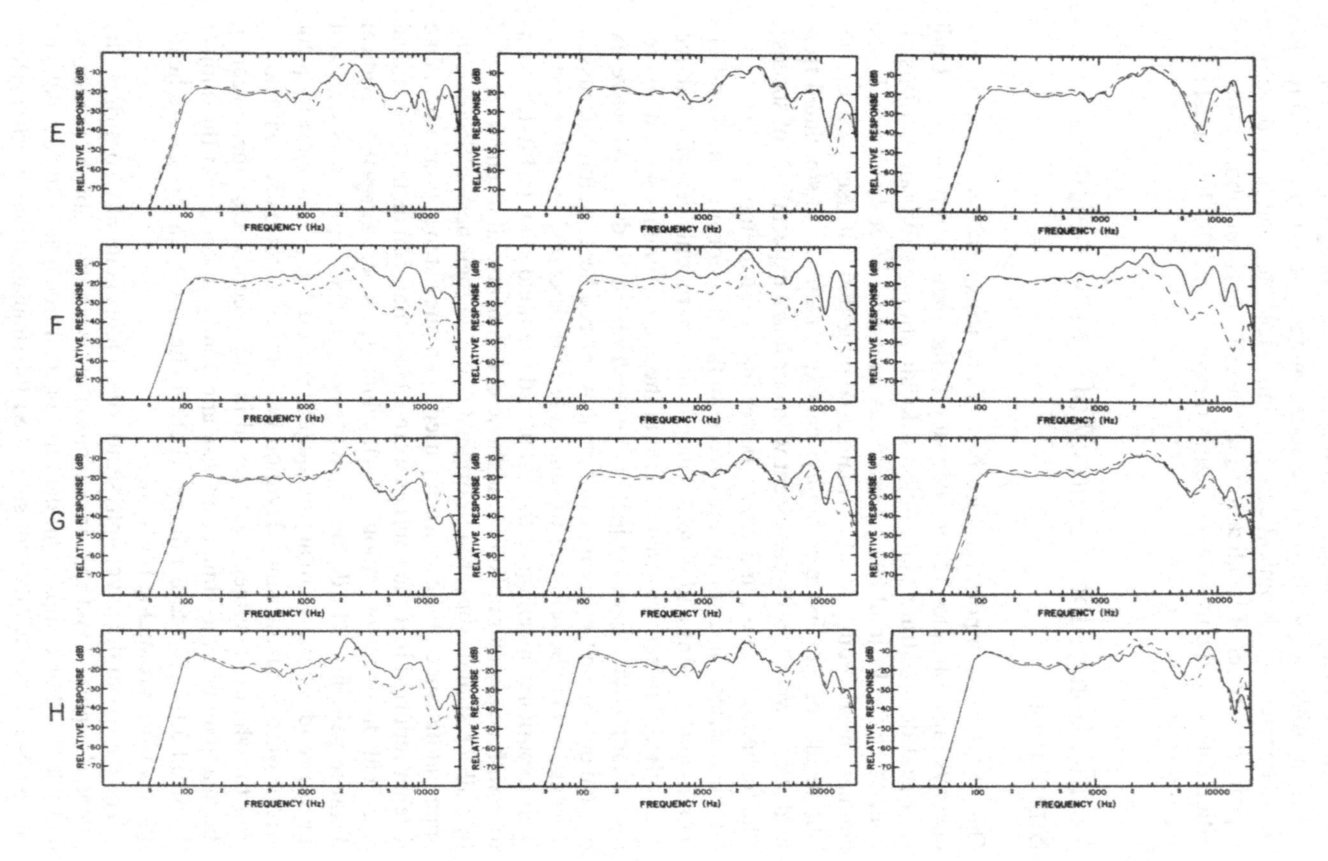
E
F
G
H
RELATIVE RESPONSE (dB)
FREQUENCY (Hz)

the large intersubject variability in the measured functions and to demonstrate the changes in the functions that accompany changes in source location. Although we have made no attempt as yet to relate the direction dependencies to anatomical details (pinna size, etc), we have developed a primitive model of localization that bases its decision about source position solely on the transfer functions (amplitude only). This model is discussed later.

Verification of the Accuracy of the Free-Field Simulation

One of the most important objectives of our research was to obtain psychophysical verification of the adequacy of the simulation procedure. Until we could demonstrate that subjects' localization performance with simulated free-field stimuli was the same as with actual free-field stimuli, the simulation procedure would remain a complicated but useless exercise. The results of our first psychophysical experiments with simulated free-field stimuli were wholly consistent with our initial expectations of success. The technique works, and works better than we had hoped.

The verification experiments were identical in form to the free-field paired-comparisons and absolute judgment experiments described above. The only difference was that stimuli in these experiments were delivered over headphones. Five subjects have been tested to date. We have extensive data from two, which will be summarized here. Before discussing the data, it may be relevant to summarize the listeners' verbal reports of their experience listening to the simulated free-field stimuli. First, in contrast with nearly all previous headphone experiments, all listeners reported that all the stimuli appeared to originate outside their heads. Second, the apparent distance of the sources varied from one listener to another. One listener reported that the sources were all less than 1-m distant (the actual free-field distance was about 2 m), while the others reported veridical distance percepts. Third, the apparent positions of the sources were not always veridical, but were always quite clear and stable. The most frequent errors were median-plane elevation errors (not front-back confusions). Fourth, when one subject was tested in the anechoic chamber (wearing the headphones), the simulation was apparently so good that the subject reported difficulty determining whether the loudspeakers or the headphones were producing the sound.

The data from the psychophysical experiments generally confirmed the listeners' reports about the simulations. Data from two subjects are shown in Table 2-1. For comparison, data from actual free-field tests are included in the table. It is clear that for subject S1 the simulation was nearly perfect. Localization performance of this subject (as indicated by the "goodness of fit" measure) was nearly the same in free-field and under headphones,

TABLE 2.1. Goodness-of-fit[a] of subjects' judgments to the actual source positions for the absolute judgment and paired comparison tasks.

Listening condition	Listener	Absolute judgment	Paired comparison
Free field	S1	.99	.96
	S2	.97	.95
Simulated free field	S1	.96	.88
	S2	.83	.81

[a]Goodness-of-fit values are correlation coefficients

whether measured in the paired-comparison or absolute judgment paradigm. Figure 2-10 compares subject S1's absolute judgment performance in free field with that in simulated free field. For the other subject, the simulation was far less successful. This subject showed a significant performance decrement in both headphone conditions.

We have carefully analyzed the possible sources of error and are confident that the relatively poor performance of some listeners on some simulations can be explained. In the worst case (subject S2 in Table 2-1) movement of the measuring microphone was almost certainly the source of the problem. This subject's ear canal and outer ear were such that while the microphone could be rigidly fixed for the free-field measure-

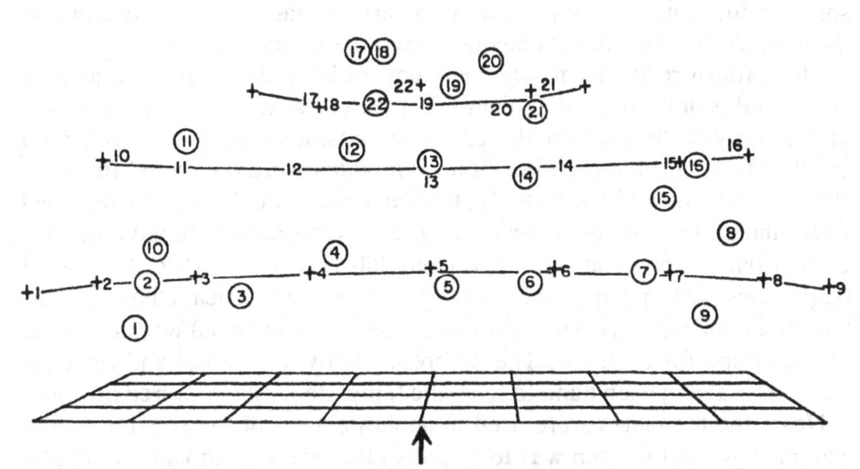

FIGURE 2.10. Perspective view (as in Fig 2.4) showing a comparison of perceived source locations obtained from one subject (S1) in freefield (numerals not enclosed by circles) and in simulated free field, under headphones (enclosed numerals). Judgments of perceived source locations were obtained in both conditions using the absolute-judment paradigm. Note the close correspondence between the positions indicated by the two sets of numerals.

ments, placement of the headphone caused the microphone to move (by S2's own report). Thus, free-field and headphone transfer functions were not measured from the same place. For the simulation to work, the same point in the ear canal must be sampled both in free field and under headphones. Small changes in microphone position cause relatively large changes in measured response, particularly at frequencies higher than about 5 kHz. With subject S1 microphone movement was not a problem. We are certain that the reason some simulations appear to be better than others is due to differences in microphone stability.

Spectral Pattern Classification and Sound Localization

Our measurements of outer-ear transfer functions (Figure 2-10) revealed large intersubject differences and substantial changes in the forms of the functions as the position of the source changed. The importance of these outer ear effects on localization has yet to be determined. One important question that we attempted to answer is whether there is sufficient information in the outer-ear transfer functions alone to allow correct identification of source locations. In an attempt to answer this question, we have followed a "pattern-recognition" approach similar to that used in automated speech recognition (Rabiner and Wilpon, 1979, 1980; Wilpon, et al, 1982). In short, we have used standard multivariate statistical techniques (Duda and Hart, 1973; Batchelor, 1974) to develop a system that can correctly localize sounds in space, based only on estimates of interaural spectral differences. Comparison of the errors made by this procedure to those made by the subjects has been extremely encouraging.

The pattern recognition scheme is described in detail by Perkins et al (1985) and is only summarized here. The outer-ear transfer functions (amplitude only) estimated for the 22 source positions were used to form a spectral template for each position. Templates were computed for three subjects, S1, S2, and S3. Initially, five measurements from each ear canal were made. The complete spectra were divided into 22 nonoverlapping critical bands spanning the range from 300 Hz to 14.2 kHz. The cutoff frequencies correspond to those beyond which the physical measurements tend to be unstable. Next, the average power (in decibels) was computed in each band for each ear. The difference between the left and right ear was computed in each band to form the template for each source position.

Next, the templates were used to generate a set of "normative" data. The purpose of this step was to establish that the 22 templates were statistically discriminable. For each subject we simulated 30 "observations" for each template. Each observation was normally distributed with its mean constrained to be the same as the average of the physical measurements. One standard deviation was chosen to be 1 dB, a value we selected as representative of the variability in the process of comparing a template to the spectrum of an incoming sound. Separate multivariate analyses of variance were performed on the data from each subject. In all cases the results indicated that sufficient information was available in the spectral measurements to distinguish the 22 source locations. The discriminant functions produced by this analysis provide an optimal linear combination of the critical band estimates. In each analysis the first discriminant function accounted for approximately 85% of the discriminating power. The coefficients of this function are nearly equal for all critical bands, with the higher-frequency bands receiving slightly greater weight.

In a manner similar to that described above, three sets of "test" data, each consisting of 15 observations, were generated. In this case, however, the "observations" were drawn from a normal distribution, with standard deviation chosen so that the proportion of errors committed by the classification procedure matched that for the subject performing the absolute judgment task in free field. These values were 1 dB for S1 and 2 dB for S2. Subject S3 did not perform the absolute judgment task; therefore, a 1-dB estimate was chosen based on his "near perfect" performance in the paired comparison task. Different sets of "test" data were used in each of three conditions; differing in the number of critical bands included wide band (full 22 bands), low-pass at 4 kHz (16 bands), and low-pass at 1 kHz (7 bands). The percentages of errors made by the classification procedure for S1 were 4%, 20%, and 59% for wide-band, 4 kHz and 1 kHz, respectively. Corresponding figures were 18%, 34%, and 71% for S2 and 0%, 2%, and 29% for S3.

For two subjects (S1 and S2) the proportion of classification errors and the source locations at which the errors occurred were compared with the errors made by the subject in the absolute judgment task. In the case of S1, 5% errors were observed in absolute judgments, and the classification procedure committed 4% errors. For S2, 20% errors were recorded in absolute judgments as compared with 18% errors in classification. A comparison of the source locations for which errors are made by the classification procedure and by the subjects themselves shows a high degree of correspondence. Of the six positions on which S1 made errors in absolute judgment, the spectral classification procedure made errors on five. The classification procedure erroneously identified 12 of the 13 source locations at which S2 made errors.

In summary, we have used a classification algorithm to extract information from the power spectrum measurements made in subjects' ear canals. This method has been quite successful both in matching the proportion of errors made by the subjects in absolute judgment and in identifying the source positions for which errors are made. All of this has been accomplished without the use of information about the phase spectra at the two ears.

Future Directions

The ability to simulate free-field listening under headphones allows many studies of human sound localization that were previously impossible. For example, it should be straightforward to study the relative importance of interaural time and intensity differences and thus directly test the duplex theory. Simple manipulations of the characteristics of the digital filters used in the stimulus synthesis algorithms could zero or randomize interaural time differences (for example) and leave interaural intensity differences unaffected. More complex variations on this theme could be used to study the perceptual relevance of the spectral patterning produced by the pinnae. However, before the simulation technique finds wide acceptance it will be important to find ways of simplifying the digital filtering algorithms. Their current complexity makes them far too time consuming for general experimental use. We are actively working on solutions to this problem.

Acknowledgments. This research was supported in part by grant BNS 8014144 from the National Science Foundation and grant NS-22060 from the National Institutes of Health (NIH)-National Institute of Neurological and Communicative Disorders and Stroke (NINCDS). MEP was supported by NIH-NINCDS training grant NS-07223. Portions of the work were presented at the 105th and 107th meetings of the Acoustical Society of America. The authors gratefully acknowledge the technical assistance in some of the research of Nancy Barker Walczak, and the editorial comments of Dr. Robert Lutfi, Mr. Larry Revit and Ms. Walczak.

References

Batchelor, B.G. (1974). Practical approach to pattern classification. New York: Plenum Press.

Batteau, D.W. (1967). The role of the pinna in human localization, Proc. Roy Soc. (Lond) B168, 1011, 158–180.

Blauert, J. (1969). Sound localization in the median plane, Acustica, 22, 205–213.

Blauert, J. (1983). Spatial hearing: The psychophysics of human sound localization, TUMIT Press, Cambridge

Bloom, P.J. (1977). Creating source elevation illusions by spectral manipulation, J. Audio Eng. Soc. 25, 560–565.

Butler, R.A. (1975). The influence of the external and middle ear on auditory discriminations. In: Handbook of Sensory Physiology, Vol. V/2; Auditory System, W. Keidel, W. Neff (eds.). New York: Springer-Verlag, pp. 247–260.

Butler, R.A., Planert, N. (1976). The influence of stimulus band-width on localization of sound in space. Percept. Psychophys. 19, 103–108.

Coleman, P.D. (1962). Failure to localize the source distance of an unfamiliar sound, J. Acoust Soc Am. 34, 345–346.

Duda, R., Hart, P. (1973). Pattern Classification and Scene Analysis. New York: John Wiley & Sons.
Feddersen, W.E., Sandel, T.T., Teas, D.C., Jeffress, L.A. (1957). Localization of high-frequency tones. J Acoust. Soc. Am. 29, 988–991.
Gardner, M.B. (1968). Proximity image effect in sound localization. J. Acoust. Soc. Am. 43, 163.
Gardner, M.B., Gardner, R.S. (1973). Problem of localization in the median plane: Effect of pinna cavity occlusion. J. Acoust. Soc. Am. 53, 400–408.
Green, D.M. (1983). Profile analysis: A different view of auditory intensity discrimination. Am. Psychol. 38, 133–142.
Hebrank, J., Wright, D. (1974a). Are two ears necessary for localization of sound sources on the median plane? J. Acoust. Soc. Am. 56, 957–962.
Hebrank, J., Wright, D. (1974b). Spectral cues used in localization of sound sources on the median plane. J. Acoust. Soc. Am. 56, 1829–1834.
Kistler, D.J., Wightman, F.L. (1985). A multidimensional scaling approach to auditory space perception. Percept. Psychophys. In press.
Lackner, J.R. (1983). Influence of posture on the spatial localization of sound. J. Aud. Eng. Soc. 31, 650–661.
Mehrgardt, S., Mellert, V. (1977). Transformation characteristics of the external human ear. J. Acoust. Soc. Am. 61, 1567–1576.
Mills, A.W. (1958). On the minimum audible angle. J. Acoust. Soc. Am. 30, 237–246.
Mills, A.W. (1960). Lateralization of high-frequency tones. J. Acoust. Soc. Am. 32, 132–134.
Morimoto, M., Ando, Y. (1982). On the simulation of sound localization. In: Localization of Sound: Theory and Applications, Gatehouse, R.W. (ed.). Groton, CT, Amphora Press.
Musicant, A.D. Butler, R.A. (1985). Influence of monaural spectral cues on binaural localization, J. Acoust. Soc. Am., 77, 202–208.
Oppenheim, A.V., Schafer, R.W. (1975). Digital Signal Processing. Englewood Cliffs, NJ, Prentice-Hall.
Perkins, M.E., Kistler, D.J., Wightman, F.L. (1985). Spectral pattern recognition and sound localization," J. Acoust. Soc. Am. In press.
Rabiner, L.R., Wilpon, J.G. (1979). Considerations in applying clustering techniques to speaker independent word recognition. J. Acoust. Soc. Am. 66, 663–673.
Rabiner, L.R., Wilpon, J.G. (1980). A simplified, robust training procedure for speaker trained, isolated word recognition systems. J. Acoust. Soc. Am. 68, 1270–1276.
Scharf, B. (1970). Critical Bands. In: Foundations of Modern Auditory Theory, Vol. I. J.V. Tobias (ed.). New York: Academic Press.
Schroeder, M.R., Atal, B.S. (1963). Computer simulation of sound transmission in rooms. IEEE Int. Conv. Rec. 11, 150–155.
Stevens, S.S., Newman, E.B. (1936). Localization of actual sources of sound. Am. J. Psychol. 48, 297–306.
Stockham, T.G. (1966). High speed convolution and correlation. Proc. AFIPS Spring Joint Comp. Conf. 28, 229–233.
Strutt, J.W. (Lord Rayleigh). (1907). On our perception of sound direction. Phil. Mag. 13, 214–232.

Thurlow, W.R., Runge, P.S. (1967). Effect of induced head movements on localization of direction of sounds. J. Acoust. Soc. Am. 42, 480–488.

Tribolet, J.M. (1977). A new phase unwrapping algorithm. IEEE Trans. Acoust. Sp. Sig. Proc. ASSP-25, 170–177.

Watkins, A.J. (1978). Psychoacoustical aspects of synthesized vertical locale cues. J. Acoust. Soc. Am. 63, 1152–1165.

Wightman, F.L., Kistler, D. (1980). A new 'look' at auditory space perception. In: Psychophysical Physiological and Behavioral Studies in Hearing Brink, G. v.d. Bilsen, F. (eds.). Delft, University Press.

Wilpon, J.G., Rabiner, L.R., Bergh, A. (1982). Speaker-independent isolated word recognition using a 129-word airline vocabulary. J. Acoust. Soc. Am. 72, 390–396.

Woodworth, R.S. (1938). Experimental Psychology, New York: Holt, Rinehart & Winston.

Wright, D., Hebrank, J., Wilson, B. (1974). Pinna reflections as cues for localization. J. Acoust. Soc. Am. 56, 957–962.

Zwislocki, J., Feldman, R.S. (1956). Just noticeable differences in dichotic phase. J. Acoust. Soc. Am. 28, 860–864.

3
Lateralization

William A. Yost and Ervin R. Hafter

In the real world sounds arrive at the two ears with interaural differences of time and intensity and with differences in the spectral pattern of information. George Kuhn, in Chapter 1, showed that these interaural variables can be complex. Wightman et al, in Chapter 2, showed that this complexity may be important for a listener's ability to localize sounds in space. Presenting these stimuli over headphones allows for exact control of these variables. For instance, a pure-tone sound source in the free field generates both an interaural intensive and a temporal difference. Thus, it is difficult to vary one interaural parameter independently of the other. If the stimulus is delivered by headphones, the tone can be presented with an interaural temporal difference independent of the interaural intensive difference. Anyone who has listened to stimuli presented dichotically (see Table 3-1 for a discription of diotic, monotic, dichotic, etc) over headphones has noticed that the sound image appears to be "inside" the head rather than "out" in the environment. When binaural hearing is studied with headphones, the task is referred to as *lateralization*. For sounds presented externally to the listener such as via a loudspeaker, the task is referred to as *localization*. In this chapter we describe some of the basic data and theories that pertain to lateralization. The first section covers lateralization of simple stimuli, primarily sinusoids. The second section describes the situations in which the stimuli are more complex.

Simple Stimuli

Lateral Position

Perhaps the most obvious question concerns the relationship between the lateral perception of sounds and the variations in interaural parameters. Studies of localization show that a sound at a particular location is represented at the ears with a certain interaural configuration of time and intensity (for sinusoids we can ignore spectral patterns). Figure 3-1 is a general description of the relationship between the location of a sinusoidal

TABLE 3.1. Definition of binaural terms.

Monotic or Monaural: Refers to a stimulus presented to only one ear.
Dichotic: Refers to stimuli that are different at the two ears.
Diotic: Refers to stimuli that are identical at the ears.
Binaural: Refers to any stimulus that is diotic or dichotic.

sound in the free field and the perceived lateral position when the sound is presented over headphones. This figure is based on work by Feddersen et al (1955) and Molino (1970). The tone is perceived toward the ear, which receives the sound first and with the greater stimulus intensity.

These studies were based on the interaction of interaural time and intensity. Figures 3-2 and 3-3 show data from a variety of studies in which either interaural intensity was varied and the interaural time difference was zero (Fig. 3-2) or interaural time was varied and the interaural intensive difference was zero (Fig. 3-3). In general, the listeners in each study were asked to indicate the lateral position of an image. A scale representing the head and locations between the ears provided the matrix for indicating the lateral judgments. The perceived lateral location is plotted as a function of the interaural variable of intensity in decibels (Fig. 3-2) or interaural phase in degrees (Fig. 3-3). The perceived lateral location is scaled as a

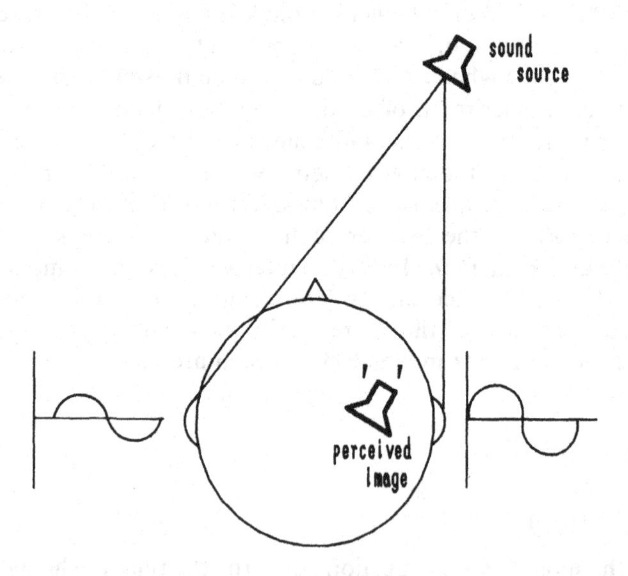

FIGURE 3.1. A schematic diagram indicating that sound sources located toward one ear produce a higher level signal that leads in time at the ear closest to the source. Presenting interaural temporal and intensive differences over headphones results in perceived locations inside the head, such that an image is perceived toward the ear with the greater intensity and which is presented the stimulus first. (Based on work by Feddersen et al, 1955, and Molino, 1970.)

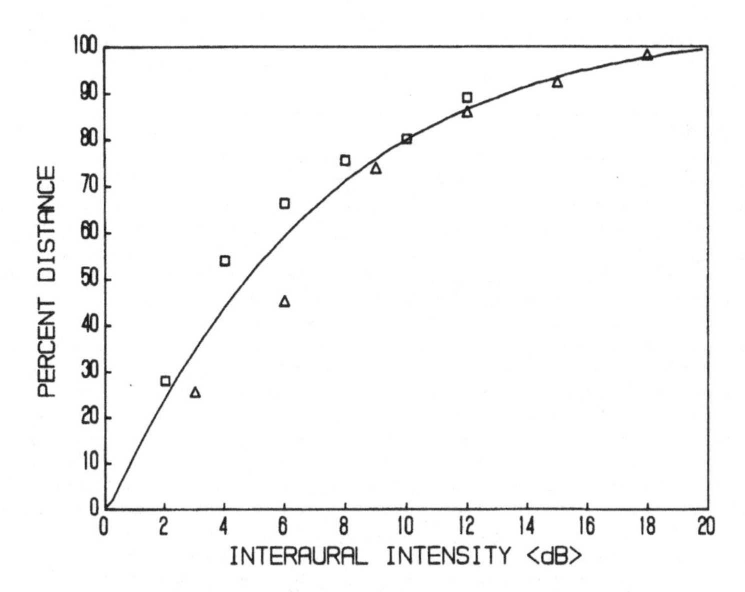

FIGURE 3.2. The perceived location of pure tones as a function of the interaural intensive difference. The perceived location is normalized across the studies in terms of the percent distance from an image lateralized at midline (0%) and one lateralized entirely toward one ear (100%). The □ data are from Watson and Mittler (1965) and the △ data from Yost (1981). The curve is an approximate fit to the data. The results are plotted for only one side of the head (as if the tone were only presented with one ear louder). Yost (1981) showed that the perceived locations were symmetric across the two sides of the head when one ear or the other was presented the more intense tone.

percentage of the distance from the middle of the head to one side. Different procedures and scales were used in the various studies, but the results allow for a normalization of the perceived location as shown in the figures. In those studies in which the images could occur on both sides of the head, the perceived locations appeared to be symmetric across the two sides. For this reason, the data for only one side of the head are shown. The figure captions list the various studies and provide additional information about the figures.

For interaural intensive differences of less than 10 to 12 dB, there is an almost linear relationship between perceived lateral location and the interaural intensive difference expressed in decibels. For interaural intensive differences greater than 12 dB, there is less change in lateral location for each additional decibel change in interaural level. Although most studies imply that interaural intensive differences of 15 to 20 dB determine the limit of lateral sensitivity to interaural level, Hafter (1977) showed that very large (greater then 30 dB) interaural intensive differences still had an effect on a listener's performance. However, for these large interaural

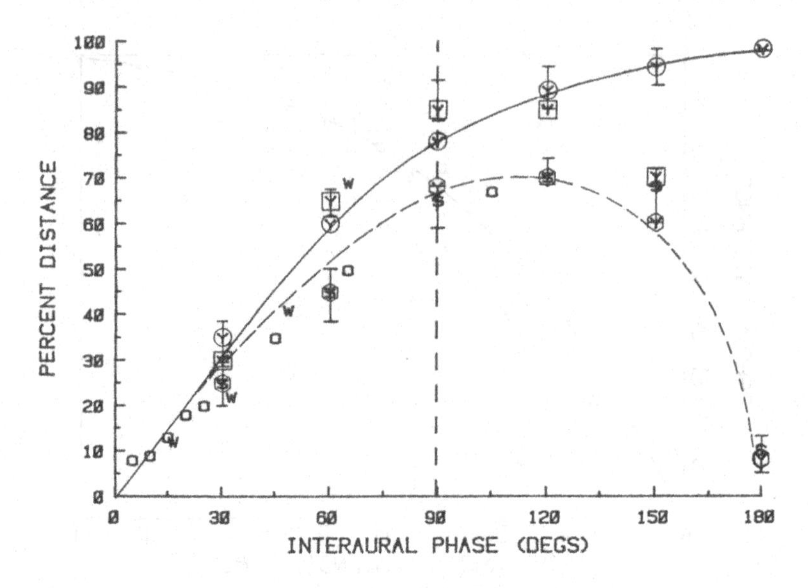

FIGURE 3.3. A similar plot as shown in Figure 3.2, except for interaural temporal differences. The O data are from Osman (1984); W from Watson and Mittler (1965); S from Sayers (1964); Y from Yost (1981) (the circled Y data are the modal data of this study and the squared Y's are the arithmetic mean results); and T from Yost, Turner, and Bergert (1974). The solid curve is a fit to those studies that report the lateral position using the mode, while the dotted curve fits those data reported as means. Only data up to 180° are shown. These tend to represent the lateral position out to one ear. As the interaural phase difference increases beyond 180°, the image appears on the other side of the head, as indicated on the lower abscissa. At or near 180°, Yost (1981) and Sayers (1964) showed that the listener often perceives two images or a diffuse image on both sides of the head (see Figure 3.4).

level differences there is an extremely small change in perceived location with changes in interaural level. Yost (1981) measured essentially the same relationship between interaural intensive difference and perceived location for a wide range of frequency, overall level, and stimulus duration. Although the overall tonal level and duration in the Watson and Mittler (1965) study were different from those used by Yost (1981), similar data were obtained. Thus, these few data suggest that the relationship between lateral position and interaural intensity is approximately the same over a considerable range of frequency, overall level, and duration.

As the data in Figure 3-3 indicate, the relationship between lateral position and interaural time appears more complex than that shown in Figure 3-2. There is a psychophysical procedural difference among the studies. For some studies the listeners were presented only one stimulus per trial for a lateral position judgment. In the other studies two stimuli were presented per trial: a standard stimulus and the stimulus to be judged. The distributions of the perceived locations shown in Figure 3-4 help explain the differences between the two curves in Figure 3-3. Each panel of Figure 3-4 shows the number of times (frequency) the listener indicated a particular lateral location for a particular stimulus condition. The lateral locations are scaled such that 0 is in the middle of the head, 10 is at one ear, and -10 is at the other ear. Each panel represents a particular interaural phase difference (0°, 60°, 120°, and 180°).

The two studies are by Yost (1981) in which two stimuli (standard and test) were presented per trial; and by Sayers (1964) in which only one stimulus (the test) was presented per trial. Notice that as the interaural phase difference increases (as the image is located further from the midline), the distribution of responses becomes bimodal, indicating lateral positions on both sides of the head. In the Sayers (1964) study the listeners sometimes indicated that the stimulus was at midline when it was presented

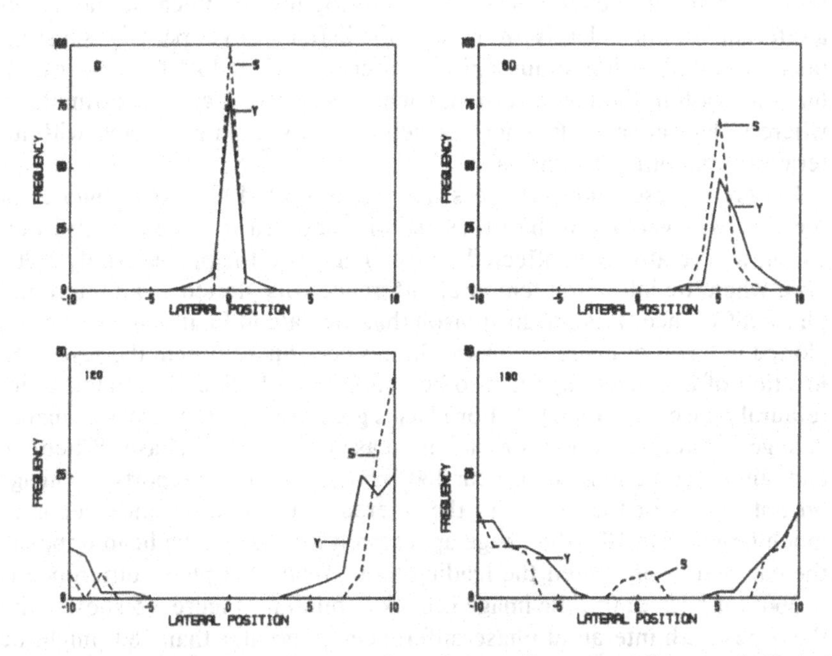

FIGURE 3.4. Distributions of the number of times (frequency) listeners indicated a particular lateral location for four different interaural phase shifts (the four different figures). 0 represents the middle of the head, 10 toward one ear, and -10 toward the other ear. The Y data are from Yost (1981) and S from Sayers (1964). The results show that for 0° the images are perceived at midline. As the interaural phase shift approaches 180°, the image is diffused, appearing on both sides of the head. Without a referent stimulus listeners sometimes perceived an interaural phase shift of 180° as being at the midline, as indicated by the results from Sayers (1964), the S data points.

with a 180° interaural phase shift. The listeners in the Yost (1981) study never made this judgment when they were given two stimuli per trial. If the arithmetic mean is used to average the data and to indicate the lateral location, then the "average" perceived location for an interaural phase difference of 180° would be near midline. In the Sayers study there were very few data indicating that the listener perceived the image at midline, and in the Yost study the listener never perceived the image at midline for these large values of interaural phase. The dashed curve in Figure 3-3 represents data that were reported using the arithmetic mean as the indicator of lateral position. The solid curve in Figure 3-3 represents data that were reported as the *modal* response. The modal lateral position is the lateral position that the listener reported *most* often. When the distributions are symmetric and unimodal, as they are for small values of interaural phase (as in the upper left-hand panel in Fig. 3-4), then the mean and the mode are very similar. However, for bimodal distributions the mean and mode can disagree by a great deal, as indicated by the difference in the dashed and solid curves in Figure 3-3. Because the use of a referent stimulus clearly shows that the listener never perceives a stimulus presented with interaural phase differences near 180° to be near midline, computing the mean response may lead to an incorrect estimate of where in lateral space listeners perceive tonal stimuli presented with interaural temporal differences.

In two of these studies (Sayers, 1964, and Yost, 1981) the frequency of the tone was varied. Both of these studies showed that the perceived location was relatively unaffected by frequency (up to approximately 1,200 Hz), when the interaural temporal difference was plotted as an interaural phase difference. It is for this reason that the data in Figure 3-3 have been plotted in terms of interaural phase. In this coordinate system the perceived location of a sinusoid appears to be approximately linearly related to interaural phases less than 90°. For phases greater than 90° there is a smaller change in lateral location for each increase in interaural phase. When the interaural phase difference is near 180°, the listener often reports the image on both sides of the head. As the interaural phase difference becomes much larger than 180° the image appears on the side of the head opposite the ear that is presented the leading tone. When the phase difference is at 360° (or back at 0°), the image is back at midline. Figure 3-5 shows why the tones with interaural phase differences of greater than 180° might be lateralized on the opposite side of the head from tones with phase differences less than 180°.

Yost (1981) measured the relationship between lateral position and interaural phase for a variety of overall levels and stimulus durations and found approximately the same results. In addition, there are differences among the studies reported in Figure 3-3 in overall level and duration and yet there is fairly good agreement in the results. Thus, it appears that the results shown in Figure 3-3 describe the approximate relationship between

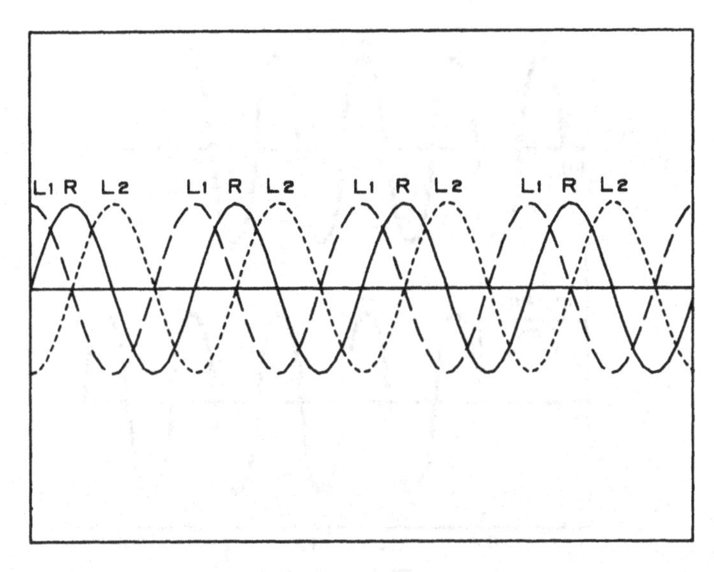

FIGURE 3.5. Two sinusoids: one with a phase shift of 90° (L1 as one possible sound going to the left ear) and one with a phase shift of 270° (L2 as another possible sound going to the left ear) are compared to a zero phase sinusoid (R for the sound going to the right ear). The R sinusoid appears to lead to the L2 sinusoid, which may cause the listener to perceive this 270° (L1) sinusoid as coming from the opposite side as the 90° (L1) sinusoid.

perceived lateral position and interaural time for a relatively wide range of frequencies (up to 1,200 Hz), overall levels, and durations.

As the tonal frequency increases beyond 1,200 Hz, interaural phase or interaural time does *not* influence the lateral position of sinusoids. Abel and Kunov (1983) demonstrated this fact in a series of recent studies. The reason for the lack of sensitivity to interaural time at frequencies above 1,200 Hz is presumably due to the location ambiguity provided by high-frequency, pure tones. Figure 3-6 describes this ambiguity, assuming the tone was presented in the free field. Because of the time-width of the head (time it takes the sound to travel from one ear to the other), the stimuli at the two ears appear identical after the first period, even though the stimulus has been presented opposite the right ear. For this frequency and any higher frequency the auditory system would not be able to use the interaural time difference for accurately determining the source of the sound. This same explanation is used to account for the fact that even over headphones (where the time-width of the head might not be a factor), one cannot perceive a change in interaural time for frequencies greater than approximately 1,200 Hz. The cause of the ambiguity is either that the nervous system is wired according to the time-width of the head or that the nervous system cannot accurately follow the time course of pure tones at frequencies higher than 1,200 Hz (see Hafter, 1977). The fact

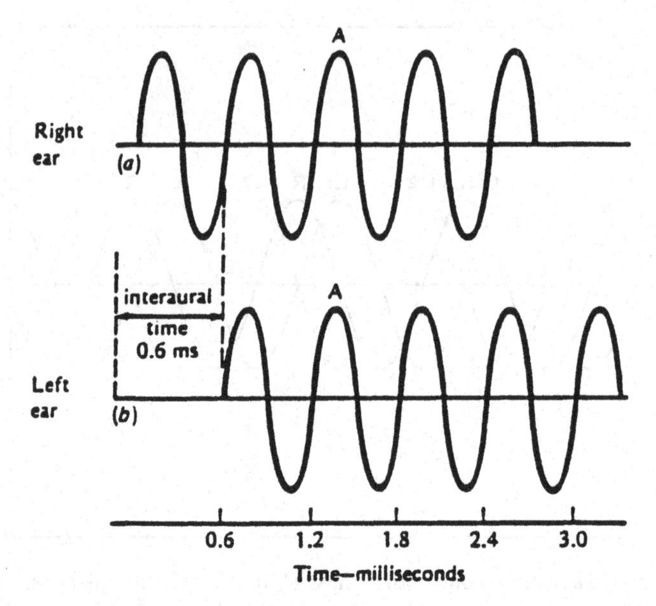

FIGURE 3.6. A 1,666 Hz sinusoid is shown at one ear and the same sinusoid interaurally delayed by 600 μ is shown at the other ear. Because 600 μs is the period of 1,666 Hz, the two sinusoids are identical after one period. If the binaural system listened at any time except during the first period it would assume the tone was at midline because there are no interaural differences. A 600-μs interaural time difference occurs when a sound source is placed opposite one ear. Therefore, the system could not localize this tone under these conditions. This is an extreme example of the ambiquity provided by interaural time differences for frequencies above 1,000 to 1,200 Hz.

that interaural time is not a binaural cue at high frequencies is a major aspect of the "duplex theory of localization," about which we will have more to say later in this chapter.

An interaural time difference may exist for the entire waveform, for just the onset of the waveform, for just the offset, or only during the tone presentation (a phase shift). Thus, there can be onset, ongoing, offset, or some combination of these interaural temporal differences. The results cited above occur for ongoing temporal differences when the onset or offset temporal differences are not more than a period or two of the pulsed sinusoid. When the onset difference becomes long, the stimulus condition becomes similar to those used to study the precedence effect, as explained in Chapter 4 by Zurek.

Trading and Equivalence Ratios

Data from Figures 3-2 and 3-3 indicate that tones appear at particular locations owing to an interaural differences of time or of intensity. For instance, according to Figure 3-2, approximately 4 dB of interaural intensive difference should place an image 50% of the distance between the midline and one ear. From Figure 3-3, a 45° interaural phase shift should place a tone at about this same lateral location. Would a listener judge these two stimuli as having the same position? The results in Figure 3-7 show some of these comparisons. The squares (□) and octagons (O) are replotted from Figures 3-2 and 3-3. For the data in Figure 3-7, the values of interaural intensity (from Fig. 3-2), which led to the same lateral location as some value of interaural time (from Fig. 3-3), are plotted as a function of the interaural temporal difference. The other data points (◇, Δ, +) represent data from studies in which listeners were directly asked to indicate the amount of interaural time it took for a lateral image to have the same location as that produced by a particular interaural intensive difference. The ◇ and + data points are two listeners from the study by Domnitz and Colburn (1977), and the Δ data points are the average data from three subjects in the Yost et al (1975) study. The two listeners from the Domnitz and Colburn (1977) study indicate the variability that can be

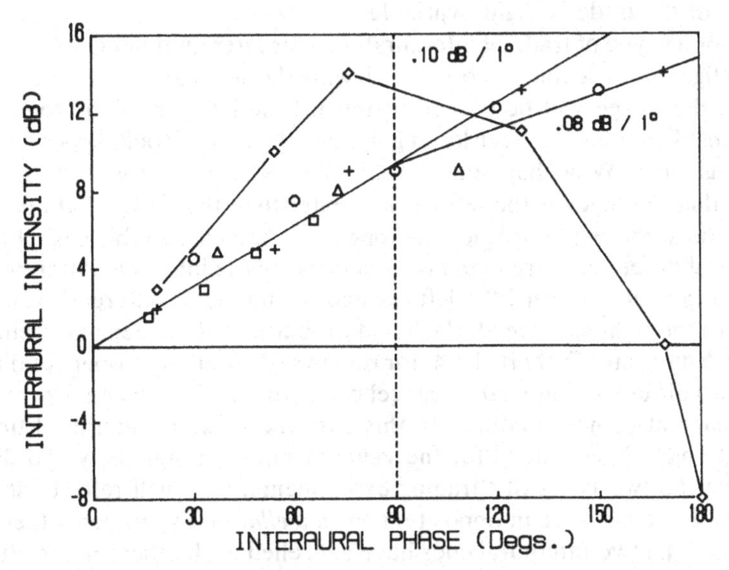

FIGURE 3.7. Various studies in which the interaural temporal difference required to place an image at one lateral position is plotted as a function of the interaural intensive difference required to place the lateral image at the same location. The squares [□] and octagon (O) data are from Figures 3.2 and 3.3 and indicate the values of interaural time and intensity required to place the images at approximately the same location. The Δ data points are from Yost et al (1975), and the + and ◇ data points are two subjects from the study by Domnitz and Colburn (1977). In these studies listeners were asked to determine if an image produced with one interaural variable was in the same location as that produced with the other variable. A variety of trading ratios are drawn as lines through the data. In these cases the trading ratio is the ratio of the interaural intensive difference to the interaural temporal difference required to place an image at the same lateral location.

obtained in these matching procedures. The difference in these two subjects' performance might be related to the observations made concerning Figure 3-4. At large values of interaural phase the image appears to be on both sides of the head (as seen in Fig. 3-4). If the listener is asked to match the lateral position of this tone to a tone presented with an interaural intensive difference (which has only one location), which image(s) does he or she use for the match? The + data might be for a listener who picked the image on only one side of his or her head. The ◇ data might be for a listener who was confused. He or she might have attempted to use the image on one side of the head part of the time and the image on the other side for the other matches. Although we do not know why these two subjects differed, the fact that lateral images (for interaural phase shifts greater than 90°) are found on both sides of the head for large values of interaural temporal differences makes the interpretation of these matching results difficult. Up to approximately 90° of interaural time the "trade" of interaural intensity for interaural phase appears to be approximately 0.10 dB/1°. Beyond 90°, where ambigious images may occur, the value of the trade is highly variable.

Another type of trade of interaural time for interaural intensity has been investigated. If a tone is presented with the left ear having the higher level, the image will be perceived toward the left ear. If the tone were presented with the right ear leading in time, the image would appear toward the right ear. What happens if a stimulus is presented with both cues, such that the level at the left ear is greater than that at the right ear *and* the tone at the right ear leads the tone at the left ear? In this case the two binaural differences are in *opposition* because the intensive difference indicates a source toward the left ear and the temporal difference a source toward the right ear (see Durlach and Colburn, 1978, Green and Henning, 1969, Young and Carhart, 1924, for reviews). Under the proper conditions the two differences appear to cancel each other and the listener perceives an image at or near midline. In this case the value of interaural time is said to have been traded for the value of interaural intensity. To differentiate the two types of "trading experiments," we will refer to studies in which the cues are in opposition as *cancellation* experiments (because it is as if the two interaural cues have cancelled each other) and to studies in which the cues are not in opposition as *equivalence* experiments (because the trade between interaural time and intensity indicates the magnitude of each cue necessary for perceiving an image at an equivalent position).

In addition to the confusions cited above, when stimuli are presented with both interaural differences, there is evidence that listeners hear two images (one associated with the interaural intensive difference and the other with the interaural temporal difference) in cancellation experiments. Even if listeners do not report two images, the data suggest that both images appear to effect their performance independently. Hafter and Jeffress (1968) provided one demonstration of the influence of both images

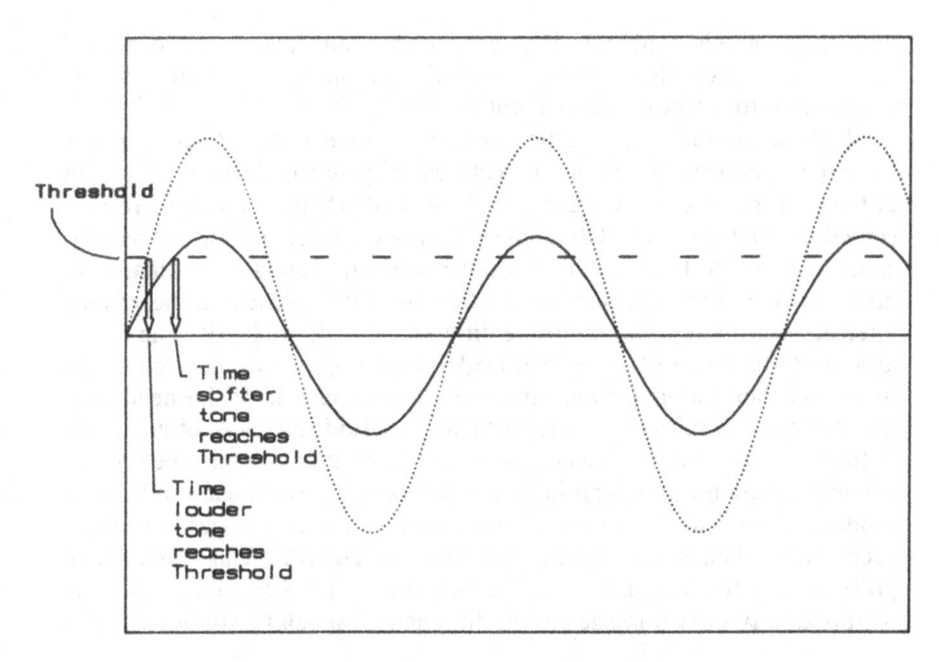

FIGURE 3.8. A schematic indicating how interaural intensity might be translated into interaural time within the nervous system. The more intense tone reaches the threshold level before the less intense tone, resulting in an apparent temporal difference between the two ears.

in these trading experiments. Listeners were asked to adjust the interaural time difference to offset a 9-dB interaural intensive difference. They were asked to determine that "trade" for the right-most image if they heard two images. On half the trials (L for left ear more intense) the left ear had the more intense tone and the right ear led in time. On the other half (R for right ear more intense) of the trials, the opposite occurred: the right ear had the greater intensity tone and the left led in time. There was a large difference between the two trading ratios obtained for the L and R trials. Because of this difference, the results substantiate the idea that listeners hear two images in these cancelling experiments. In addition to these problems, Trahiotis and Kappant (1978) described other potential procedural problems in measuring binaural trading ratios.

For both the cancellation and the equivalence trading experiments, a trade might be possible because the two interaural values are encoded by the same process at some level within the nervous system. For more than 100 years (see Green and Henning, 1969) it has been suggested that interaural intensive differences may be coded in terms of interaural temporal differences, as shown in Figure 3-8. That is, the more intense signal stimulates the nervous system sooner relative to the weaker stimulus. Thus an interaural level difference might appear to the nervous system as an interaural temporal difference (see Hafter and Carrier, 1972, for additional discussion of this concept). The two independent images, cited above, appear to be in conflict with such encoding in that both interaural differences seem to affect lateral perception.

All these lateralization studies indicate a great deal between subject variability and usually require a great deal of practice. Listeners have no difficulty reporting an image inside their head when pure tones are presented over headphones. Obtaining accurate estimates of that image's location has proved to be a very difficult psychophysical task. Qualitatively, however, the image appears toward the ear that is presented the greater intensity up to interaural intensive differences of 15 to 20 dB. The image appears to be toward the ear that leads in time up to half a period of the tone, and then the image appears on the opposite side of the head. For interaural temporal values near one half a period, listeners often report diffuse images on both sides of the head. Attempts to determine a trade between interaural time and intensity, either in terms of each difference producing the same perceived position or the two differences cancelling each other when in opposition, have proved difficult. This difficulty is probably due to ambiguity about the position of the lateral images and/or the possibility of two images in the trading or cancelling studies.

Lateral Discriminations

The results described above are analogous to asking where a listener perceives a sound in space. In addition to these results, there are excellent data on the acuity of the binaural system in discriminating between two different sound source locations (the minimal audible angle, MAA, as explained in Chapter 2 by Wightman et al). Over headphones the analogous question concerns the discriminability of the interaural differences of time and intensity. In localization, humans can detect a 1° angular separation between sound sources when the sources are directly in front and a 5° to 7° separation when the sound sources are off to one side of the listener. Figures 3-2 and 3-3 indicate that the perceived lateral location of a sound can be moved from midline toward one ear by introducing an interaural intensive or temporal difference. Thus, in a manner similar to the MAA experiments, one can determine the just discriminable change in interaural time from some referent interaural temporal difference or the just discriminable interaural intensive difference from some referent interaural intensive difference. The referent interaural differences are used to place the lateral image at different locations within the head.

Figure 3-9 displays the just discriminable interaural intensive differences and Figure 3-10 the interaural temporal differences as a function of frequency, with the parameter on each curve being either the referent interaural intensive difference (Fig. 3-9) or the referent interaural temporal difference (Fig. 3-10). The lines are fit to the data of Yost (1981), and the various data points represent data from other studies.

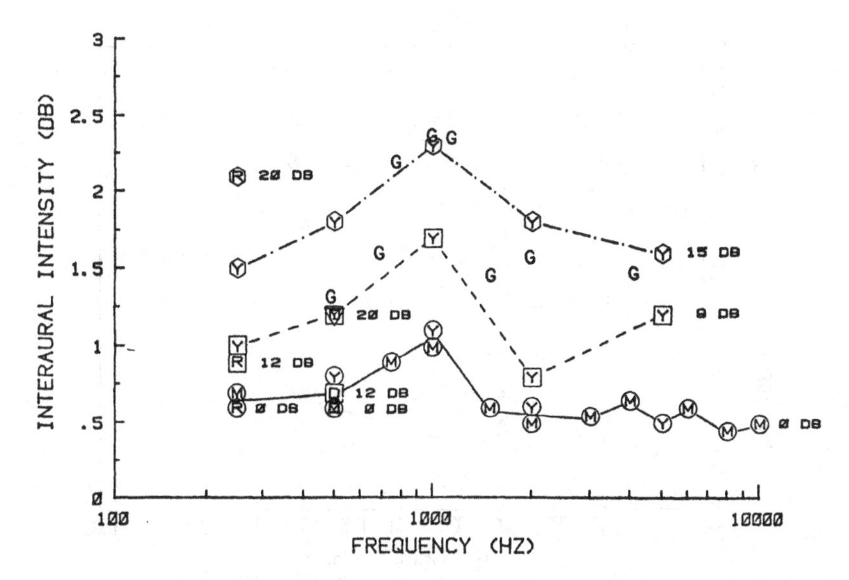

FIGURE 3.9. Interaural intensive difference thresholds as a function of the frequency of the tone. The parameter is the interaural intensive difference of the referent tone which serves to mark a position in lateral space. The Y data are from Yost and Dye (1987), the M from Mills (1960), the R from Rowland and Tobias (1967), the D from Herskowitz and Durlach (1969), and the G from Grantham (1984). The thresholds are relatively constant across frequency except at 1,000 Hz, and the thresholds increase as the interaural intensive difference of the referent increases (i.e., as the tone is lateralized further from midline).

As can be seen in Figure 3-9, the interaural intensive difference required for threshold increases near 1,000 Hz, when the referent interaural intensive difference is 0 dB. That is, when the image is in the middle of the head, the just discriminable interaural intensive difference is between 0.5 and 1.0 dB (except near 1,000 Hz). As the referent interaural intensive difference increases (as the referent is moved from midline toward one ear), the just discriminable interaural intensive difference increases. This means that as the image is moved off midline, the binaural system is less sensitive to interaural intensity.

Grantham (1984) has suggested that the increase in interaural intensive difference threshold near 1,000 Hz is due to an interaction of two underlying processes: one that operates well at high frequencies but not at low frequencies and another that operates well at low but not high frequencies. The spectral region near 1,000 Hz is the crossover region where neither process is functioning best. Grantham (1984) suggests two sets of processes that might function in this manner.

Although there are procedural and stimulus differences among the various studies, there is fairly good agreement among the results. A major concern in studies of interaural intensive discrimination is the possibility

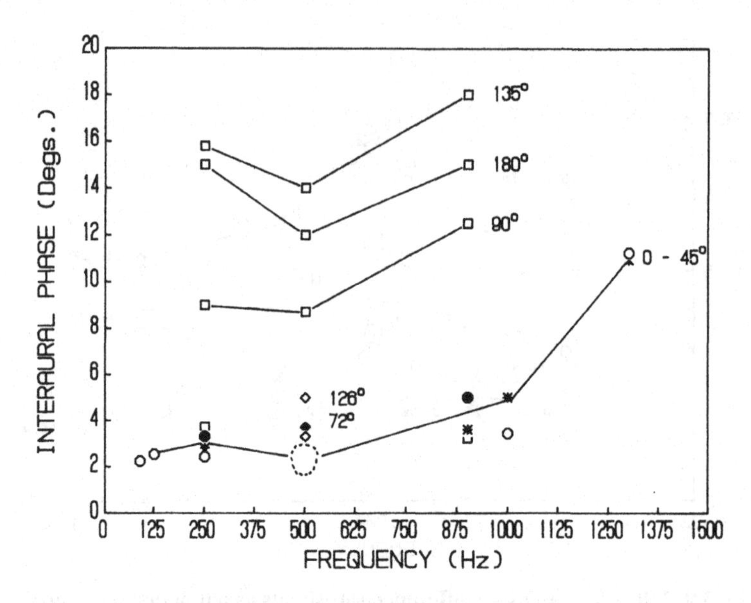

FIGURE 3.10. Interaural temporal difference thresholds (plotted as interaural phase differences) as a function of the frequency of the tone. The parameter is the interaural temporal difference of the tone. The parameter is the interaural temporal difference of the standard, which serves to mark a position in lateral space. The squares (□) are from Yost (1974), the closed circles (●) from Mills (1960), the diamonds (◇) from Herskowitz and Durlach (1969), the stars (⋆) from Zwislocki and Feldman (1956), and the open circles (○) from Klump and Eady (1956). The dotted large circle at 500 Hz contains data from Yost, Herskowitz and Durlach, Zwislocki and Feldman, Mills, and Klump and Eady. Data from the line through the data points represent data for standards at 135°, 180°, 90°' and all conditions between 0° and 45°. Interaural temporal difference thresholds, when expressed in units of interaural phase, remain approximately the same from 125 Hz to 1000 Hz. At 1000 Hz the interaural temporal difference thresholds increase for higher frequencies and interaural temporal differences thresholds are almost impossible to measure for frequencies above 1500 Hz. The interaural difference thresholds tend to increase as the interaural temporal difference of the standard increases, although there is considerable differences among the various studies.

that monaural level differences or binaural loudness cues might mediate the subject's performance. Imagine that stimulus level L is presented to both ears, resulting in a 0.0-dB interaural intensive difference for the referent stimulus. For the test stimulus the level at one ear is also L while the level at the other ear is $L + \Delta L$, generating an interaural intensive difference of ΔL. The listener is asked to discriminate between the test and referent stimulus with the magnitude of ΔL required for some criterion performance determining threshold. The subject might be able to make the discrimination by listening only to the ear that was presented $L + \Delta L$ as the test stimulus. The difference between the test and referent stimulus at this ear is a monaural intensity change. In addition, the overall loudness of the combined test stimuli from the two ears (L and L + Δ L) might be louder than the referent stimulus (L at one ear and L at the other ear). Thus, the obtained threshold (size of Δ L required for threshold performance) might be due to a monaural intensity cue, a binaural loudness cue, or the interaural intensive difference (the desired cue). By randomizing the overall level (L) at the ears, the monaural intensity cue and binaural loudness cue will not be correlated with the interaural intensive difference. Thus the listener must monitor the interaural difference in order to perform the task. Some of the discrepancy among the results from the various studies might be due to the fact that only Yost (1981), Grantham (1984), and Yost and Dye (1987), randomized overall level.

The data in Figure 3-10 show that the interaural temporal difference threshold is approximately 2° at low frequencies and increases as frequency increases. Above 1,200 to 1,500 Hz the binaural system is insensitive to changes in interaural time as explained earlier in this chapter. As the interaural temporal difference of the referent increases (as the image is moved off midline), the interaural temporal difference thresholds also increase (although there is very little change until the interaural phase difference of the referent is more than 45°). This increase in interaural temporal discrimination for tones lateralized off midline indicates that the binaural system is less sensitive to interaural temporal differences off midline. There is no clear explanation for the discrepancy among the various studies in terms of the magnitude of the shift in threshold when the interaural temporal difference of the referent is varied. As explained in the section on lateral position, images associated with interaural phase differences greater than 90° are diffuse and are often perceived in more than one position. This confusion might lead to a difficult discrimination for some subjects.

These data indicate that the binaural system is sensitive to one half of a decibel of interaural intensity and 2° of interaural phase. For some frequencies a 2° interaural phase shift corresponds to 10 μs of interaural delay. These small values demonstrate the high sensitivity of the binaural system to variations in the interaural configuration of intensity and time.

Lateralization Models

As Colburn and Durlach (1978) describe in their review chapter, most models or theories of lateralization (especially for sinusoids) may be viewed as various cross-correlation schemes. Rather than providing a review of the many models, we will explain the general concept of cross-correlation and then indicate some of the problems this type of model must overcome to account for the data cited in the first part of this chapter.

The basic concept of these binaural models is that the inputs at the two ears are cross-correlated: If $X_L(t)$ is the input to the left ear, $X_R(t,\tau)$ is the input to the right ear, τ is the interaural delay, and $R(T,\tau)$ is the cross-correlation:

$$R(T,\tau) = \int X_L(t)*X_R(t-T,\tau)\,dt \tag{1}$$

If X(t) is a sinusoid, then $R(T,\tau)$ is a sinusoidal function, as shown in Figure 3-11. The location of the peaks in $R(T,\tau)$ indicates the interaural delay (τ) introduced between the two ears.

Jeffress (1948) described a coincidence network, as shown in Figure 3-12, as a possible way the nervous system could perform such an analysis. The place in the network of coincidence corresponds to the peaks in the function $R(T,\tau)$, assuming T is mapped onto neural place. Thus the location of the peaks in the cross-correlation function, $R(T,\tau)$, or the place of maximal coincidence in the neural network, encodes the value of the interaural delay. The value of this derived interaural delay is used to indicate the location of the stimulus such that values near zero are near midline and large values indicate sound sources toward one side.

The following aspects of the cross-correlator or the coincidence network must be considered in order for these models to account for the data cited in the first part of this chapter:

(1) As already mentioned, T in the function $R(T,\tau)$ must be assumed to map a decision axis or neural location.

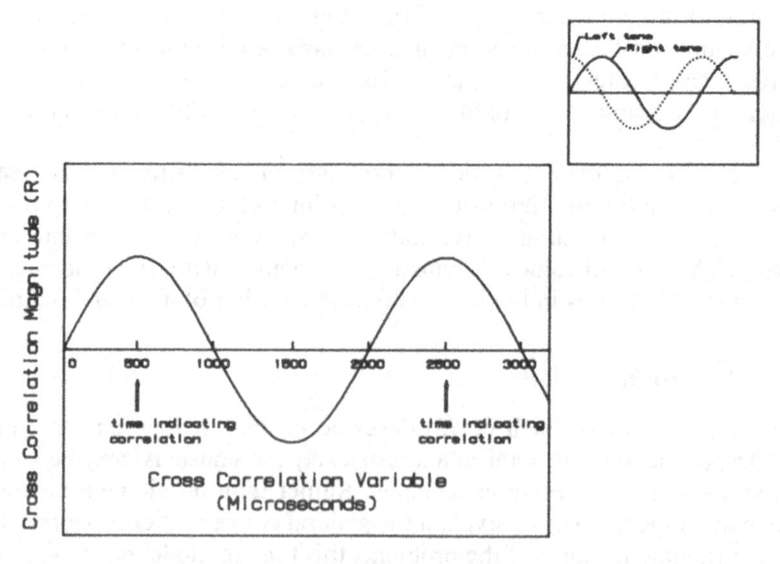

FIGURE 3.11. The cross-correlation function of two 500-Hz sinusoids presented with a 500 μs (90°) interaural delay. The peak at 500 μs in the cross-correlation function can be used to code stimulus location.

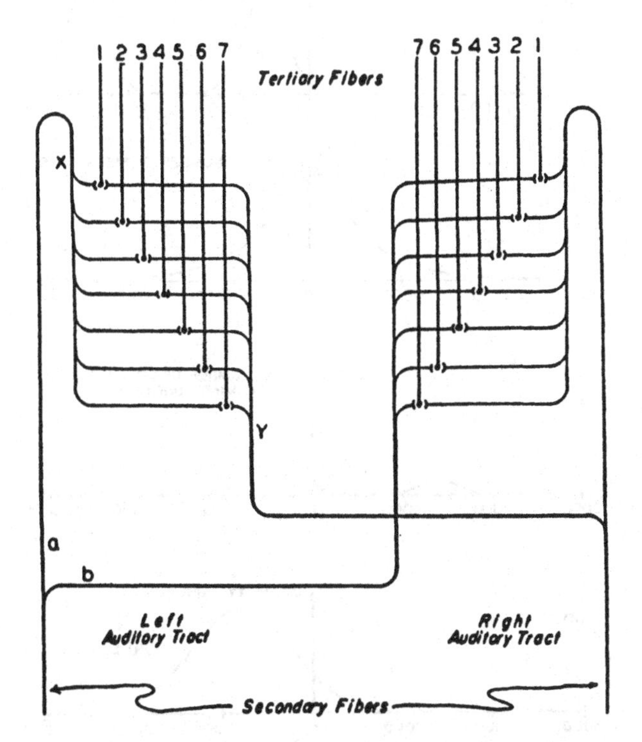

FIGURE 3.12. A coincidence network that might perform the type of cross-correlation described in Figure 3.11. The points (1–7) of coincidence occur at locations in the network which could decode the location of the sound source fed to the two auditory tracks. (From Jeffress, 1948, with permission.)

(2) Cross-correlation accounts only for interaural time; some mechanism is required to account for interaural intensity.
(3) The cross-correlation function has multiple peaks (indicating multiple possible delays and therefore multiple possible stimulus locations), yet the data suggest that often only one image is heard. Some assumptions must be added so that for cases in which there is only one delay and only one stimulus location, there is only one peak in the derived cross-correlation function or coincidence network.
(4) The data (see Fig. 3-3 and 3-10) suggest that stimulus location and interaural time discrimination are frequency dependent. Cross-correlation predicts that the same interaural time delay will produce the same lateral location for all frequencies. Because lateral location based on interaural time appears to be frequency dependent, the cross-correlation models must be similarly frequency dependent.
(5) The cross-correlators or coincidence networks should account for the insensitivity of the binaural system to interaural time for frequencies above 1,200 to 1,500 Hz.

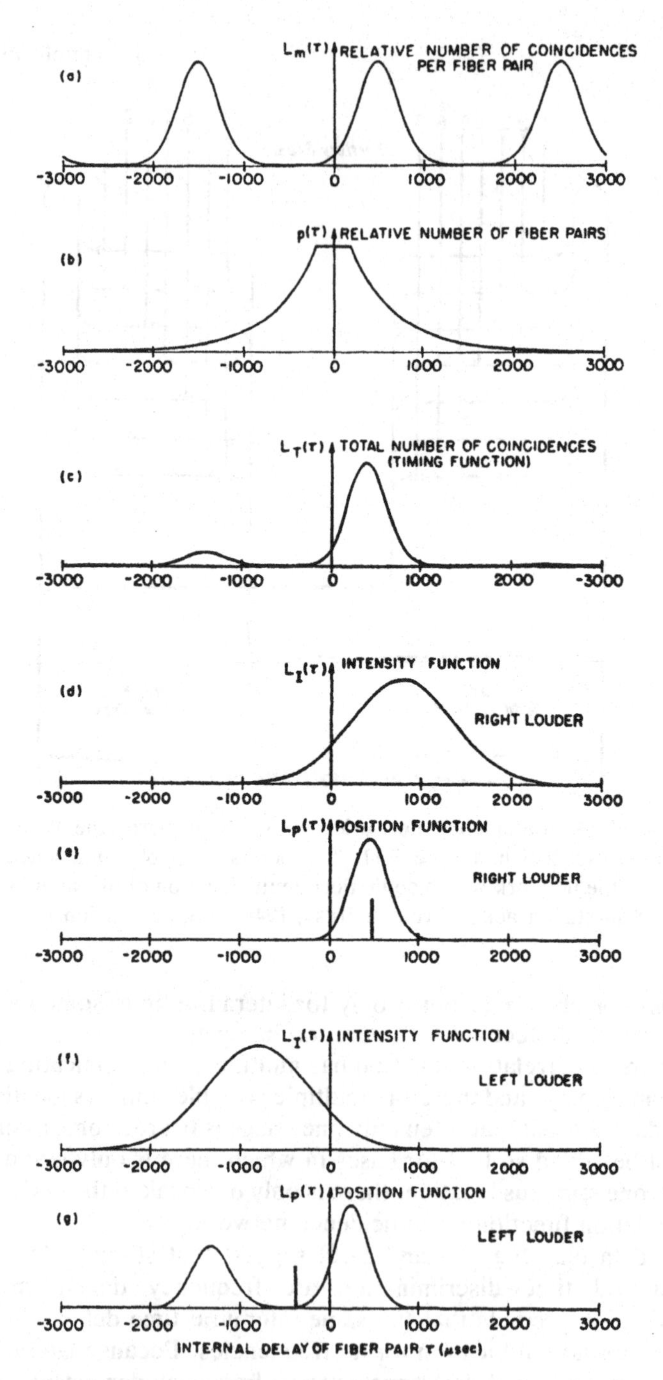

FIGURE 3.13. The model proposed by Stern and Colburn (1978, with permission), which is based on cross-correlation. See text for an explanation of the weighting functions that reduce multiple peaks (panel b) and account for interaural intensity (panels d and f). The resulting patterns (panels e and g) are used to predict the location and diffusness of the lateral image.

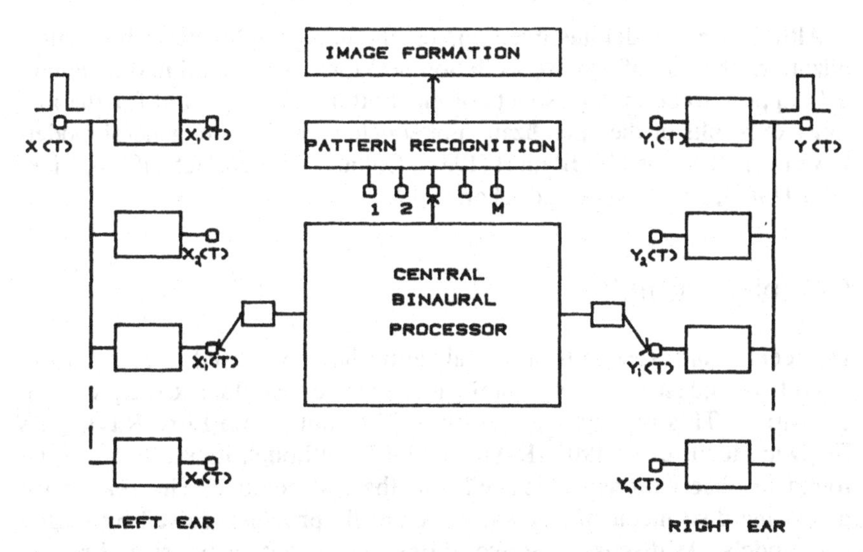

FIGURE 3.14. An overall model of binaural processing in which the signals, x (T) and y (T), at each ear are divided into frequency channels, X_m(T) Y_m(T), whose outputs are cross-correlated in the central processor. The cross-correlation functions are weighted by interaural intensity variables. The information is combined to form a pattern of interaural differences as a function of the channels (1,2,...M).

Figure 3-13 shows the result of the model by Stern and Colburn (1978), which attempts to deal with many of these five problems. Panel a shows the type of coincidences that result from cross-correlation of a 500-Hz pure tone with a 500-μs interaural delay (the function in panel a is similar, but not identical, to the cross-correlation function shown in Fig. 3-11). Panel b indicates a weighting function that reduces the number of coincidence peaks as seen in panel c. Intensity weighting functions are then assumed (panels d and f) for cases in which the right ear or the left ear is presented the more intense stimulus. The resultant functions (panels e and g) indicate the assumed distribution of activity across the network, such that for panel g the stimulus might be perceived as having two locations (as appears to be the case for interaural delays near one-half period of the sinusoid), whereas for panel e there is only one location to the right of midline.

This model and others like it (see Coburn and Durlach, 1978) can account for the qualitative aspects of the results at any one frequency. Additional assumptions are required to account for the effects across frequency, especially the two mentioned in points 4 and 5 above. For instance, one might assume that each frequency channel has its own coincidence or cross-correlation network (this was part of Jeffress', 1948, earlier description). As such, a global model for binaural processing, especially as it pertains to lateralization, might be characterized as in Figure 3-14.

Although no model has been universally accepted for describing lateralization, the use of cross-correlation and the scheme outlined in Figure 3-14 appear to be major aspects of most attempts to account for the majority of results of the lateralization of simple stimuli. More general models have been described by Blauert (1983), Colburn and Durlach (1978), Licklider (1962), and Sayers and Cherry (1957).

Complex Stimuli

In recent years, research in lateralization has been used to reformulate one of the oldest theories in biological science, the duplex theory of localization. This theory is generally said to date from Lord Rayleigh's Sidgwick lectures of 1907 (Rayleigh, 1907), although its seeds are to be found in observations made well into the last century. The reason for proposing dual mechanisms was to reconcile problems raised by single-cue models. As discussed above, differences of intensity created by the head's shadow on the distal ear become vanishingly small in the lower frequency regions; yet localization with these frequencies is quite good. Conversely, an appeal to temporal cues must deal with the fact that for signals as low as 700 Hz, the potential for interaural phases in excess of 180° makes these differences untrustworthy. Rayleigh's solution included separate mechanisms, each responsible for a different portion of the spectrum. Accordingly, localization of low frequencies is based on interaural differences of time or phase, while localization of high frequencies relies on differences of intensity.

The duplex theory received a major thrust from Stevens and Newman (1936) in a classic study of localization in the free field. An echoic stimulation was achieved by perching listeners on a chair in the open air, and observers were asked to name the directions of tones originating from a circle of speakers placed in the horizontal plane. A plot of the errors of localization as a function of frequency showed that performance was worst for the midfrequency range, a fact taken as evidence of two independent mechanisms, with the low-frequency portion of the function presumably mediated by time and the high-frequency portion by intensity. More direct support for the duplex theory, or at least for the contention that time or phase cannot be used for high frequencies, has come from observations with pure tones presented through headphones. As shown in Figure 3-3, an interaural difference in phase in a low-frequency tone moves the fused image closer to the leading ear, but for frequencies above 1,200 to 1,500 Hz, there is no sensation of sidedness. For these tones, the image stays centered and the binaural system can be said to be truly phase-blind.

Why then should one question the duplex theory? The rest of this chapter attempts to show that its contention regarding high frequencies is generally invalid. (Hafter, 1984) Thus, while it certainly is true that we cannot detect interaural delays in narrow-band signals such as tones, the situation is quite different with wider bandwidth complex stimuli. Given the general acceptance of the more traditional model outside of the specialties of binaural hearing, it is surprising to note that sensitivity to interaural differences of time with transients and noise was demonstrated many years ago (e.g., Klumpp and Eady, 1956; Harris, 1960; Yost et al, 1971). Indeed, the most popular of recent tools, the amplitude modulated (AM) high-frequency tone, was used to study localization by Leakey, Sayers, and Cherry in 1958. However, the work that has had the largest impact on the direction of this research has come from Henning and his colleagues. In a series of papers (e.g., 1974; 1980; Henning and Ashton, 1981), Henning examined the interaural phase relations of components in sinusoidal amplitude modulation (SAM), asking how they affect the laterality of dichotic stimulation. SAM with a modulation frequency of f_m consists of a carrier frequency, f_c, and two half-power sidebands, one at $f_m + f_c$ and one at $f_c - f_m$. Figure 3-15 shows samples of stimuli that have been used in various experiments with SAM. In each panel, the two waveforms are for the right and left ears, respectively. The three types of interaural delay depicted are: in (a), a delay of just the carrier frequency; (b), a delay of just the envelope of modulation; and (c), a delay of the entire waveform. The inserts within each panel are from Henning (1980). They show the interaural phase spectra for each type of delay. Note that for a delay of the carrier only, as in (a), all components are delayed by the same phase. For (b), phases of the upper and lower sidebands move in opposite directions, with the upper sideband being delayed while the lower is advanced. This produces a so-called group delay, defined as the local slope of the continuous phase spectrum. Finally, for a delay of the entire waveform, sometimes called a pure phase delay, there are both carrier and group delays. A major thrust of Henning's work has been to show that lateralization with an interaural delay in the carrier affects lateralization with low-frequency carriers, whereas with high frequencies the envelope is critical. Obviously, a delay of the entire waveform produces some of each, and this works well at any carrier frequency. Another way of explaining these results is to say that purely high-frequency sounds can indeed be lateralized and, one presumes, localized, on the basis of interaural differences of time as long as the envelope is delayed.

Henning's fundamental paper was soon followed by experiments from McFadden and Pasanen (1976) and McFadden and Moffitt (1977) using as signals pairs of high-frequency tones and narrow bands of noise centered at high frequencies. Similarly, Nuetzel and Hafter (1976; 1981) used SAM to study limitations of bandwidth and duration. In nearly all cases, the basic phenomenon, that is, the ability to lateralize high frequencies on the basis of interaural time, has been repeated. Indeed, listeners can be found whose interaural thresholds are as small as those achieved with low-frequency pure tones at the frequency of the modulator. Finally, it

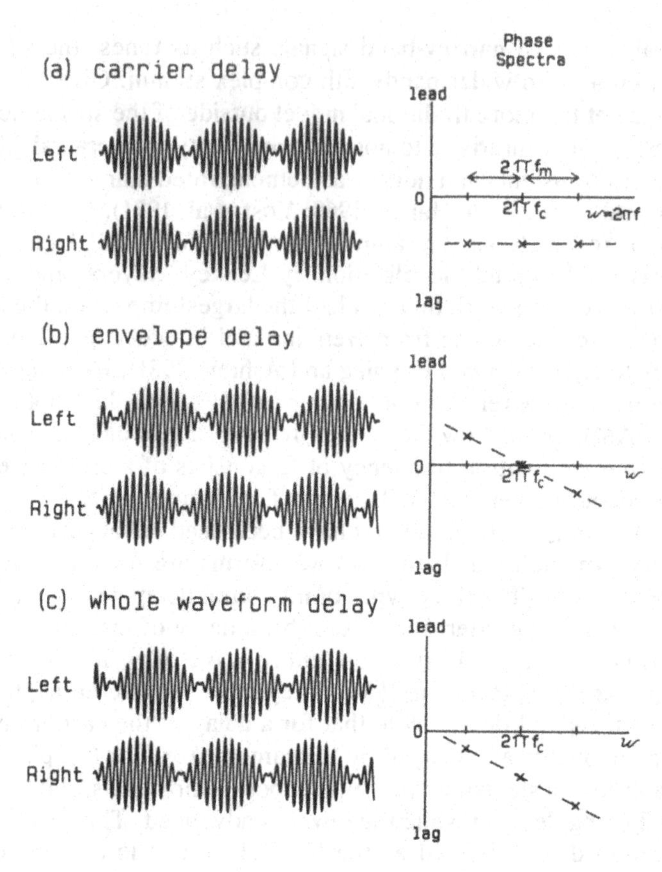

FIGURE 3.15. The time domain waveforms and the phase spectra for sinusoid amplitude modulated (SAM) tones. The modulation frequency is f_m and the carrier frequency is fc. The three types of delays—carrier only, envolpe only, and the whole waveform—are depicted. Notice that in each case there are important differences in the phase of the individual components of the three component SAM stimulus. (After Henning, 1980)

has been common to support the idea of insensitivity to temporal features of high-frequency signals by pointing to the inability of auditory neurons to entrain to the phases of very high-frequency tones. In Chapter 6, Kuwada and Yin show neurophysiological data on high-frequency timing, demonstrating as would be expected from the psychophysics, that single binaural units can encode interaural differences in the envelopes of AM.

One of the two basic messages to be derived from this chapter is that the ability to detect the interaural delay at high frequencies is neither ephemeral nor in doubt. Rather, the long life and continued existence of the duplex theory simply points to the dangers of building complex theories on the basis of simple stimuli. Now that it is clear that time is indeed a useful cue for complex high-frequency signals, the focus of the field has turned to questions about the nature of the underlying mechanisms. The work to be discussed here is restricted to studies using psychophysics (see Chapter 6 for discussions of physiology).

Information Conveyed by the Ongoing Stimulus

A useful trick for understanding the mechanisms of high-frequency timing is to reverse the question, asking what it is about high-frequency tones that prevents their lateralization by time. Hafter and his colleagues (e.g., Hafter and Dye, 1983; Hafter, Dye, and Wenzel, 1983; Dye and Hafter, 1984) have approached this issue by studying the lateralization of high-frequency AM as a simultaneous function of the rate of modulation (f_m) and of the number of peaks in the envelope. First, why rate? Consider any form of AM whose rate is so high that the carrier frequency and the sidebands, i.e., the components at f_c + integer multiples of f_m, fall into separate auditory filters. In this case, the depth of modulation in each filter goes to zero, and the stimulus becomes effectively a series of unlateralizable high-frequency tones. In a sense, then, the transition from low rates of modulation to high defines the change from AM to tones. As such, the relation between thresholds for interaural differences of time and the frequency of modulation helps to define an upper limit of temporal coding based on envelopes. However, it has been argued that these functions of performance versus rate are insufficient to give a full understanding of the underlying limitations with tones. In the sections that follow it will be argued that a more complete answer can be found by paying attention to an additional factor, the stimulus duration.

As noted earlier, most current models of interaural-time sensitivity call for a cross-correlation of inputs to like-frequency channels, much like that shown in Figures 3-12, 3-13, and 3-14. If the monaural channels leading to the point of binaural comparison were free of temporal noise or "jitter," there would be activity only at the point in the correlator corresponding to Δ t. Of course, this could not be the case in any real situation since the so-called inter-arrival times of neural impulses at the point of binaural interaction would be distributed over a set of elements whose central value was defined by the interaural delay but whose variance would be related to the sums of the two jitter-variances in monaural channels. The probability-density distributions in Figure 3-16 are meant to represent the binaural responses in the two intervals of a hypothetical forced-choice detection task. In this experiment, the dichotic stimulus leads to the left ear in one interval and to the right ear in the other. Thus, the means of the two distributions are at the mirror-image values. It is between samples drawn from such distributions that the listener must discriminate.

The information transmitted by a stimulus is inversely proportional to its variance. Where performance is defined as the value that produces a constant percentage correct and, by implication, a constant mean-to-sigma

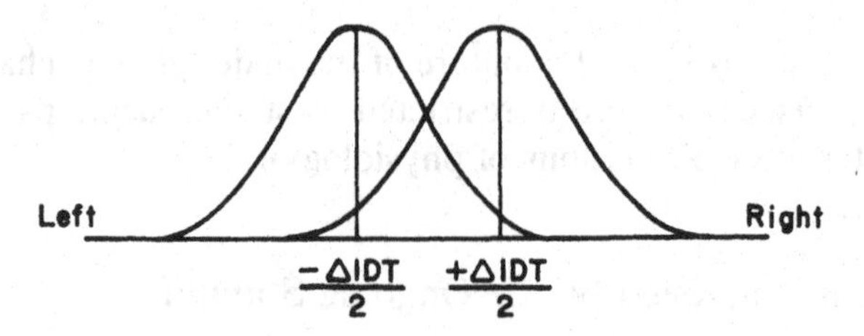

FIGURE 3.16. Theoretical distributions of interaural temporal differences for two values of interaural time (-ΔIDT and +ΔIDT), assuming that the binaural system encodes these variables with some normally distributed random jitter. The subjects' task is to discriminate between these two distributions.

ratio, the threshold values become a direct measure of the standard deviation of the internal, binaural noise. As such, a ratio of squared thresholds tells about the relative amount of information transmitted in each case. For example, if a stimulus transmits twice the information of another, the ratio of the two thresholds should be equal to the square root of 2. This idea was applied effectively by Houtgast and Plomp (1968) to a study of the detectability of an interaural delay in a low-frequency noise as a function of its duration. They argued that for a system capable of deriving information uniformly throughout its duration, the lengthening of the signal should produce a proportional reduction in the temporal variance; thus, the prediction of a reduction in threshold proportional to the square root of duration. Their tests were made against a background of unrelated noise at several values of signal-to-noise ratio, S/N. What they basically found was that the square-root relation held for low values of S/N, but that for signals presented in the quiet, the data took on a different form (a graphical representation of these data is presented later in the chapter).

The interaction between interaural thresholds and duration has also been applied to studies of high-frequency, complex stimuli (Nuetzel and Hafter, 1976; McFadden and Pasanen, 1976; McFadden and Moffitt, 1977; Hafter, Dye, and Nuetzel, 1980; Hafter and Dye, 1983; Hafter, Dye, and Wenzel, 1983; Hafter and Wenzel, 1983, Bernstein and Trahiotis, 1985). In one such paradigm, the stimuli are trains of high-frequency clicks, that is, clicks whose energy is restricted to the higher spectral regions. Figure 3-17 depicts the logic of these experiments with a schematic representation of the trains presented in a two-alternative, forced-choice paradigm. Simplified drawings show unfiltered impulses rather than the actual waveforms presented to the two ears. The information to be detected is interaural delay. Five parameters are needed to describe these trains: 1) the shape and 2) center frequencies of the filter used to produce each click; 3) the modulation rate, denoted here by its inverse, the interclick interval, (ICI); 4) the length of each train in number of dichotic clicks, (n); 5) the interaural delay between waveforms to the two ears (Δt). By applying the logic of Houtgast and Plomp (1968), one can predict the performance of an observer

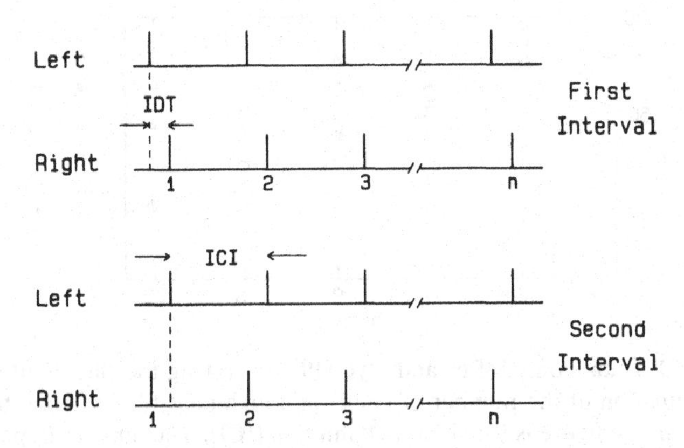

FIGURE 3.17. A schematic diagram of the click trains presented to subjects in a two-interval task. Δt is the interaural temporal difference, ICI is the interclick interval, and n is the number of clicks.

who is able to process and average the information conveyed by successive samples of the internal signal-plus-noise (Δt + jitter). Independent treatment of each click in the train should lead to a $\sqrt{n}$ decline in the threshold. Equation (2) takes the argument one step further, showing the relation between Δt_1, the threshold for only a single dichotic click, and Δt_n, the threshold for a train of length *n*. Obviously, detection must be based on the neural representation of the signal, not on the clicks themselves. Note then that a new parameter, N, has replaced the variable n. It is defined as the number of neural events evoked by the train of n clicks; it is the number on which the observer's judgment of "right-left" must be made.

$$\Delta t_n = \frac{\Delta t_1}{N^{0.5}} \tag{2}$$

Equation (3) is simply the logarithm of Equation (2). Its purpose is to show that for integration of information transmitted uniformly throughout the train, one should obtain a straight line in the log plots, with an intercept of Δt. An exponent (0.5) was used in lieu of a square-root sign in order to show that the slopes of the optimal functions in log-space should be -0.5.

$$\log \Delta t_n = \log \Delta t_1 - 0.5 \log N \tag{3}$$

According to this line of reasoning, the hypothetical number N measures the efficiency with which the nervous system encodes interaural information; as such, its joint relation to both n and the ICI can be used to help illuminate the limitation placed by high rates of modulation. While these relations cannot be measured directly, they can be inferred from application of Equation (3) to actual data. Figure 3-18 shows the results from such an experiment (Hafter and Dye, 1983), plotting the logarithm of the ratios $\Delta t_1 / \Delta t_n$ as functions of log n. The dashed line depicts the

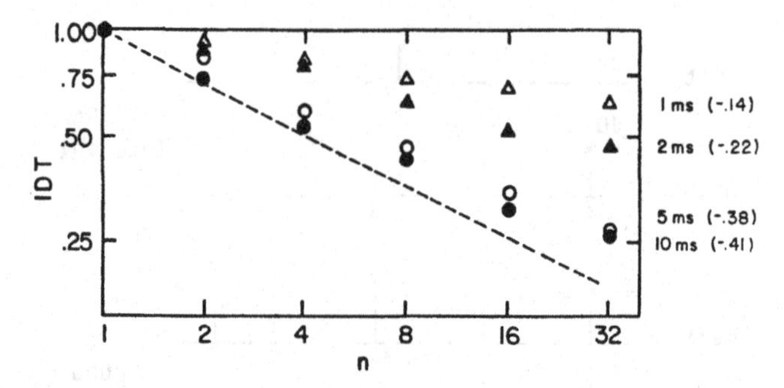

FIGURE 3.18. Data from Hafter and Dye (1983) showing the threshold values of IDT as a function of the number of clicks in a high-pass filtered click train. The parameter on the figure is the interclick interval (ICI). The number in parenthesis next to use ICI is the slope of the line that could be fit to the data. The dashed line has a slope of -0.5.

hypothetical complete-information slope of -0.5. These data show that while the log-threshold log-n functions are indeed linear, the slopes grow more shallow with smaller ICIs, that is, higher rates of modulation. Variation in the slopes can be described by inclusion of a multiplicative constant, k, whose values run from 0.0 to 1.0, depending on the ICI. This is shown in Equation (4). Taking the inverse logarithm produces Equation (5). It suggests that the number of neural events evoked by the n clicks is proportional to a compressive (exponent less than 1.0) power function of the input.

$$\log \Delta t_n = \log \Delta t_1 - 0.5k \log N \tag{4}$$

$$N \alpha\, n^k \qquad 0.0 < k \leqslant 1.0 \qquad k = f(\mathrm{ICI}) \tag{5}$$

The power function in Equation (5) can be expanded to a form that shows that each click in the train reduces the effectiveness of the one that follows. This is shown in Equation (6) where what is calculated is the probability that the jth click in a train will evoke an event within a single neural channel; the summation of these hypothetical probabilities over n clicks is proportional to n^k. From this formulation, one can see that for values of k less than 1.0, the probability of evoking a neural event with the jth click is greater than on the j plus 1st click. The legitimacy of the argument derives from the fact that the increment in k afforded by the jth click can be determined from parametric data, subtracting the performance for $j - 1$ clicks from that for j.

$$\text{prob}\,(j) = j^k - (j-1)^k \tag{6}$$

Elsewhere it has been shown that this nonstationary process cannot be accounted for by a lack of independence of the successive samples of internal noise or by reductions in the depths of modulation imposed by selective filtering of the sidebands (Hafter and Dye, 1983). Rather, there is evidence that the high-frequency binaural centers are fed by channels (Hafter and Wenzel, 1983) that saturate after the onset, with a rate of saturation determined by the rapidity of stimulation. For values of k less than 1.0, this is like a gentle form of "binaural precedence" (see Chapter 4 by P.M. Zurek) but with the proviso that the relative extra importance depends upon the modulation frequency.

The remainder of this chapter discusses lateralization from a variety of binaural experiments that bear on the generality of the saturation hypothesis. It has long been known that interaural information in a signal's onset can be especially effective, even if it conflicts with that later information (Tobias and Schubert, 1959; Abel and Kunov, 1983). Houtgast and Plomp's (1968) paper contains an excellent discussion of the role of onsets in localization. Even with high-frequency tones, delays in the onset can be detected. The question is, what is the nature of the process whereby more binaural information is provided by the onset than by rest of the signal? Can the compressive power function suggested in Equation (5) be widely applied? Figure 3-19 shows results from Tobias and Zerlin (1959). Their stimulus was a wide-band noise. There is no way to determine the number of envelope peaks in the noise without defining the width or widths of the listening bands. However, within each band, the number of peaks

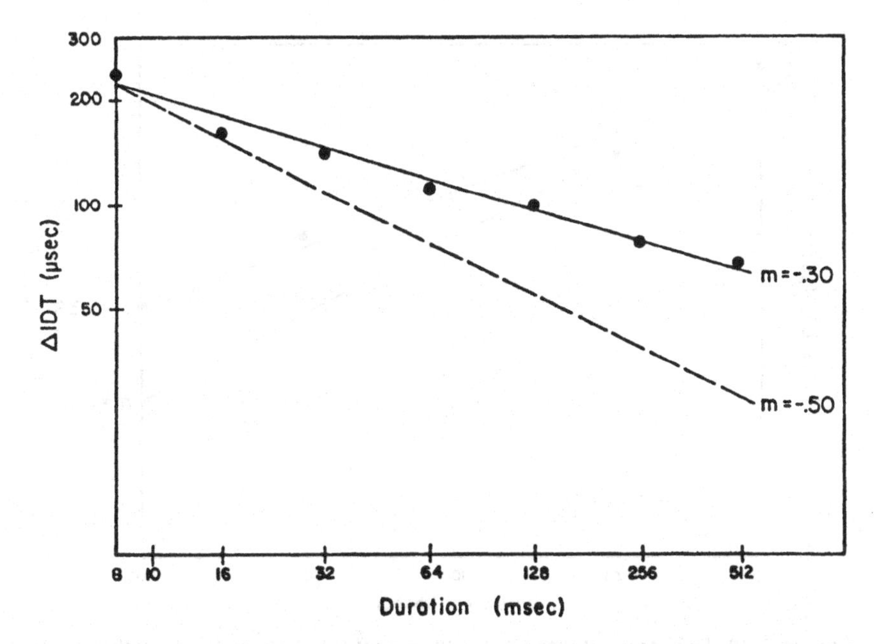

FIGURE 3.19. Data from Tobias and Zerlin (1959) showing that the change in IDT threshold has a function of the duration of a high-frequency noise. The data are fit with a slope of -.3 not -.5, which is indicated by the dashed line.

grows linearly with duration, so log duration was chosen for the abscissa. Note then that the points seem well fit by a straight line with absolute slope of less than 0.5.

Similarly, Figure 3-20 shows data from Houtgast and Plomp (1968). Although the stimulus was primarily low frequency, much the same basic result can be seen to have prevailed. That is, for the signals lateralized in the quiet, the absolute slope of the log-threshold, log-duration function was diminished.

For high-carrier frequencies, the slopes of the functions seem invariably to show the effect of modulation frequency. Where modulation is relatively slow, there is no saturation and the steeper slopes predicted by optimal summation are achieved despite the high-carrier frequency. Figure 3-21 shows data from Yost et al (1971) for lateralizations of high-pass clicks. There the ICI was 20 ms. In Figure 3-22 we see that with a higher rate, the shallower slopes prevailed. These are data from Nuetzel and Hafter (1976) obtained with 300-Hz sinusoidal modulation (ICI = 3.3).

Figure 3-23 compares data from Yost (1976) for clicks high-passed at 2 kHz. In this case, the slopes remain reasonably steep despite a relatively short ICI of 2.2 ms. Perhaps the difference here stems from the extensive low-frequency energy in the lower skirts of the low-pass filters. Finally, Figure 3-24 is taken from a study by McFadden and Moffitt (1977). Here the stimuli were pairs of tones centered at 3,500 Hz. The parameter is

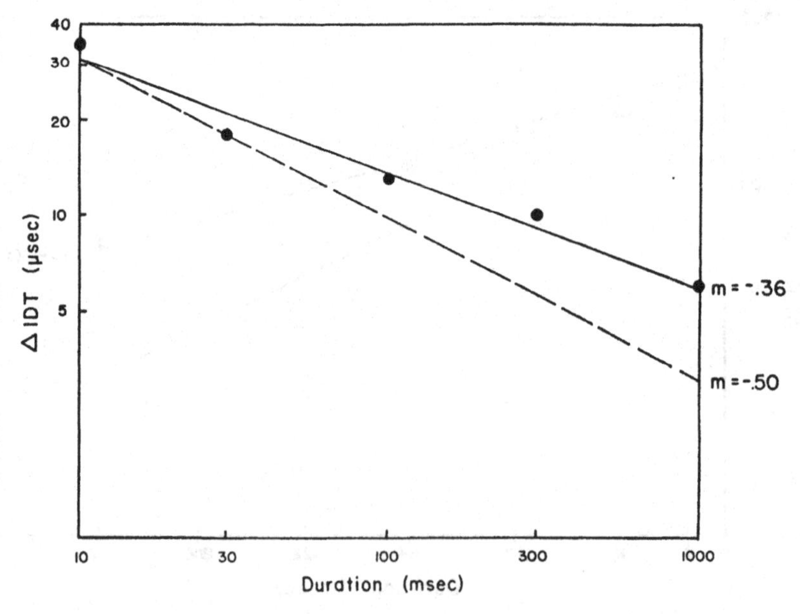

FIGURE 3.20. Data from Houtgast and Plomp (1968) showing the IDT thresholds as a function of the duration of low-frequency noise. Again the slope of the data (-.36) is less than -.5 (the dashed line).

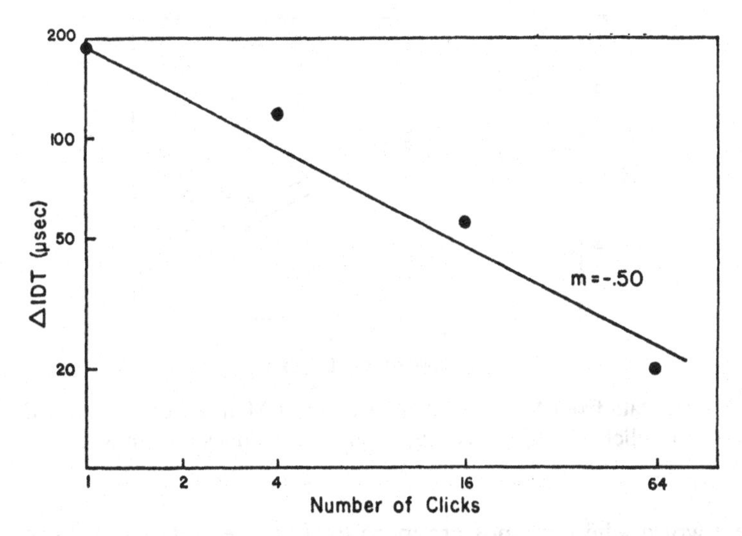

FIGURE 3.21. Data from Yost et al (1971) showing the IDT thresholds as a function of the number of high-pass clicks presented with a 20 ms ICI. For this low click rate the data are fairly well fit by a line with a slope -.5.

beat frequency, i.e., the frequency of the envelope. Again, the data are fit well by straight lines whose absolute slopes diminish with higher frequencies of modulation.

Our attempt has been to make two points. First, listeners are clearly able to detect even very small differences of interaural rime in high-frequency signals as long as there is a modulation of the waveform at a sufficiently slow rate. This fact has great significance for localization in

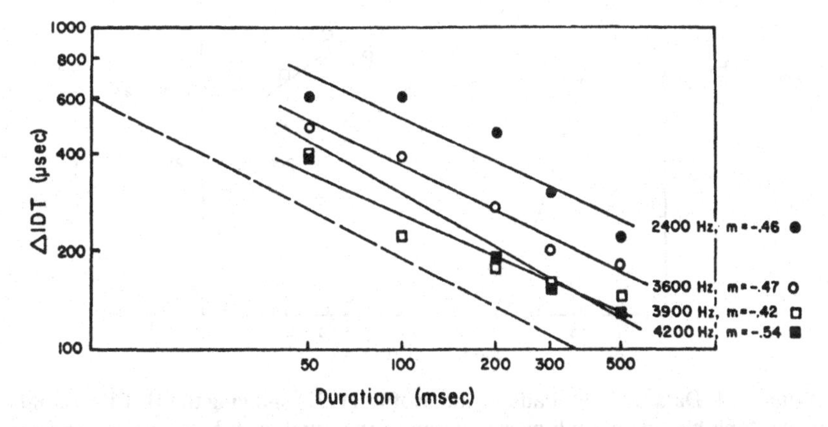

FIGURE 3.22. Data from Nuetzel and Hafter (1976) showing the IDT thresholds as a function of the duration of SAM tone. The parameter is the carrier frequency for the 300-Hz modulated tones. Again the dashed line represents a slope of -.5.

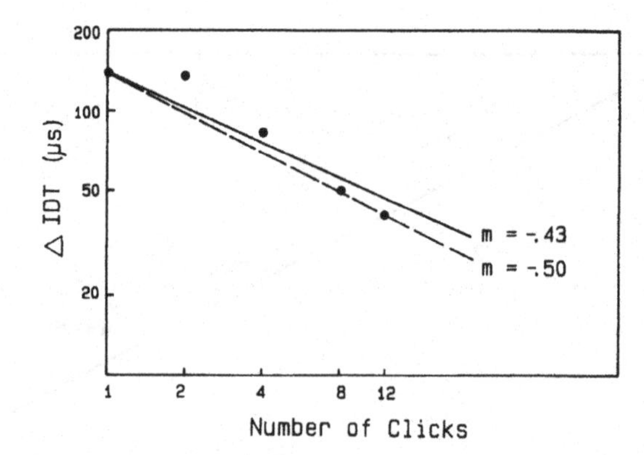

FIGURE 3.23. Data from Yost (1976) showing the IDT threshold as a function of the number of clicks for high-pass click train. Two slopes are shown.

the real world where sounds are more likely to be wide band. Second is that the process that prevents lateralization with very narrow bands such as with pure tones is a kind of neural saturation that takes the form of a compressive power function. By showing rapid adaptation to information presented at too high a rate, this mechanism enhances the importance of

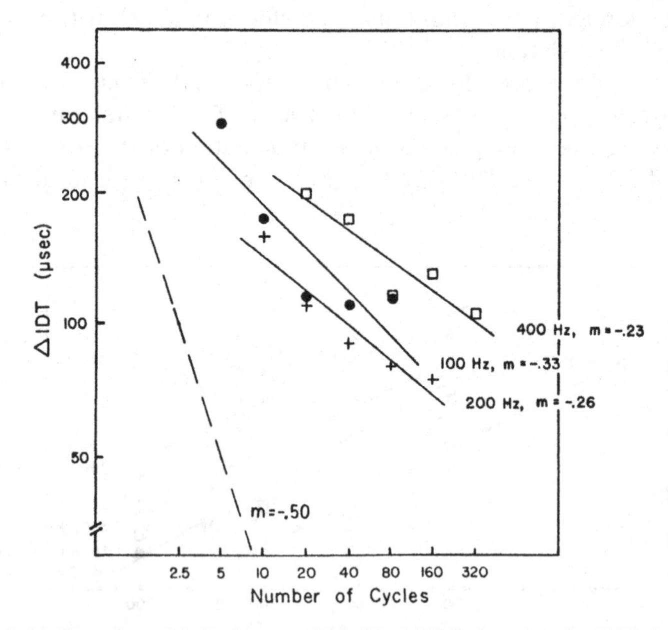

FIGURE 3.24. Data from McFadden and Moffitt (1977) showing the IDT thresholds for two-tone beating stimuli as a function of the number of beat cycles. The parameter on the figure is the difference in the frequency between the two tones whose center frequency was 3,200 Hz.

onsets relative to the rest of the signal, perhaps supplying the mechanism of binaural "precedence" (see Chapter 4).

A final note on lateralization of envelopes must deal with distinctions between high- and low-frequency carriers. Here, the answers are still unclear. For example, one might suspect that the timing code elicited by the individual peaks of a low-frequency tone would be subject to a saturation signal similar to that found with envelopes of high-frequency AM; and this fairly well describes the data which were obtained by Richard and Hafter (1973). However, comparable measures with tones by Hafter and Dye (unpublished) did not reveal a significant flattening of the log-threshold, log-duration slopes; and more data will be needed before the case with tones is fully understood. The issue is equally complicated when the stimulus is AM with a low-frequency carrier. Henning and Ashton (1981) have evidence that suggests that listeners can discriminate on the basis of interaural timing in the envelopes only for carriers above approximately 1,500 Hz. On the contrary, Bernstein and Trahiotis (1983), measuring the extent of laterality, suggested that envelope timing at low frequencies is an important factor.

Lateralization Across Frequency Bands

The previous discussion described the lateral perceptions associated with the interaural variables of time and intensity. The binaural system operates over a wide spectral region. The diagram in Figure 3-14 indicates a way to view binaural processing across this spectral region. That is, the information is processed at the output of frequency selective channels. What has not been discussed is how the binaural system processes information across these frequency channels, especially for complex stimuli that might contain different interaural information in different channels. Such complexity exits frequently in the real world. We have little trouble knowing the location of two different people talking at the same time, even if they are at a noisy party. In this case the interaural variables are different in different frequency bands. Despite the ability of the binaural system to efficiently process these cues, little research has been directed at understanding how the binaural system combines interaural information across frequency bands. Lateralization provides an excellent paradigm for investigating these issues.

Figure 3-25 shows the results from a study by Dye and Baumann (1984), which attempts to begin to understand these interactions. The experiment determined the just discriminable interaural temporal difference for different tones in a three-tone complex. One, two, or all three tones had their interaural temporal difference varied, while the remaining tones were presented diotically. As can be seen when two or three of the tones had the interaural temporal shift, interaural temporal discrimination was the same as that obtained for each tone presented in isolation. However, when

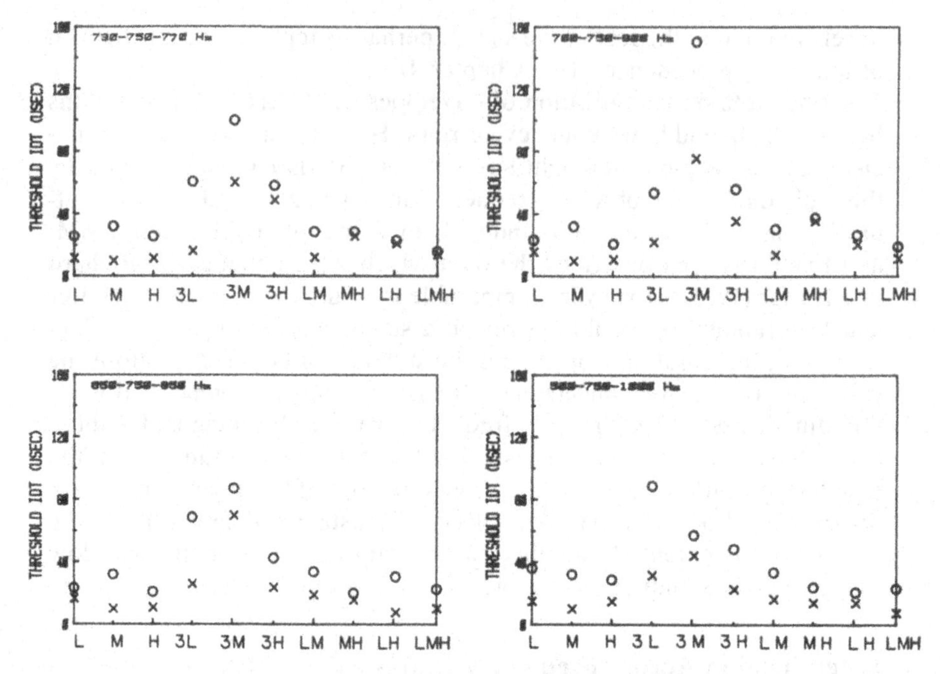

FIGURE 3.25. Data from Dye and Baumann (1984) in which the IDT thresholds are plotted for different combinations of a three-tone complex when the three tones were 300, 750, and 1,200 Hz. L, M, and H refer to the 300, 750, and 1000 Hz tones, respectively, presented alone; 3L, 3M, 3H refers to all three tones presented together, but only the L, M, or H tone had its interaural temporal difference threshold determined; LM, MH, and LH refers to all three tones being presented with two of the tones (LM, MH, or LH) having their interaural temporal differences threshold determine together; and LMH refers to all three tones presented together with all three tones having their interaural temporal difference threshold determined. As can be seen, the threshold of the 300-Hz tone is greatly increased when it is combined with the other two tones (the 3L condition), indicating that tones 450 Hz and 900 Hz removed from the 300 Hz tone can still affect the 300-Hz tones IDT threshold.

only one tone of the three had the interaural temporal difference, interaural temporal difference thresholds were elevated. These data indicate that even when tones are far apart in the spectrum, the two diotic tones still influence the lateral judgments of the third spectral component. Put another way, there appears to be a wide band of spectral integration for these binaural tasks.

These data suggest that the binaural schematic shown in Figure 3-14 might have to be modified. Rather than independent channels leading to a decision about stimulus location, this general model might have to assume some sort of integration across frequency bands before a decision about stimulus location is determined. Although the data of Figure 3-25 are suggestive of an interaction among frequency channels, more research investigating lateralization across frequency bands is needed before the details of such a system can be described.

Conclusions

This chapter describes some research concerning lateralization and the role interaural time and intensity play in determining the source of a sound. There are many other studies that involve headphone presentation of dichotic stimuli: For instance, studies of binaural masking (see Green and Yost, 1973, or McFadden, 1978, for reviews), dichotic listening, and hemispheric dominance (see Lauter, 1983, for a recent review), etc. (see Tobais, 1974). These and other studies do not *directly* investigate the properties of the binaural system responsible for directional hearing and, thus, were not discussed in this chapter.

The data we have described show that interaural time and intensity are powerful cues for determining the source of a sound. Cross-correlation is the common mode used to describe the type of mechanisms which might underlie these perceptions for simple stimuli. Rate of stimulus modulation rather than only stimulus frequency is a major variable in determining the influence of interaural time on directional sensitivity. Finally, we suggest that there might be integration of binaural information across frequency channels underlying our ability to lateralize or localize complex stimuli with different interaural information in different frequency bands.

Acknowledgments. The sections of this chapter on Simple Stimuli were written primarily by W.A. Yost and taken from his presentation at the Directional Hearing Symposium. The section on Complex Stimuli was written primarily by E.R. Hafter and was taken from his presentation at the symposium. We would like to thank Beth Wenzel for her assistance and her presentation of the complex stimulus paper at the Directional Hearing Symposium. Toby Dye's comments were very valuable. We would like to acknowledge the National Science Foundation, the National Institutes of Health (National Institute of Neurological and Communicative Disorders and Stroke) and the Air Force Office of Scientific Research for their support of our research.

References

Abel, S.M., Kunov, H. (1983). Lateralization based on interaural phase differences: Effects of frequency, amplitude, duration, and shape of rise/decay. J. Acoust. Soc. Am. 73, 955–961.

Bernstein, L.R., Trahiotis, C. (1983). Envelope based lateralization of low-frequency waveforms. J. Acoust. Soc. Am. 74, 585.

Bernstein, L.R., Trahiotis, C. (1985). Lateralization of sinusoidally amplitude-modulated tones: Effects of spectral locus and temporal variation—J. Acoust. Soc. Am., 78, 514–523.

Blauert, J. (1983). Spatial Hearing: The Psychophysics of Human Sound Localization. Cambridge, MA: The MIT Press.

Colburn, H.S., Durlach, W.I. (1978). Models of binaural interaction: In: Handbook of Perception, Vol. IV. Hearing. Carterette, E.C., Friedman, M.P. (eds.). New York: Academic Press.

Domnitz, R.H., Colburn, H.S. (1977). Lateral position and interaural discrimination. J. Acoust. Soc. Am. 61, 1586–1598.

Durlach, N.I., Colburn, H.S. (1978). Binaural phenomena. In: Handbook of Perception. Vol. IV. Hearing. Carterette, E.C., Friedman, M.P. (eds.). New York: Academic Press.

Dye, R.H. and Baumann, J. (1984). The combination of interaural information across frequencies, J. Acoust. Soc. Am. 75, 588.

Feddersen, W.E., Sandel, T.T., Teas, D.C., Jeffress, L.A. (1955). Measurements of interaural time and intensity difference. J. Acoust. Soc. Am. 27, 1008.

Grantham, D. W. (1984). Interaural intensity discrimination: insensitivity at 1000 Hz. J. Acoust. Soc. Am. 75 (4), 1191–1194.

Green, D.M., Henning, G.W. (1969). Audition. In: Annual Review of Psychology. Vol. 20. pp. 105–128. Palo Alto, CA: Annual Reviews Inc.

Green, D.M., Yost, W.A. (1975). Binaural Analysis. In: Handbook of sensory physiology (edited W. Keidel and D. Wineff). Springer-Verlag, Netherlands.

Hafter, E.R. (1977). Lateralization model and the role of time-intensity trading in binaural masking: Can the data be explained by a time-only hypothesis? J. Acoust. Soc. Am. 62, 633–636.

Hafter, E.R. (1984). Spatial hearing and the duplex theory: How viable? In: Dynamic Aspects of Neocortical Function. Edelman, G.M., Gall, W.E., Cowan, W.M. (eds.). New York: Wiley.

Hafter, E.R., Carrier, S.C. (1972). Binaural interaction in low-frequency stimuli: The inability to trade time and intensity completely. J. Acoust. Soc. Am. 51, 1852–1862.

Hafter, E.R., Dye, R.H. Jr. (1983). Detection of interaural differences of time in trains of high-frequency clicks as a function of interclick time interval and number. J. Acoust. Soc. Am. 73, 644–651.

Hafter, E.R., Dye, R.H., Nuetzel, J.M. (1980). Lateralization of high-frequency stimuli on the basis of time and intensity. In: Psychophysical, Physiological and Behavioral Studies in Hearing. van den Brink, G., Bilsen, F.A. (eds.). Delft, The Netherlands: Delft University Press.

Hafter, E.R., Dye, R.H. Jr., Wenzel, E.W. (1983). Detection of interaural differences of intensity in trains of high-frequency clicks as a function of interclick interval and number. J. Acoust. Soc. Am. 75, 1708–1713.

Hafter, E.R., Jeffress, L.A. (1968). Two-image lateralization of tones and clicks. J. Acoust. Soc. Am. 44, 563–569.

Hafter, E.R., Wenzel, E. (1983). Lateralization of transients presented at high rates: site of the saturation effect. In: Hearing — Physiological Basis and Psychophysics. Klinke, R., Hartmann, R. (eds.). Berlin/Heidelberg: Springer-Verlag.

Harris, C.G. (1960). Binaural interactions of impulsive stimuli and pure tones. J. Acoust. Soc. Am. 32, 685–692.

Henning, G.B. (1974). Detectability of interaural delay in high-frequency complex waveforms. J. Acoust. Soc. Am. 55, 84–90.
Henning, G.B. (1980). Some observations on the lateralization of complex waveforms. J. Acoust. Soc. Am. 68, 446–454.
Henning, G.B., Ashton, J. (1981). The effect of carrier and modulation frequency on lateralization based on interaural phase and interaural group delay. Hear. Res. 4, 185–194.
Hershkowitz, R.M., Durlach, N.I. (1969). Interaural time and amplitude jnd's for a 500 Hz tone. J. Acoust. Soc. Am. 46, 1464–1467.
Houtgast, T., Plomp, R. (1968). Lateralization threshold of a signal in noise. J. Acoust. Soc. Am. 4, 807–812.
Jeffress, L.A. (1948). A place theory of sound localization. J. Comp. Physiol. Psychol. 41, 35–39.
Klumpp, R.G., Eady, H.R. (1956). Some measurement of interaural time difference thresholds. J. Acoust. Soc. Am. 28, 859–860.
Lamter, J.L. (1983). Stimulus characteristics and relative ear advantage: A new look at old data, J. Acoust. Soc. Am. 74, 1–17.
Leakey, D.M., Sayers, B. McA., Cherry, C. (1958). Binaural fusion of low- and high-frequency sounds. J. Acoust. Soc. Am. 30, 222–223.
Licklider, J.C.R. (1962). Auditory-process models. J. Acoust. Soc. Am. 34, 713.
McFadden, D.M. (1975). Masking and the Binaural System. In: The Nervous System, Vol. 3 (edited E.L. Eagles), Raven Press, New York.
McFadden, D.M., Moffitt, C.M. (1977). Acoustic integration for lateralization at high frequencies. J. Acoust. Soc. Am. 61, 1604–1608.
McFadden, D., Pasanen, E. (1976). Lateralization at high frequencies based on interaural time differences. J. Acoust. Soc. Am. 59, 634–639.
Mills, A.W. (1960). Lateralization of high-frequency tones, J. Acoust. Soc. Am. 32, 132–134.
Molino, J.A. (1970). Simulation of the localization of distant sound sources by earphones. J. Acoust. Soc. Am. 48, 85.
Nuetzel, J.M., Hafter, E.R. (1976). Lateralization of complex waveforms: Effects of fine structure, amplitude, and duration. J. Acoust. Soc. Am. 60, 1339–1346.
Nuetzel, J.M., Hafter, E.R. (1981). Lateralization of complex waveforms: Spectral effects. J. Acoust. Soc. Am. 69, 1112–1118.
Osman, E. Personal communication.
Raleigh, Lord (1907). On our perception of sound direction. Phil. Mag. 13, 214–232.
Ricard, G.L., Hafter, E.R. (1973). Detection of interaural time differences in short-duration low-frequency tones. J. Acoust. Soc. Am. 53, 335(A).
Rowland, R.C. and Tobias, J.F. (1967). Interaural Intensity difference linen, J. Speech and Hearing Res. 10, 745–756.
Sayers, B. (1964). Acoustic-image lateralization judgments with binaural tones. J. Acoust. Soc. Am. 36, 923.
Sayers, B., Cherry, E.C. (1957). Mechanism of binaural fusion in the hearing of speech. J. Acoust. Soc. Am. 29, 973–986.
Stern, R.M., Colburn, H.S. (1978). Theory of binaural interaction based on auditory-nerve data. IV. A model for subjective lateral position. J. Acoust. Soc. Am. 64 (1), 127–140.
Stevens, S.S., Newman, E.B. (1936). The localization of actual sources of sound. Am. J. Psychol. 48, 297–306.

Tobias, J.V. (1972) Curious binaural phenomena in Foundations of Modern Auditory Theory (edited J.V. Tobias), Academic Press, New York.

Tobias, J.V., Schubert, E.D. (1959). Effective onset duration of auditory stimuli. J. Acoust. Soc. Am. 31, 1595–1603.

Trahiotis, C., Kappant, H. (1978). Regression interpretation of differences in time-intensity trading ratios obtained in studies of laterality using method of adjustment. J. Acoust. Soc. Am. 64 (4), 1041–1048.

Watson, C.S., Mittler, B.T. (1965). Time-intensity equivalence in auditory lateralization: A graphical method. Psychon. Science 2, 219–220.

Yost, W.A. (1974). Discrimination of interaural phase differences. J. Acoust. Soc. Am. 55, 1299–1303.

Yost, W.A. (1976). Lateralization of repeated filtered transients, J. Acoust. Soc. Am. 60, 178–181.

Yost, W.A. (1981). Lateral position of sinusoids presented with interaural intensive and temporal differences. J. Acoust. Soc. Am. 70 (2), 397–409.

Yost, W.A., Dye, R.H. (1987). Discrimination of Interaural Differences of Level as a function of Frequency, J. Acoust. Soc. Am., in press.

Yost, W.A., Tanis, D.C., Nielsen, D.W., Bergert, B. (1975). Interaural time versus interaural intensity in a lateralization paradigm. Percept. Psychophys. 18, 433–440.

Yost, W.A., Wightman, F.L., Green, D.M. (1971). Lateralization of filtered clicks. J. Acoust. Soc. Am. 50, 1526–1531.

Young, I.L., Jr., Carhart, R. (1974). Time-intensity trading functions for pure tone and a high-frequency AM signal. J. Acoust. Soc. Am. 56, 605–611.

Zwislocki, J.J. and Feldman, R.S. (1956). Just noticeable dichotic phase difference, J. Acoust. Soc. Am 28, 152–154.

4
The Precedence Effect

P.M. ZUREK

When two binaural sounds are presented with a brief delay between them, and are perceived as a single auditory event, the localization of that event is determined largely by the directional cues carried by the earlier sound. This observation is known as the precedence effect in sound localization.

The precedence effect is of interest for several reasons, both practical and theoretical. The most often-cited example of the practical importance of the precedence effect arises when a listener must localize a sound source in a reverberant environment. If the listener is in line-of-sight of the sound source, the first-arriving, or direct, wavefront will arrive from the direction of the source. Waves that are reflected off nearby surfaces will arrive later and from other directions. Because the directional cues associated with the direct wave are given more weight perceptually than those of the reflections, localization is usually quite accurate. Although the ability to localize in reverberant surroundings has always seemed apparent from casual observation, this belief has only recently been formally tested and confirmed (Hartmann, 1983).

The precedence effect has received considerable attention in the fields of architectural acoustics and stereophony where an important concern is the localization of an auditory event induced by multiple coherent sound sources. Investigations into the design of sound-reinforcement systems for lecture halls (Parkin and Humphreys, 1958) and sound-reproduction systems for motion pictures (Snow, 1953) have either exploited or otherwise tried to take into account the strong influence on localization of the sound arriving earlier at a listener's ears.

In addition to being of practical importance, the precedence effect has been a topic of continuing theoretical interest in psychoacoustics because of what it reveals about the sound-localization process. The effect can be viewed as a reflection of the temporal weighting of incoming localization information, a weighting that can change greatly over the course of milliseconds. Changes in perception of localization, however, occur over time intervals on the order of tens or hundreds of milliseconds (Grantham and Wightman, 1978). Thus, the fact that the short-time sequence of cues is highly influential means that localization in a multiple-source sound field will not necessarily be predictable from long-time-averaged variables.

Despite the large number of studies of the precedence effect, understanding of the phenomenon has not yet been woven into existing theories of binaural hearing. This state of affairs may result partly from the development of theories largely around steady-state stimuli. In part it may also result from the fact that many studies of the precedence effect have been motivated by applications rather than analysis and thus have had relatively little impact on theories. In an attempt to improve upon this state, the emphasis in the present review is on analytic understanding of the precedence effect. The goals of this chapter, which are pursued in the following three sections, are to familiarize the reader with the extensive work that has been done on the precedence effect, to point out the relation of it to other binaural phenomena, and to outline a theoretical framework incorporating the precedence effect into models of directional hearing.

Measurements of the Precedence Effect

There have been several independent discoveries of the precedence effect dating back to the mid-19th century. The history of these discoveries has been described by Gardner (1968) and need not be repeated here. The reader should be aware, however, that as a consequence of the multiple discoveries, and because of interest in relatively nonoverlapping disciplines, the precedence effect is known by various aliases: "the Haas effect," "the law of the first wavefront," "the first-arrival effect," and "the auditory suppression effect." Blauert (1983) distinguishes between "summing localization" and "the law of the first wavefront" on the basis of somewhat different phenomena when the intersource delay is less than or greater than approximately 1 ms. The term "precedence effect" is used here because, generally, it seems to have become the most common title in the psychoacoustic literature. In audio and architectural acoustics the terms "Haas effect" and "law of the first wavefront" seem to be more widely used (e.g., Alkin, 1973; Parkin and Humphreys, 1958) and also take on slightly different meanings in different contexts.

The Precedence Effect With Loudspeakers

Some of the early descriptions of the precedence effect were concerned with the utility of the effect for sound-reinforcement systems. Fay (1936) and Hall (1936) described a system that induced source localization at the talker's mouth rather than at the loudspeaker by delaying the amplified version of the talker's voice. They found that within limits of time delay and relative intensity of the loudspeaker signal, localization at the talker could be achieved. Other implementations of such systems are described by Parkin and Humphreys (1958) and Gardner (1969).

At about the same time as the Fay and Hall reports, Snow was working on stereophonic sound-reproduction systems that also exploited the precedence effect (see Snow, 1954). To control and measure localization, Snow employed a paired-loudspeaker configuration like that shown in Figure 4-1. The two loudspeakers were driven by the same source except that the signal from one loudspeaker could be delayed while the signal from the other loudspeaker could be attenuated. The precedence effect, which results in localization toward the left loudspeaker when τ is positive, is measured by the attenuation of the signal from the leading loudspeaker required to bring the image to the midpoint between the loudspeakers.

Snow (1954) found that with τ between about 0.5 and 10 ms, the leading source had to be reduced 5 to 8 dB to center the image. Figure 4-2 shows Snow's results along with those from other studies (Haas, 1951; Leakey and Cherry, 1957; Leakey, 1959) in which the same configuration, stimulus (running speech), and method were used. Also shown is a curve from deBoer (1940), which was measured using a different procedure that involved two steps. Listeners first judged the azimuth, phi, of the image, as τ was varied with g = 0 dB; they then judged phi as a function of g with $\tau = 0$. The curve in Figure 4-2 was obtained by eliminating the common variable phi, resulting in a relation between g and τ. This procedure does not necessarily give the same results as when τ and g are both nonzero and adjusted for center localization. Despite the assumptions needed to get from one type of measurement to the other, the results are in general

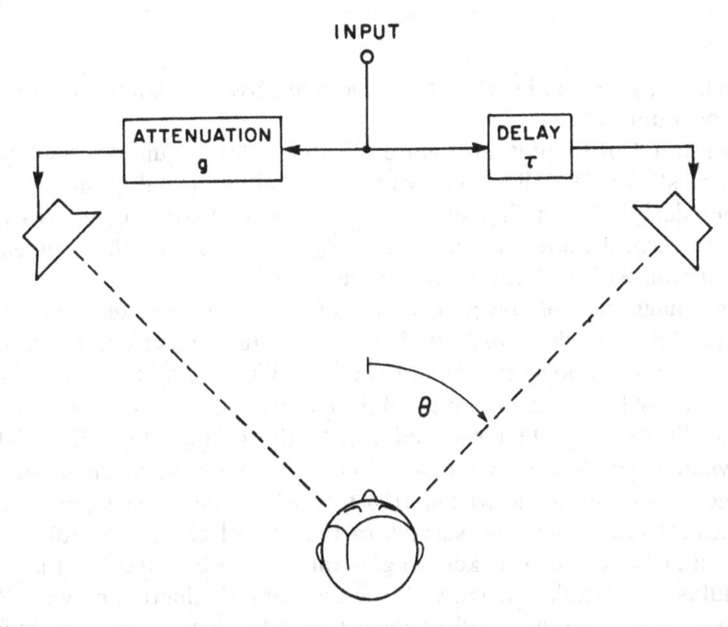

FIGURE 4.1. The two-loudspeaker configuration used in demonstrations and measurements of the precedence effect.

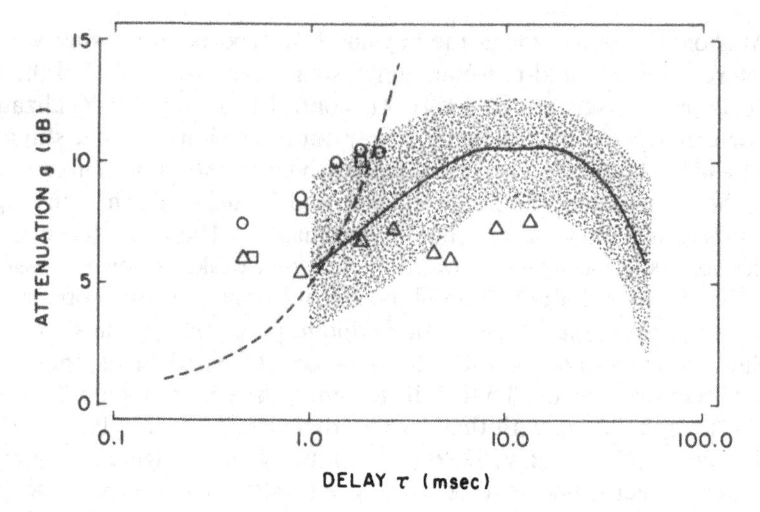

FIGURE 4.2. Measurements of the precedence effect in the two-loudspeaker configuration from studies employing running speech as the stimulus and angular displacements (Θ) of the loudspeakers between 22.5 and 45 degrees. The signal from the leading loudspeaker was attenuated (ordinate) to balance the shift in localization caused by the inter-loudspeaker delay (abscissa). The solid curve and stipling give the mean and range of attenuation adjustments by 15 subjects (Haas, 1951). Triangles are average data from at least four listeners (Snow, 1954). Circles are averages from six subjects (Leakey and Cherry, 1957). Squared are averages from six subjects (Leakey, 1959). The dashed curve (deBoer, 1940) was obtained with an indirect method explained in the text.

agreement, particularly when the large interobserver differences are taken into account.

The trend of the data in Figure 4-2 shows the required attenuation to be largest, 5 to 12 dB, at delays between about 1 and 20 ms. At much longer delays the trailing sound is heard as a separate echo. Also, as τ goes to zero, the attenuation must go to zero. However, the course of the attenuation with $\tau < 1.0$ ms is not clear.

The magnitude of the precedence effect is partly determined by the nature of the stimulus. Haas (1951) made measurements like those in Figure 4-2 with filtered noise in place of speech. With octave bands centered at either a low (.28 kHz) or a high (4.5 kHz) frequency and delays between 1 and 40 ms, the effect was reduced to the range 3 to 8 dB. Wallach, Newman, and Rosenzweig (1949) found that with either clicks or piano music the sound image was shifted toward the closer speaker, but with orchestral music only the scratch of the record player's needle was localized at the closer speaker. In general, it is to be expected that as the stimulus bandwidth is narrowed, the temporal distinction between "preceding" and "trailing" will become blurred. Also, there will obviously be a perceptual limit on the delay between two temporally punctate stimuli for fusion into a single auditory event. Wallach et al (1949) estimated this limit to be about 5 ms for clicks. Thus, the curves in Figure 4-2 would be expected to return to zero at shorter delays with clicks as stimuli than they do with wideband speech.

Leakey and Cherry (1957) found that adding broadband noise disrupted the precedence effect. Their experiment again used the same two-loudspeaker configuration, but included a third loudspeaker emitting broadband noise directly in front of the observer. As the level of the noise relative to the continuous speech was raised, the effect diminished systematically. For example, with speech 10 dB above masked threshold, the maximum increase in the level of the delayed sound required to center the image was only 3.5 dB. The disrupting effect of noise on the precedence effect was also observed by Thurlow and Parks (1961) using click stimuli. Thurlow and Parks also determined that the precedence effect is obtained when the loudspeakers are asymmetrically located with respect to the observer (e.g., 10° right and 50° right).

Blauert (1971) showed that localization is determined primarily by the earlier sound not only for left-right (horizontal azimuth) judgments, but also for front-back judgments. This study used one loudspeaker in front and another behind the observer, who was asked whether the sound emanated from front, above, or rear. When the inter-loudspeaker delay was greater than about 500 μs, responses tended to follow the precedence effect in that most of the responses were "front" when the front sound led, and vice versa. For smaller delays this effect was not observed; listeners tended to give "rear" responses for delays in the range −500 to +500 μs. In particular, when the test signal was music and the delay was zero, nearly all of the responses were "rear." The significance of Blauert's finding is that the precedence effect is obtained even though the stimulus is essentially identical at the two ears.

It has been noted (e.g., Wallach et al, 1949) that transients seem to be necessary to induce the precedence effect. That initial transients have a strong influence on the localization of subsequent sounds is demonstrated by the Franssen effect (see Blauert, 1983). Suppose a pulsed sine wave is split into two signals, one composed of the transient onset and offsets and the other composed of the steady-state portion. These two signals are then presented via separate loudspeakers. The entire signal is localized at the transient loudspeaker even when the steady-state duration is a few seconds. Similar results with transients from one loudspeaker influencing the localization of subsequent tones have been reported by Thurlow, Marten, and Bhatt (1968) and Scharf (1974). Note that these findings demonstrate that the trailing sound need not be a replica of the preceding sound in order for the precedence effect to be observed.

Thus far, the only nonhuman animals to have been tested in the two-loudspeaker configuration are rats (Kelly, 1974) and cats (e.g., Whitfield et al, 1972), and both these species exhibited the precedence effect (i.e., responded to the leading loudspeaker as though it was a single source).

A series of investigations with animals, human infants, and special clinical populations has attempted to assess the importance of various brain structures for the precedence effect. Initial ablation-behavior studies (Cranford et al, 1971; Whitfield et al, 1972) indicated that unilateral ablation of auditory cortex in cats severely disrupts the precedence effect while leaving unimpaired the ability to localize single sound sources. The nature of the disruption was such that animals did not exhibit the precedence effect when the leading loudspeaker was contralateral to the side of the ablation. For example, if the left auditory cortex was removed, left-right stimuli (23-ms 1-kHz tone pulses) were localized at the leading source but right-left stimuli were not. Results of later studies (Cranford and Oberholtzer, 1976; Whitfield, Diamond, Chiveralls, and Williamson, 1978) are more equivocal in that some cats did not show a deficit in the precedence effect after unilateral cortical ablation. Factors such as testing procedures, extent of the lesion, and learning of new strategies have been suggested to account for the differences.

On the basis of these animal studies, Clifton et al (1981a) predicted that newborn human infants, in whom cortical development normally lags behind that of peripheral and brainstem structures, would not exhibit the precedence effect. This prediction was testable because it had been shown by Muir and Field (1979) that newborn infants are capable of orienting to sounds. Thus, if newborns exhibit the precedence effect, they should respond equivalently to a single sound source and the leading source in the paired-source configuration. The results of the Clifton et al (1981a) study were quite clear. A single source elicited a head turn on 42 of 72 trials (four trials from each of 18 infants), and 40 of those 42 were in the correct direction (where direction is either right or left). Paired loudspeakers, with a 7-ms delay between them, elicited a head turn on eight of the 72 trials, only three of which were in the direction of the leading source.

At some point in development, of course, the precedence effect must appear. Subsequent studies using the same experimental procedures (Clifton et al 1981b; Clifton, Morrongiello, and Dowd, 1984) determined that it is roughly at 5 to 6 months of age that the precedence effect can first be measured. This development is proposed to result from the rapid maturation of the cortex during the first few months of life.

Hochster and Kelly (1981), also motivated by the ablation-behavior studies of the precedence effect in cats, studied the effect in children, 6 to 16 years of age, with temporal-lobe epilepsy. These subjects could localize (i.e., indicate which of two loudspeakers produced a single click) as well as normal children. In the paired-source condition, in which the normal children always identified the leading loudspeaker, the epileptic subjects exhibited a deficit that depended on the intersource delay. With delays of 8 and 16 ms the epileptic subjects performed at 80% and 50% (chance) correct, respectively. With 1 and 4 ms delays, however, the clinical group performed at nearly normal (100%) levels. This pattern of results is difficult to interpret solely as a deficit in the precedence effect, since the effect is strongest at delays of 1 to 4 ms where the clinical group was unaffected (unless a ceiling effect prevented a difference). If it is assumed that the click stimuli with longer delays were heard separately by all subjects (Wallach et al, 1949), then a deficit in the ability to judge spatial-temporal order, not in the precedence effect, would seem to be suggested.

The Precedence Effect With Earphones

As with other topics in directional hearing, studies of the precedence effect have often been conducted with sounds presented via earphones to achieve precise and independent control of the stimuli to the two ears (see Chapter 2). The auditory perceptions evoked under earphone stimulation are, to a large extent, internalized versions of the images aroused by free-field stimulation. If there is no difference between the sounds at the two ears, a well-defined image is heard intracranially in the median plane. An intensity increase or a time lead of the sound to one ear tends to shift, or "lateralize," the image toward that ear.

The timing of stimuli used in many earphone studies of the precedence effect is shown in Figure 4-3. In terms of their correspondence to events in a free field, the first pair of pulses represents a direct wave arriving from a direction that determines the first interaural delay T1 while the second pair of pulses represents a reflected wave, delayed τ seconds, arriving from a different direction that determines T2. The pulses shown in Figure 4-3 display only the temporal pattern of stimulation that would be produced at the ears of an observer in a sound field; direction- and frequency-dependent amplitude effects are not represented. These latter effects have typically not been part of earphone studies of the precedence effect.

With direct control of stimulus parameters it is possible to assess the

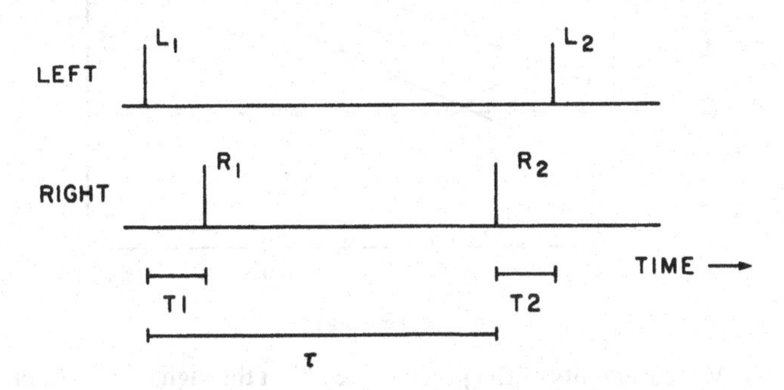

FIGURE 4.3. The impulses represent the timing of brief, impulsive signals used in earphone studies of the precedence effect. Note that T1 and T2 are defined to be positive when their respective senses of delay (left re right) conflict.

precedence effect using a number of possible dependent measures. Von Bekesy (1930), for example, measured lateral position with either T1 or T2 equal to zero and the other nonzero. He noted that the position of the image was determined by the time difference on the first pair when τ was greater than 900 to 1,200 μs but that with τ less than 900 μs, T2 also influenced lateral position. Similar observations were made by Langmuir et al (1944) whose measure of the directional strength of the initial sound was its level relative to the delayed sound for a criterion effect on lateralization.

Wallach et al (1949) employed pulses of 1-ms duration presented in two pairs, as in Figure 4-3. With T2 fixed, subjects judged whether the fused image was left or right of center as T1 was varied. The value of T1 that resulted in 50% "left" judgments was taken to be the value that just balanced the effect of a given T2. Their results from two subjects with τ = 2.0 ms are shown in Figure 4-4. The slope of the line for T2 < 400 μs indicates that a unit time delay for the first pair requires approximately seven units of delay in the opposite direction for the second pair to produce a centered image.

Yost and Soderquist (1984) recently replicated the results of Wallach et al (1949), but have questioned the interpretation that the lateralization effects of the first and second pulse pairs literally cancel. They found that when presented with T2 fixed at rather large values (600 or 800 μs) and T1 at a value that produces 50% "left" judgments (the criterion for equal effect used by Wallach et al), subjects rarely judged that stimulus to be in the same lateral position as a diotic stimulus. This may be partly explained by the width of the image produced by the dichotic stimulus. If

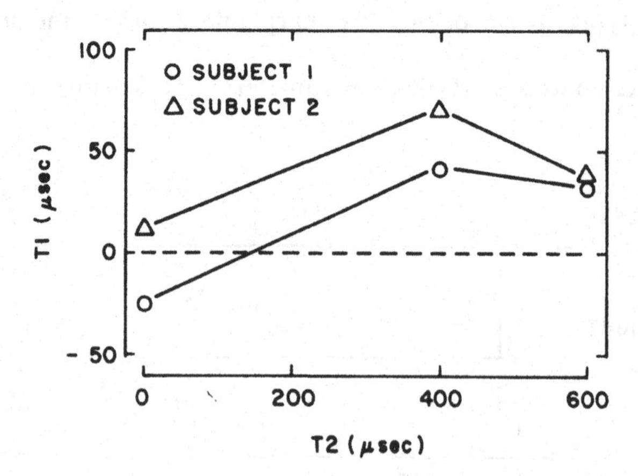

FIGURE 4.4. Measurements of the precedence effect through central lateralization. The data points indicate the values of opposing interaural delays, T1 and T2, on successive pairs of clicks that result in a central lateralization. (Adapted from Wallach, Newman, and Rosenzweig, 1949.)

that image is broader and more diffuse than the diotic stimulus, then greater variance in same/different judgments of lateral position would be expected.

Yost and Soderquist (1984) made two other noteworthy points. First, they demonstrated that the precedence effect can be observed when the lateral position of the four-pulse stimulus is equated to that of an off-midline stimulus (rather than a centered stimulus). This demonstration in lateralization is analogous to Thurlow and Parks' (1961) demonstration with asymmetrically positioned loudspeakers. Second, Yost and Soderquist showed that detection threshold for the binaural four-click complex (as in Fig. 4-3) in a diotic white noise is lower if the first pair, as opposed to the second pair, is presented with an interaural delay.

The uncertainties inherent in subjective judgments of lateral position can be avoided by measuring just-noticeable differences (JND) in a forced-choice paradigm. In such studies (Zurek, 1980; Gaskell, 1983) the precedence effect is evidenced as an increase in the JND (decrease in resolution) in the interaural parameters of the trailing sound relative to that measured on the leading sound. Both Zurek (1980) and Gaskell (1983) measured $\Delta T1$ and $\Delta T2$, the JNDs from zero interaural delay on the first and second sounds, using transient stimuli (1-ms noise bursts in the former study and 20-μs clicks in the latter study). When the JND on the first burst was measured, the second burst was diotic, and vice versa. The parameter of interest was the delay τ between leading and trailing sounds. Results of these two studies, plotted in Figure 4-5, show $\Delta T1$ to be independent of τ and $\Delta T2$ to depend strongly on τ. The point of maximal increase in $\Delta T2$ is at a smaller τ in Gaskell's data ($\approx$1.0 ms) than in Zurek's (2–4 ms). This difference may be due to the difference in stimulus duration; intuitively, an earlier return to smaller JNDs as τ increases are expected with briefer stimuli.

Several other findings of the Zurek (1980) and Gaskell (1983) studies are of note. First, as was noted in earlier loudspeaker studies, it was found that the second transient need not be a repetition of the first to obtain the precedence effect. A pattern essentially the same as that shown in Figure 4-5 was obtained when the trailing noise burst was statistically independent of the first (Zurek, 1980). Second, the form of the dependence of interaural time JNDs on τ was found also for interaural amplitude JNDs (Zurek, 1980). However, when an interaural amplitude difference is imposed on the second pair of sounds and τ is on the order of a millisecond, the lateral image is moved in the direction opposite that indicated by the amplitude difference (Gaskell, 1983). With reference to Figure 4-3, imagine that L_1 and R_1 are coincident, as are L_2 and R_2 τ seconds later. Set L_1, R_1, and R_2 to unit amplitude and L_2 to a smaller amplitude. On the basis of the amplitude difference between L_2 and R_2, an image on the right would be expected, but it is found that the fused image is on the left. A possible explanation for this effect is seen by considering the case of extreme attenuation of L_2 so that there are essentially only three pulses, two forming

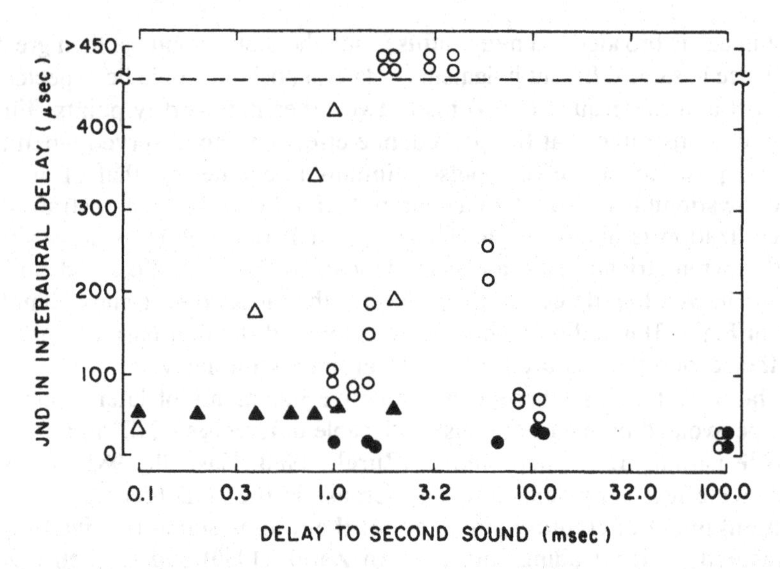

FIGURE 4.5. Measurements of the precedence effect through interaural resolution. Plotted is the just-noticeable difference (JND) in interaural delay imposed on either the preceding sound (filled symbols) or the trailing sound (open symbols) as a function of the delay to the trailing sound. Circles indicate data from a study by Zurek (1980) in which the stimuli were 1-ms noise bursts. Triangles indicate data from a study by Gaskell (1983) in which the stimuli were 20-μs clicks.

a coincident pair and a monaural trailing pulse at the right ear. If the delay τ is short enough, there will be two binaural pulse-pulse delays to be considered—one equal to zero and the other, leading to the *left*, of τ seconds. Apparently this long-delay interaction is incorporated with the zero-delay interaction to produce, in this case, leftward lateralization.

The Precedence Effect in Relation to Other Binaural Phenomena

From the demonstrations and measurements that have just been summarized, it is evident that the precedence effect is a fairly robust phenomenon. We should expect then that the effect would be manifested in a variety of binaural studies that may or may not have been conducted, or interpreted, with the precedence effect in mind. In this section we point out some of the binaural phenomena that seem to be related to the precedence effect even though the stimulus structure may have been quite different from that considered thus far.

On the basis of the precedence effect it might be expected that increasing stimulus duration would have little effect on the resolution of interaural delay. Since the perception of lateral position is dominated by cues present at onset, increasing duration should contribute relatively little. A specific alternative prediction, which is based on a statistical decision model (Houtgast and Plomp, 1968), is that the JND should decrease as the square root of duration, at least up to some maximum temporal integration time. Results from three studies (Tobias and Schubert, 1959; Houtgast and Plomp, 1968; and Ricard, 1974), presented in Figure 4-6, show that there is a modest improvement in the interaural delay JND as the duration of the stimulus is increased. [Similar dependencies on stimulus duration have been shown for interaural delay JNDs and interaural amplitude JNDs in high-frequency click trains (Hafter and Dye, 1983; Hafter, Dye, and Wenzel, 1983).] The results clearly deviate from the square-root dependence on duration; whether they are in accord with expectations from the precedence effect is questionable, since that prediction was not precisely stated. If one believed the precedence effect to be absolute, then no improvement with duration could be accepted. If, as is more reasonable in

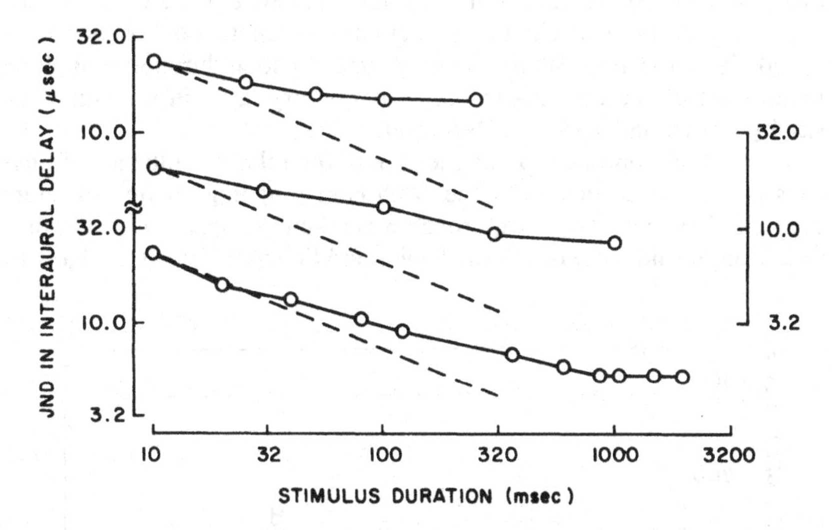

FIGURE 4.6. Just-noticeable differences (JNDs) in interaural delay as a function of the duration of the stimulus. The lower set of data points are from a study by Tobias and Zerlin (1959) in which the stimulus was noise low-pass filtered at 5 kHz. Stimuli longer than 100 ms were presented at 65 dB SPL whereas those less than 100 ms were presented at greater intensity in an attempt to equate loudness. The middle set of data points (refer to right ordinate) are from a study by Houtgast and Plomp (1968) in which the stimulus was an octave band of noise centered at 500 Hz, presented at a constant level that was 50 dB above threshold for the long-duration signals. The upper set of data points are from a study by Ricard (1974) in which the stimulus was a 500-Hz tone presented at 70 dB SPL with a 2.5 ms rise-fall time. In this study interaural delay was present only in the phase of the tone, not in its envelope. The dashed lines represent the theoretical prediction that the JND is inversely proportional to the square root of duration.

light of the results described above, the precedence effect is viewed as a matter of degree depending on time after onset, then the prediction should be somewhere between no dependence and the square-root dependence on duration.

In a different approach to the question of how resolution changes after onset, sensitivity was probed at various times after the onset of a stimulus of fixed duration (Zurek, 1980). The stimulus was a 50-ms burst of noise that was diotic except for a 5-ms segment that was interaurally delayed. JNDs in the interaural delay of this segment were measured for various placements of the segment after onset. The results of these measurements, presented in Figure 4-7, show a pattern of 1) highest sensitivity near onset, 2) a brief period of extreme insensitivity occurring 1 to 3 ms after onset, and 3) a gradual recovery of interaural sensitivity at longer postonset delays. [A temporal dependence similar to that of the interaural delay JNDs in Figure 4-7 was also observed for interaural amplitude JNDs (Zurek, 1980).] It is clear from the dependence of interaural sensitivity on postonset time that if the stimuli are transients and τ is between 1 and 3 ms, there will be a gross insensitivity to the interaural parameters of the succeeding sound. It can also be seen, however, that for long-duration stimuli the precedence effect amounts to only a temporary lapse in interaural sensitivity occurring 1 to 5 ms after onset.

Yet another approach to the question of the relative influence of onset cues places those cues in conflict with later (ongoing, or fine-structure) cues. Studies in which this approach was taken are apparently at odds as to the importance of onset cues. Tobias and Schubert (1959) used a wide-

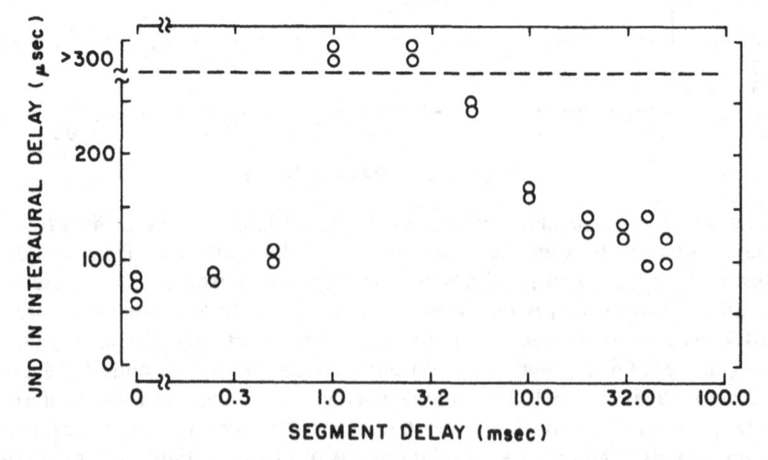

FIGURE 4.7. Just-noticeable differences (JNDs) in the interaural delay of a 5-ms segment of a 50-ms noise burst. JND measurements from a single subject are plotted as a function of the delay of the 5-ms segment from the onset of the 50-ms burst. (Adapted from Zurek, 1980.)

band noise burst in which envelope onset occurred earlier at the ear to which the fine structure of the noise was delayed. When the stimulus duration was greater than about 200 ms, the fine-structure delay dominated the judgments and onset time differences as large as 400 μs were ineffective in altering lateral position. With shorter stimulus durations the onset was relatively more important. However, even with a stimulus duration of only 10 ms (the shortest duration employed), an onset difference of 400 μs was balanced by an ongoing delay of 95 μs, indicating that the ongoing difference was four times as influential as the onset cue. Later investigations (Kunov and Abel, 1981; Abel and Kunov, 1983) placed the two types of cues in opposition by interaurally delaying tone bursts by an amount greater than one-half period (but less than a full period). Thus, the envelope cue led at the nondelayed side while the phase cue (the interaural phase angle that is smaller in magnitude than 180°) led on the delayed side. By controlling the strength of the onset cue through the rise time of the stimulus, Kunov and Abel (1981) found that in order for position judgments of long-duration tones to be free of the effects of onset cues, the rise time had to be about 200 ms. This indicates a very powerful influence of the onset.

The seeming discrepancy between the results of Tobias and Schubert (1959) and those of others (e.g., Wallach et al, 1949; Kunov and Abel, 1981; Abel and Kunov, 1983) may lie in the many differences among the stimuli used in those studies. However, Tobias and Schubert's findings can also be interpreted in at least two other ways. First, they may indicate that a gross envelope disparity is not sufficient to elicit the precedence effect independent of the waveform fine structure. In other words, in addition to the abrupt onset, interaural coherence may be necessary to elicit the precedence effect. Second, the two cues in Tobias and Schubert's study were confounded in that the fine-structure cue could also make a contribution during the onset. Other studies (von Bekesy, 1930; Zurek, 1980; Gaskell, 1983) indicate that interaural differences within approximately the first 800 μs contribute to lateral position. Thus, during that period (after the onset of the delayed envelope) the fine-structure delay could contribute to the lateralization judgment, thereby decreasing the apparent effectiveness of the envelope cue.

The precedence effect can also be seen to be operative in lateralization studies that have employed the "three-click paradigm" (Guttman, van Bergeijk, and David, 1960; Guttman, 1962; Harris, Flanagan, and Watson, 1963; Guttman, 1965). In these studies two clicks, separated in time by dt, are presented to one ear, and a third click is presented to the opposite ear. Usually the parameters of the monaural click pair are fixed, and the subject adjusts the time of occurrence of the contralateral "probe" click to center an image (there can be multiple images and hence more than one probe placement resulting in an image being centered). Of the many interesting results of these studies, the most relevant here is that reported by Harris et al (1963), who found that subjects obtained centered images by adjusting the probe to be coincident with the succeeding click when dt was 0.5, 1.0, 4.0, and 8.0 ms, but *not* when dt was equal to 2.0 ms. These results (and a similar result observed by Sayers and Toole, 1964, with a slightly different paradigm) provide further evidence that there is a temporary lapse in binaural sensitivity after an abrupt onset or transient.

Models of the Precedence Effect

In this section we examine explanations that have been offered for at least some aspects of the precedence effect and conclude with what seems to be, on the basis of the reviewed data, the minimal necessary structure of a model.

Forward Masking

One of the first notions that comes to mind in thinking about the precedence effect is that the delayed stimulus has less influence in localization and lateralization because it is subject to forward masking that is stronger than the backward masking of the leading sound. This hypothesis was advanced, in neurophysiological terms, by David (1959) to explain results such as those in Figure 4-2 obtained in the two-loudspeaker configuration. David's explanation of the precedence effect consisted of two parts, only the second of which involved forward masking: at short delays, where τ is less than the travel time around the head, subjects could compensate for the delay with attenuation of the leading source (in the manner of a time-intensity trade); at longer delays, the trailing sound is underrepresented in its neural response relative to that of the leading sound (i.e., forward masking plays a role).

A testable feature of the forward-masking hypothesis is that the essential disrupting phenomena occur separately in the two monaural channels. Thus, the effects observed in binaural discrimination tasks should have counterparts in monaural discrimination tasks. Two such comparisons lead us to discount the role of forward masking in the precedence effect. For the case of transient stimuli we can look for asymmetries between backward and forward masking of clicks by clicks. According to Ronken (1970), at interclick delays of 1 and 10 ms there are approximately equal amounts of forward and backward masking. Whereas at intermediate delays there is greater forward than backward masking by 5 to 8 dB (Table 4-1). Now consider the interaural JND data of Figure 4-5. Clearly, masking is not the primary determinant of interaural sensitivity if, as delay is varied, backward and forward masking both vary over the same range of more than 30 dB, yet interaural performance on the first sound is independent of delay whereas that on the second sound is strongly dependent on delay.

The second comparison between monaural and binaural performance

TABLE 4-1. Thresholds for a target click masked by another click separated in time (from Ronken, 1970)[a].

	Target-masker separation (ms)			
	1	2	5	10
Forward	−5.5	−16.5	−32.5	−41.5
Backward	−7.0	−21.5	−40.5	−40.5

[a]Data from one subject. In the forward condition the masker precedes the target; the reverse for the backward condition. Thresholds are given as target-to-masker ratio in dB.

involves the data of Figure 4-7. On the basis of the masking hypothesis we would expect monaural (or diotic) detection of a brief noise burst embedded in a longer burst to follow a dependence on delay similar to that seen in the interaural JND data. Such a configuration was investigated by Zwicker (1965) and Zurek (1976), both of whom found detection threshold for a brief noise probe to be independent of its temporal placement within a longer noise burst.

Cross-Correlation

A revision of the forward-masking portion of David's explanation came about as a result of experiments by David and Hanson (1962). They presented to subjects a diotic pulse followed by a single monaural pulse and found that the temporal position of the single pulse influenced the lateral position of the combined image. For example, if the single delayed pulse was added to the right channel, the image moved from the center to the left. (This phenomenon was discussed in Section I-B; see Gaskell, 1983). In Figure 4-3 we see that although the intended leftward delay is T1, and the intended rightward delay is T2, there is an additional leftward delay, between pulses L_1 and R_2, of τ seconds. There is also an additional rightward delay, between pulses R_1 and L_2, of ($\tau - T1 + T2$) seconds. David and Hanson proposed that all the interaural delays between pulses are averaged to give the lateral position of the single image.

This notion of multiple interactions can be expressed more generally by the concept of cross-correlation, a concept that has been used to model many phenomena in binaural hearing (see Chapter 3; for surveys see Colburn and Durlach, 1978, and Blauert, 1983). However, it is not difficult to see (from the definition of the cross-correlation function) that cross-correlation fails to emphasize early interaural delays. If it is assumed that the centroid of the cross-correlation function is the correlate of lateral position (e.g., Stern and Colburn, 1978), one finds that the predicted lateral position, in terms of an equivalent interaural delay in a simple stimulus, is the average of all the interaural delays that can be found between the pulses taken in pairs. Alternatively, if lateral position is defined in terms of local maxima in the cross-correlation function, then the problem of choosing among multiple peaks arises (Blauert and Cobben, 1978). In general, cross-correlation models will fail to account for the precedence effect unless some aspect of the processing is made time dependent.

Funnelling and Inhibition

Several authors have proposed that the precedence effect results from a control process that acts to alter the interaural effects of stimuli following soon after an onset. In some discussions (von Bekesy, 1930; Thurlow et al, 1965) a funnelling process is proposed in which the localization of succeeding sounds is biased toward that established by the first-arriving sound. In others (Haas, 1951; Harris et al, 1963; Zurek, 1980) the onset of a sound triggers a delayed reaction in which cues are momentarily ineffective in contributing to localization. It is implicit in these speculations that the funnelling or inhibition occurs centrally, at least after the first locus of possible binaural interaction.

Despite their vagueness, the inhibitory theories make some testable predictions. First, since it is assumed that the inhibitory effect takes place in neural centers devoted to directionality, there should be no necessary counterpart to the precedence effect in monaural phenomena. This prediction was discussed above in conjunction with the forward-masking hypothesis. Second, since the precedence effect is proposed to result from a time-varying process, the effect should not be predictable from steady-state stimulus cues. This prediction was examined in conjunction with cross-correlation theories. Finally, although no author has made the prediction explicitly, it would not contradict the spirit of these theories if the precedence effect were shown to be separable from steady-state sound localization. While the case is not yet firmly established and alternative explanations could be given, the simplest single account of the animal and infant studies described above proposes that the precedence effect can be affected separately from steady-state localization.

It would be parsimonious if some form of inhibitory theory could be merged with other postulated mechanisms of binaural hearing to account for steady-state phenomena as well as the precedence effect. Lindemann (1986a,b) made use of the structure proposed by Jeffress (1948) for cross-correlation analysis based on neural delays and coincidence elements and then augmented this model by proposing that a coincidence at a given delay triggers inhibition of coincidence detectors at neighboring delays, an arrangement similar to that proposed by McFadden (1973). The model not only predicts the precedence effect (at least qualitatively) but also implies that responses to steady-state stimuli are sharpened and that interaural level differences are converted to interaural delays. This last effect is not predicted from the Jeffress model without assuming a peripheral neural conversion of interaural intensity differences to interaural delays.

According to Lindemann's model the precedence effect is closely coupled to steady-state localization. It would seem, however, that the model could be recast into a form that is more suggestive of separable inhibition without losing its essential features.

Overview

An overview of the precedence effect is offered in the form of a model (Fig. 4-8) that appears to satisfy the constraints imposed by the data. In the upper path of the model, which can be considered separately for steady-state stimuli, all sources of localization information, both interaural and monaural, are evaluated to form L(t), the instantaneous localization/lateralization estimate. Perceived localization is a running average of L(t) over the present and past. This part of the model can be thought of as corresponding to any theory of directional hearing that accounts for results with steady-state stimulus cues. For example, for localization in the horizontal plane, L(t) could be formed by a combination of interaural delay and interaural amplitude difference. It would be necessary, of course, to postulate interaural noise along the upper branch of the model to account for performance in detection and discrimination tasks.

The lower branch of the model comes into play when the stimulus has a rapid onset. An "onset detector" triggers a momentary inhibitory reaction I(t) that multiplies L(t), attenuating the latter for a period on the order of 5 ms after onset. The precedence effect is proposed to be triggered by rapid onsets because of the many studies, reviewed above, indicating the importance of transients and rapid onsets in obtaining the effect. The inhibition process is included on the basis of results indicating a temporary loss in interaural sensitivity for a few milliseconds after onset. This inhibition takes effect after a delay of about 800 μs.

As for the anatomical locus of these operations, it is intended that they be thought of as central for two reasons. First, the evidence suggests that the precedence effect is obtained whether the localization/lateralization cues are interaural delay, interaural level difference, or diotic spectral

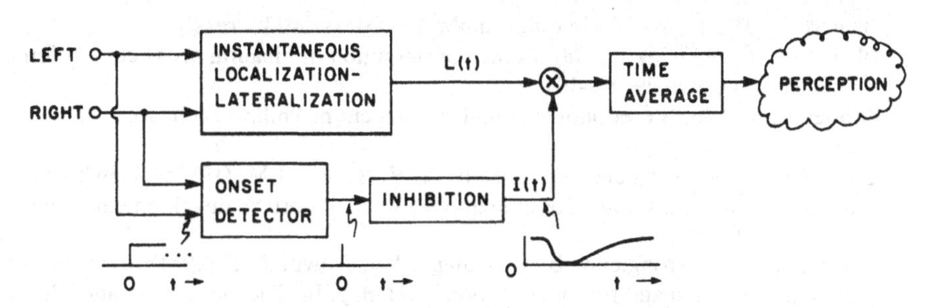

FIGURE 4.8. General structure for a model of the precedence effect.

cues. This implies that either the precedence effect occurs for the different cues separately or it operates on a more abstract feature. For simplicity we opt for the latter. The second motivation for a high-level locus comes from the animal and infant studies, which suggest that the precedence effect is separable from simple localization and that the mechanism underlying the effect may reside in the cortex.

There are obvious but necessary omissions in the model. For example, it is known that to a first approximation the right and left acoustic signals are subjected to peripheral (critical-band) spectral analysis and that binaural interaction occurs primarily between associated bands from the two ears. Filters are not pictured in the model because, first of all, precedence effect aside, there is still a poor understanding of how information is combined across bands to form a unitary localization decision. Second, there is no information on the spectral extent of the precedence effect—whether, for example, an onset in one spectral region inhibits succeeding localization information in other spectral regions. This is only one of the unspecified aspects of the model, but one that at least can be addressed empirically.

Acknowledgments. Preparation of this chapter was supported by grant R01-NS10916 from the National Institute of Neurological and Communicative Disorders and Stroke. I wish to thank R. Clifton, W. Lindemann, and several colleagues in the Auditory Perception Group at the Massachusetts Institute of Technology for their comments on a draft of this review.

References

Abel, S.M., Kunov, H. (1983). Lateralization based on interaural phase differences: Effects of frequency, amplitude, duration, and shape of rise/decay, J. Acoust. Soc. Am. 73, 955–960.

Alkin, L.M. (1973). Sound with Vision. New York: Crane, Reusak & Co.

Blauert, J. (1971). Localization and the law of the first wavefront in the median plane. J. Acoust. Soc. Am. 50, 466–470.

Blauert, J. (1983). Spatial Hearing Cambridge, Mass: MIT Press.

Blauert, J., Cobben, W. (1978). Some consideration of binaural cross correlation analysis. Acustica 39, 96–104.

deBoer, K. (1940). Stereophonic sound reproduction. Philips Tech. Rev. 5, 107–114.

Clifton, R.K., Morrongiello, B.A., Kulig, J.W., Dowd, J.M. (1981a). Newborns' orientation toward sound: Possible implications for cortical development. Child Dev. 52, 833–838.

Clifton, R.K., Morrongiello, B.A., Kulig, J.W., Dowd, J.M. (1981b). Developmental changes in auditory localization in infancy. In: The Development of Perception. Vol. I. Audition, Somatic Perception and the Chemical Senses. Aslin, R., Alberts, J., Petersen, M. (eds.). New York: Academic Press.

Clifton, R.K., Morrongiello, B.A., Dowd, J.M. (1984). A developmental look at an auditory illusion: The precedence effect. Dev. Psychobiol. 17, 519–536.
Colburn, H.S., Durlach, N.I. (1978). Models of binaural interaction. In: Handbook of Perception. Vol. IV—Hearing. Carterette, E.C., Friedman, M.P. (eds.). New York: Academic Press.
Cranford, J., Oberholtzer, M. (1976). Role of neocortex in binaural hearing in the cat. II: The 'precedence effect' in sound localization. Brain Res. 111, 225–239.
Cranford, J., Ravizza, R., Diamond, I., Whitfield, I. (1971). Unilateral ablation of the auditory cortex in the cat impairs complex sound localization. Science 172, 286–288.
David, E.E. (1959). Comment on the precedence effect. Proc. 3rd Int. Congress Acoustics, Amsterdam.
David, E.E., Hanson, R.L. (1962). Binaural hearing and free field effects. Proc. 4th Int. Congress Acoustics, Copenhagen.
Fay, R.D. (1936). A method for obtaining natural directional effects in a public address system. J. Acoust. Soc. Am. 7, 131–132.
Gardner, M.B. (1968). Historical background of the Haas and/or precedence effect. J. Acoust. Soc. Am. 43, 1243–1248.
Gardner, M.B. (1969). Image fusion, broadening, and displacement in sound location. J. Acoust. Soc. Am. 46, 339–349.
Gaskell, H. (1983). The precedence effect. Hear. Res. 11, 277–303.
Grantham, D.W., Wightman, F.L. (1978). Detectability of varying interaural temporal differences. J. Acoust. Soc. Am. 63, 511–523.
Guttman, N. (1962). A mapping of binaural click lateralization. J. Acoust. Soc. Am. 34, 87–92.
Guttman, N. (1965). Binaural interactions of three clicks. J. Acoust. Soc. Am. 37, 145–150.
Guttman, N., van Bergeijk, W.A., David, E.E. (1960). Monaural temporal masking investigated by binaural interaction. J. Acoust. Soc. Am. 32, 1329–1336.
Haas, H. (1951). Uber den Einfluss eines Einfachechos auf die Horsamkeit von Sprache. Acustica 1, 49–58. English translation in: Haas, H. (1972). The influence of a single echo on the audibility of speech. J. Audio Eng. Soc. 20, 146–159.
Hafter, E.R., Dye, R.H. (1983). Detection of interaural differences of time in trains of high-frequency clicks as a function of interclick interval and number. J. Acoust. Soc. Am. 73, 644–651.
Hafter, E.R., Dye, R.H., Wenzel, E. (1983). Detection of interaural differences of intensity in trains of high-frequency clicks as a function of interclick interval and number. J. Acoust. Soc. Am. 73, 1708–1713.
Hall, W.M. (1936). A method for maintaining in a public address system the illusion that the sound comes from the speaker's mouth. J. Acoust. Soc. Am. 7, 239.
Harris, G.G., Flanagan, J.L., Watson, B.J. (1963). Binaural interaction of a click with a click pair. J. Acoust. Soc. Am. 35, 672–678.
Hartmann, W.M. (1983). Localization of sound in rooms, J. Acoust. Soc. Am. 74, 1380–1391.
Hochster, M.E., Kelly, J.B. (1981). The precedence effect and sound localization by children with temporal lobe epilepsy. Neuropsychologia 19, 49–55.
Houtgast, T., Plomp, R. (1968). Lateralization threshold of a signal in noise. J. Acoust. Soc. Am. 44, 807–812.
Jeffress, L.A. (1948). A place theory of sound localization, J. Comp. Physiol. Psych. 41, 35–39.

Kelly, J.B. (1974). Localization of paired sound sources in the rat: Small time differences. J. Acoust. Soc. Am. 55, 1277–1284.

Kunov, H., Abel, S.M. (1981). Effects of rise-decay time on the lateralization of interaurally delayed 1-kHz tones. J. Acoust. Soc. Am. 69, 769–773.

Langmuir, I., Schaefer, V.J., Ferguson, C.V., Hennelly, E.F. (1944). A study of binaural perception of the direction of a sound source. OSRD Report No. 4079.

Leakey, D.M. (1959). Some measurements on the effects of interchannel intensity and time differences in two channel sound systems. J. Acoust. Soc. Am. 31, 977–986.

Leakey, A.M., Cherry, E.C. (1957). Influence of noise upon the equivalence of intensity differences and small time delays in two-loudspeaker systems. J. Acoust. Soc. Am. 29, 284–286.

Lindemann, W. (1986a). Extension of a binaural cross-correlation model by contralateral inhibition. I. Simulation of lateralization for stationary signals. J. Acoust. Soc. Am. 80, 1608–1622.

Lindemann, W. (1986b). Extension of a binaural cross-correlation model by contralateral inhibition. II. The law of the first wavefront. J. Acoust. Soc. Am. 80, 1623–1630.

McFadden, D. (1973). Precedence effects and auditory cells with long characteristic delays. J. Acoust. Soc. Am. 54, 528–530.

Morrongiello, B.A., Kulig, J.W., Clifton, R.K. (1984). Developmental changes in auditory temporal perception. Child Dev. 55, 461–471.

Muir, D., Field, J. (1979). Newborn infants orient to sounds. Child Dev. 50, 431–436.

Parkin, P.H., Humphreys, H.R. (1958). Acoustics, Noise and Buildings. London: Faber & Faber.

Ricard, G.L. (1974). Lateralization of short-duration tones. Ph.D. Thesis, University of California, Berkeley.

Ronken, D.A. (1970). Monaural detection of a phase difference between clicks. J. Acoust. Soc. Am. 47, 1091–1099.

Sayers, B.McA., Toole, F.E. (1964). Acoustic-image lateralization judgments with binaural tansients. J. Acoust. Soc. Am. 36, 1199–1205.

Scharf, B. (1974). Localization of unlike tones from two loudspeakers. In: Sensation and Measurement: Papers in Honor of S.S. Stevens. Moskowitz, H.R., Scharf, B., Stevens, J.C. (eds.). Dordrecht, Holland: Reidel.

Snow, W.B. (1953). Basic principles of stereophonic sound. J. Soc. Motion Pict. Telev. Eng. 61, 567–587.

Snow, W.B. (1954). Effect of arrival time on stereophonic localization. J. Acoust. Soc. Am. 26, 1071–1074.

Stern, R.M., Colburn, H.S. (1978). Theory of binaural interaction based on auditory nerve data. IV. A model of subjective lateral position. J. Acoust. Soc. Am. 64, 127–140.

Thurlow, W.R., Marten, A.E., Bhatt, B.J. (1965). Localization aftereffects with pulse-tone and pulse-pulse stimuli. J. Acoust. Soc. Am. 37, 837–842.

Thurlow, W.R., Parks, T.E. (1961). Precedence-suppression effects for two click sources. Percept. Mot. Skills 13, 7–12.

Tobias, J.V., Schubert, E.R. (1959). Effective onset duration of auditory stimuli. J. Acoust. Soc. Am. 31, 1595–1605.

Tobias, J.V., Zerlin, S. (1959). Lateralization threshold as a function of stimulus duration. J. Acoust. Soc. Am. 31, 1591–1594.

von Bekesy, G. (1930). Zur Theorie des Horens: Uber das Richtungshoren bei einer Zeitdifferenz oder Lautstarkenungleichheit der beiderseitigen Schalleinwirkungen. Physik. Z. 31, 824–835, 857–868. English translation in: von Bekesy, G. (1960). Experiments in Hearing pp. 272–301. New York: McGraw Hill.
Wallach, H., Newman, E.B., Rosenzweig, M.R. (1949). The precedence effect in sound localization. Am. J. Psychol. 52, 315–336.
Whitfield, I.C., Cranford, J., Ravizza, R., Diamond, I. (1972). Effect of unilateral ablation of auditory cortex in cat on complex sound localization. J. Neurophysiol. 35, 718–731.
Whitfield, I., Diamond, I., Chiveralls, K., Williamson, T. (1978). Some further observations on the effects of unilateral cortical ablation on sound localization in the cat, Experimental Brain Research 31, 221–234.
Yost, W.A., Soderquist, D.R. (1984). The precedence effect: Revisited. J. Acoust. Soc. Am. 76, 1377–1383.
Zurek, P.M. (1976). An investigation of the binaural perception of echoed sound. Ph.D. Thesis, Arizona State University, Tempe.
Zurek, P.M. (1980). The precedence effect and its possible role in the avoidance of interaural ambiguities. J. Acoust. Soc. Am. 67, 952–964.
Zwicker, E. (1965). Temporal effects in simultaneous masking by white-noise bursts. J. Acoust. Soc. Am. 37, 653–663.

von Békésy, G. (1930). Zur Theorie des Hörens. Über das Richtungshören bei einer Zeitdifferenz oder Lautstärkenungleichheit der beiderseitigen Schalleinwirkungen. Physikal. Z. 31, 824–835, 857–868. English translation in von Békésy, G. (1960). Experiments in Hearing (pp. 272–301). New York: McGraw-Hill.
Wallach, H., Newman, E. B., Rosenzweig, M. R. (1949). The precedence effect in sound localization. Am. J. Psychol. 62, 315–336.
Whitfield, I. C., Cranford, J., Ravizza, R., Diamond, I. (1972). Effects of unilateral ablation of auditory cortex in cat on complex sound localization. J. Neurophysiol. 35, 718–731.
Whitfield, I. C., Diamond, I. T., Chiveralls, K., Williamson, T. G. (1978). Some further observations on the effects of unilateral cortical ablation on sound localization in the cat. Exp. Brain Res. 31, 221–236.
Yost, W. A., Soderquist, D. R. (1984). The precedence effect: revisited. J. Acoust. Soc. Am. 76, 1377–1383.
[illegible] (1976). [illegible] Arizona State [illegible] Tempe.
[illegible] M. (19[illegible]). The [illegible] and [illegible] its possible role in [illegible]
[illegible] (19[illegible]). [illegible] Am. 7[illegible]

Part II Structure and Function of the Mammalian Binaural Nervous System

One of the tasks for understanding directional hearing is to unravel the anatomical and functional complexities of the binaural nervous system. The two chapters in this section present mammalian anatomy and neurophysiology that are believed to be essential for processing variables pertinent to the localization of sound.

The departure point for the anatomical analysis is beyond the cochlea and the eighth nerve, since only ipsilateral information is processed by these structures. The analysis begins with the cochlear nucleus, which is the source for a group of neurons that interact at the following neural level (superior olivary complex) with corresponsing neurons from the contralateral side, and which extends to neural structures at higher levels of the CNS where further processing of binaural information takes place.

Although directional hearing is primarily binaural, it can also be monaural. Anatomical analyses are also being applied to this aspect of directional hearing and suggest that separate pathways exist for processing monaural directional information.

Our knowledge of the neurophysiology of directional hearing is concentrated on the processing of interaural time and interaural intensity cues—that is, on the mechanisms underlying the duplex theory. In this vain, recent research has advanced considerably our understanding of how neural structures respond to these cues, particularly in the brain stem. Progress has also been made in another aspect of the neural mechanisms of directional hearing, that is, in establishing the existence of neural representation of auditory space.

In many respects, however, our comprehension of the neurophysiology of directional hearing is less than that of the neuroanatomy of directional hearing. Little is known about the neural activity that extracts directional information from the binaural cues of "natural" auditory signals; for example, spectral differences and time differences in high-frequency complex waveforms. Similarly, there is a dearth of information on neural events underlying the precedence effect.

The material in this section should provide an understanding of the anatomical and neurophysiological processing of interaural information and help the reader to gauge some of the material in the remaining sections of the book. It should also give the reader an appreciation of the problems that remain for finding exact correspondences between directional hearing observations made in behavioral studies and those made in neural studies.

5
Central Auditory Pathways in Directional Hearing

JOHN H. CASSEDAY AND ELLEN COVEY

Introduction

The central auditory system consists of several pathways that parallel one another at some levels of the nervous system and converge or diverge at other levels. The purpose of this chapter is to examine the structure of these pathways for clues to their function in directional hearing. Figure 5-1 introduces the main components, up to the inferior colliculus, of pathways that may be involved in processing cues for directional hearing. The analysis of these pathways will begin at the cochlear nucleus, where all fibers of the auditory nerve terminate (cf Jones and Casseday, 1979). The anteroventral division of the cochlear nucleus is particularly important because it is the origin of pathways to the medial and lateral superior olives, structures that have connections ideally arranged to function as centers for binaural hearing. Other pathways originate in the cochlear nucleus but bypass the superior olivary complex and cross the brain stem to terminate in the nuclei of the lateral lemniscus and inferior colliculus (Fig. 5-1A). These are also considered in this analysis, even though they probably have little to do with binaural interaction at this first level of processing. It will be proposed that the ascending auditory pathways may be divided into binaural and monaural channels. The binaural channels are those pathways in which information from both ears is integrated at the superior olives. Each monaural channel transmits information from the contralateral ear without integration at the superior olives. Evidence reviewed in a later section suggests that this separation of binaural and monaural channels persists to the level of the auditory cortex.

The inferior colliculus is of special interest because it is the target of pathways from the cochlear nuclei, medial and lateral superior olives, and nuclei of the lateral lemniscus (Fig. 5-1A,B). In fact, most ascending fibers in the brain stem auditory system ultimately terminate in the inferior colliculus (Goldberg and Moore, 1967; Aitkin and Phillips, 1984). Thus the question arises of how all these pathways converge with or diverge from one another at the inferior colliculus. For the purpose of this chapter it

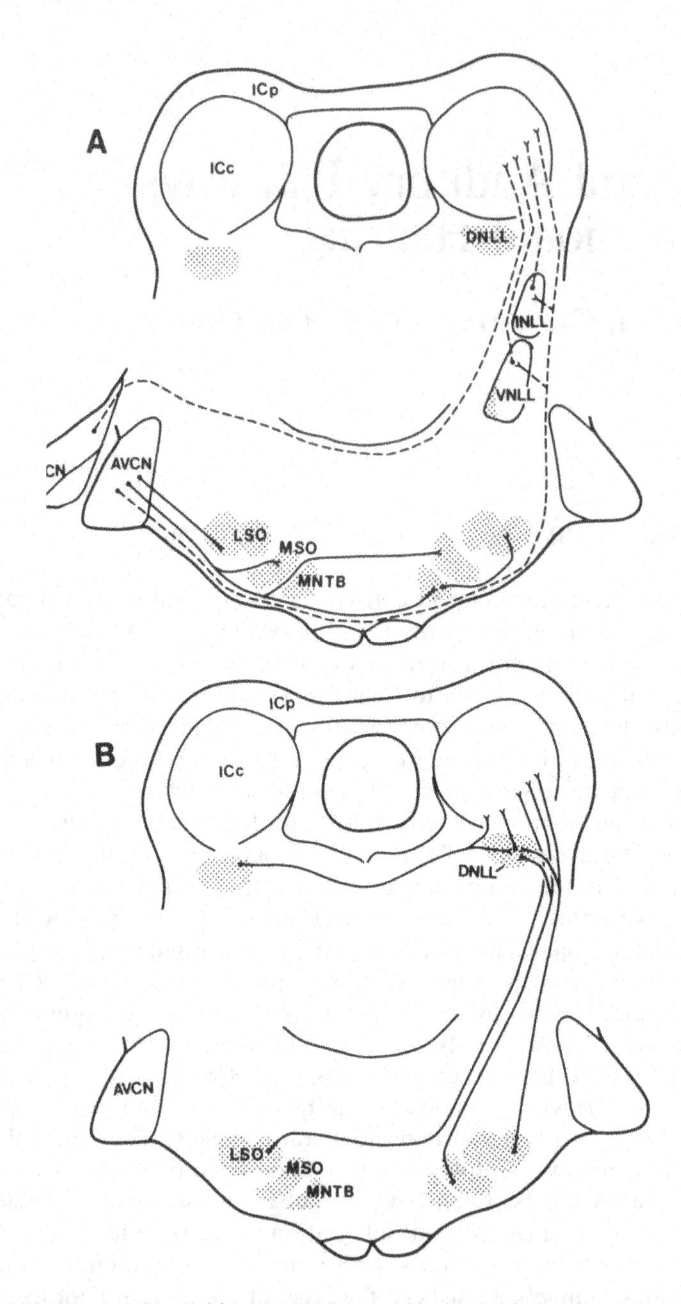

FIGURE 5.1. Monaural and binaural pathways in the brain stem of the cat. **A:** Pathways that originate in AVCN and DCN. Some of these go directly to the central nucleus of the inferior colliculus (dashed lines) while others terminate in the superior olivary complex and nuclei of the lateral lemniscus (solid lines). Major binaural centers below the inferior colliculus are indicated by stippling. Note that direct projections from AVCN innervate MSO bilaterally, whereas LSO receives direct projections from the ipsilateral side and indirect contralateral projections via MNTB. **B:** Pathways to the inferior colliculus from the major binaural centers in the brain stem. For simplicity, pathways to only one inferior colliculus are shown.

would be useful to know which other pathways converge with those from the medial and lateral superior olives. The answer is far from complete, but as is shown later, an important clue may lie in the arrangement of sheets of projections to the inferior colliculus.

The analysis of pathways in auditory thalamus and auditory cortex depends to a large extent on how the above question is answered. At present we know that there are multiple pathways at both levels, i.e., the lower brainstem and the forebrain. How are the two linked at the inferior colliculus? The solution to this puzzle would provide clues to functional differences among the various thalamic pathways to auditory cortex. Because the answer is not at hand, this chapter is weighted toward the role played in directional hearing by auditory pathways in the lower brain stem. Nevertheless, it is important to attempt to extend this analysis to the auditory thalamus and auditory cortex because these structures are important for localization of sound (Strominger, 1959; Diamond et al, 1969; Neff and Casseday, 1977; Jenkins and Merzenich, 1984), even though the primary cues for localization are encoded in the brain stem. Described are a few recent results that relate the binaural characteristics of strips of primary auditory cortex, to individual laminae in the medial geniculate body.

Divisions of the Cochlear Nucleus

The cochlear nucleus can be divided into three main structures, each of which differs from the others in afferent and efferent innervation as well as in internal architecture. Incoming fibers of the auditory nerve form a basis for subdivision in that most fibers bifurcate so as to send an ascending branch to the anteroventral cochlear nucleus and a descending branch to the posteroventral and dorsal cochlear nuclei (Fig. 5-2A; Held, 1893; Cajal, 1909; Lorente de No, 1933; Fekete et al, 1984). It is important to note that many cells in the anterior part of the anteroventral cochlear nucleus are innervated by large endings, called calyces of Held, from the auditory nerve (Fig. 5-2). The large size of these endings provides relays that are both rapid and reliable (Pfeiffer, 1966; Morest, 1968) and in this way may be responsible for an important feature of stimulus coding. Neurons in this part of the cochlear nucleus follow the input of the auditory nerve fibers with "primary-like" responses and so preserve the temporal pattern of the waveform of the auditory stimulus (Kiang et al, 1973). This timing information is essential for later encoding of cues for sound localization at binaural centers. A second important point, which will be discussed later, is that each fiber provides several endings along its path through the anteroventral cochlear nucleus (Fig. 5-2B).

Although this chapter deals primarily with the mammalian auditory system, one of the best examples of separation of binaural function is found in the barn owl (Takahashi et al, 1984). In experiments to record spatial

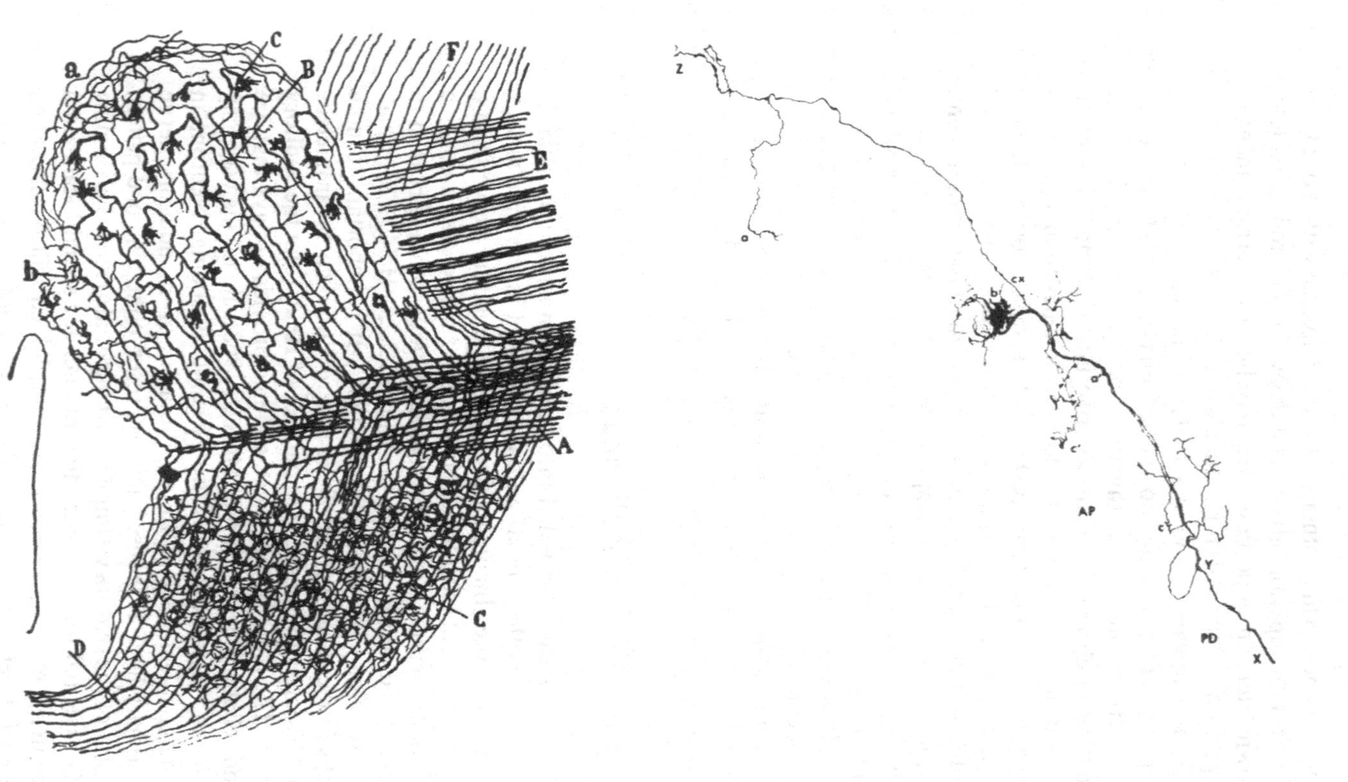

FIGURE 5.2. Termination of auditory nerve fibers in the cochlear nucleus. *Left:* Entering fibers bifurcate to send ascending branches to AVCN (top) and descending branches to PVCN and DCN. DCN is not included in the drawing (from Cajal, 1909). Note the numerous large endings in AVCN. A, auditory nerve; B, AVCN; C, PVCN; D, continuing fibers of the descending root destined for DCN; a, dorsal marginal plexus; b, terminal arborization of the ascending root; c, calyx formed by collaterals of the ascending root. *Right:* Calyx of Held (b) and other endings of one fiber in AVCN (from Brawer and Morest, 1975). The fiber is shown from its cut end (X) in posterior AVCN (dorsal part, PD) to course into the anterior division (posterior part, AP) at point Y; a,a′, types of axonal appendages; b, axonal end bulb; CX, terminal part of axon, which ends at Z.

tuning of neurons in the inferior colliculus, different divisions of the cochlear nucleus were anesthetized. The results show that the nucleus magnocellularis supplies information for analysis of interaural time differences and the nucleus angularis supplies information for analysis of interaural intensity differences.

Three Main Pathways From the Cochlear Nucleus

Each of the three main divisions of the cochlear nucleus is the source of at least one distinct pathway. As will be described in detail later, the anteroventral cochlear nucleus (AVCN) may be regarded as the origin of binaural pathways in that it is the main source of bilateral projections to the medial and lateral superior olives via the trapezoid body. The dorsal cochlear nucleus (DCN) is the source of pathways which largely bypass the superior olivary complex and project directly, via the dorsal acoustic stria, to the contralateral inferior colliculus (Fernandez and Karapas, 1967; van Noort, 1969; Adams, 1979; Oliver, 1984a). Thus, DCN is a candidate for the source of a monaural pathway. The posteroventral cochlear nucleus (PVCN) is a major source of pathways to the periolivary cell groups; pathways from PVCN also project across the brain stem in the intermediate acoustic stria and ascend to the ventral and intermediate nuclei of the lateral lemniscus and directly to the inferior colliculus (van Noort, 1969; Warr, 1982; Jones 1979). Neither DCN nor PVCN is the source of a major projection to the binaural centers of the superior olivary complex. The pathways from the cochlear nucleus, especially those in the trapezoid body, are essential for normal directional hearing. Even partial interruption of the trapezoid body disrupts an animal's ability to localize sound in space (Moore et al, 1974; Casseday and Neff, 1975; Jenkins and Masterton, 1982) or to detect binaural time differences (Masterton et al, 1967), and interruption of all three pathways can completely eliminate an animal's ability to localize sound (Casseday and Neff, 1975). In contrast, if the auditory commissures are interrupted at higher levels, at the commissure of the inferior colliculus or at the corpus callosum, for example, animals are unaffected in their ability to localize sound (Moore et al, 1974). This summary has accounted for the major pathways that are capable of transmitting the large amounts of information necessary for precise encoding of binaural differences in time or intensity.

Binaural Versus Monaural Pathways

All three divisions of the cochlear nucleus are sources of some pathways that probably have little to do with the initial stages of binaural interaction. These pathways cross the brain stem and bypass the superior olivary complex to reach the nuclei of the lateral lemniscus or the inferior colliculus or both. For this reason these pathways may be considered as "monaural" channels of information, at least until they reach the midbrain. Later, the question will be addressed of whether or not the "monaural" channels interact with "binaural" channels at the inferior colliculus. First it is necessary to have a picture of the structure and innervation of binaural centers in the superior olivary complex.

Binaural Centers in the Superior Olivary Complex

Structure of the Superior Olivary Complex

The study of the superior olivary complex has revealed a striking example of the relationship between structure and function. This section will show that the medial and lateral superior olives are ideally suited to process the two main binaural cues for sound localization—time differences and intensity differences at the two ears.

In most mammals the major components of the superior olivary complex are the medial superior olive (MSO), lateral superior olive (LSO), and

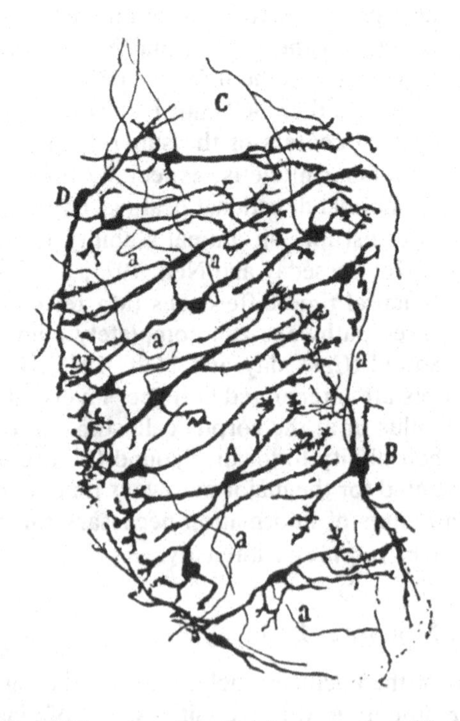

FIGURE 5.3. Cells in MSO as seen in a Golgi preparation. Note the long dendrites that extend in a medial-lateral direction to span almost the entire width of the nucleus (from Cajal, 1909). A, central cells; a, axons; B and D, marginal cells; C, fiber plexus.

medial nucleus of the trapezoid body (MNTB). Viewed in the frontal plane, MSO contains elongate cells with dendrites that extend medially and laterally (Fig. 5-3). The LSO is similar in structure, but in most mammals it has a convoluted shape, and in the cat it is S-shaped (Fig. 5-4). The cell dendrites in the LSO span the width of each fold of the nucleus. Seen in the sagittal plane, the dendrites of most cells in both LSO and MSO extend for a considerable distance along the anterior-posterior dimension (Scheibel and Scheibel, 1974; Schwartz, 1977). The arborizations of incoming axons form sheets that in the frontal plane are seen in cross-section (Fig. 5-5); the sheets extend throughout the anterior-posterior dimension of LSO and MSO (Scheibel and Scheibel, 1974; Schwartz, 1984). The third main structure in the superior olivary complex, MNTB, is quite different from the other two structures in type of cell and in the type of endings its cells receive from the cochlear nucleus. The principal type of neuron in MNTB is spherical or oval in shape, with short, branched dendrites. Axonal endings from the cochlear nucleus form calyces (Fig. 5-5) that are similar to, but larger than, the type of endings found around spherical cells in the anteroventral cochlear nucleus (Morest, 1968). The function of these endings as fast relays is probably also similar (Pfeiffer, 1966).

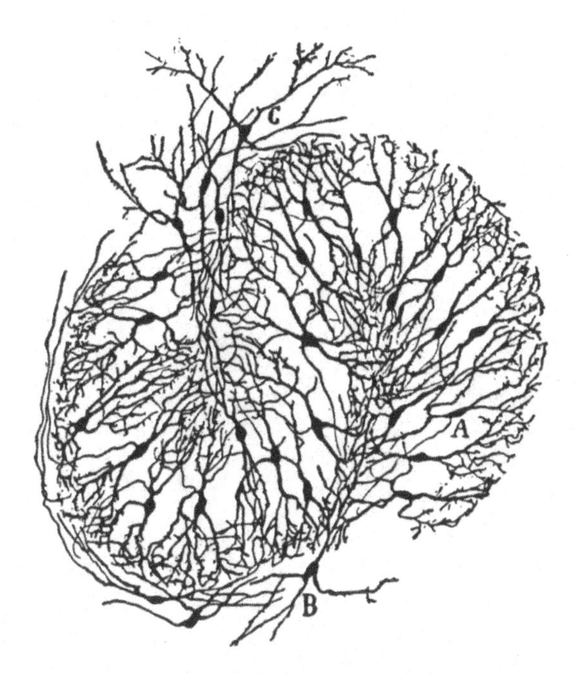

FIGURE 5.4. Cells in LSO as seen in a Golgi preparation. Note the similarity of these cells to those in MSO (Fig. 5-3). The dendrites span the width of LSO in an orientation perpendicular to the folds of the nucleus (from Cajal, 1909). A, intralaminar fusiform cells; B and C, marginal cells.

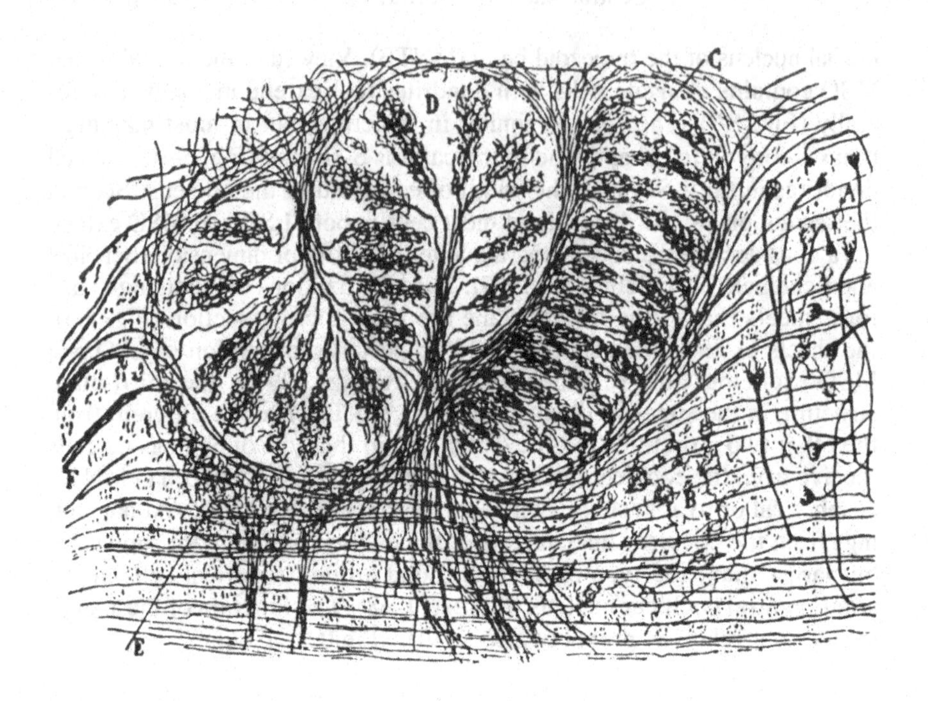
D
C
B
E

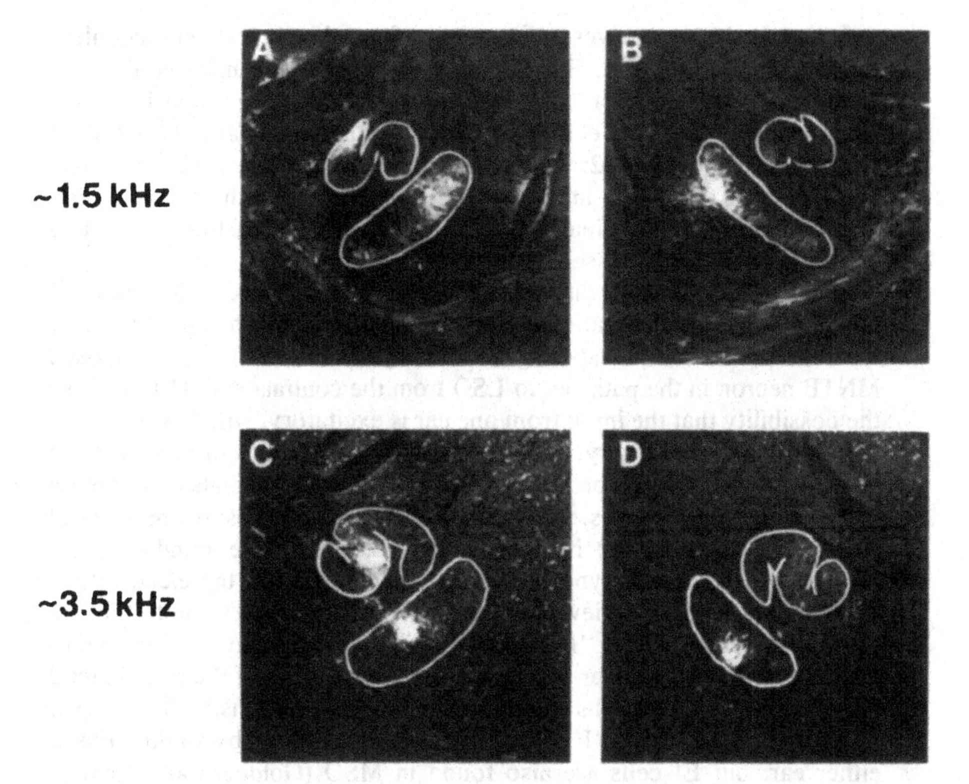

FIGURE 5.6. Photomicrographs illustrating divergence of the pathway from AVCN into separate tonotopically organized nuclei in the superior olivary complex. All photomicrographs are from frontal sections through the brain stem of cats following injections of WGA-HRP in AVCN. All were taken with dark-field illumination so that the reaction product appears as a light patch against a dark background. LSO and MSO have been outlined in white. **A**, **B**: Respectively, the ipsilateral and contralateral superior olives of a cat with a small injection of WGA-HRP in AVCN. The best frequencies of cells recorded at the injection site were about 1.5 kHz. **C**, **D**: Corresponding sections in a second animal in which the best frequencies of cells at the injection site were about 3.5 kHz. Note that in both cases there is transport to both LSO and MSO on the ipsilateral side, but only to MSO contralaterally.

of cells at the injection sites. Several features of the innervation of MSO and LSO demonstrate that these nuclei must represent a further subdivision of the pathway. First, the pathway from AVCN diverges so that there is one complete tonotopic representation at MSO and a second at LSO. Second, AVCN projects only to the ipsilateral LSO, but projects to MSO of both sides. The projection to each MSO is most dense in the half of the nucleus that is ipsilateral to the injection. Although LSO receives direct input only from the ipsilateral cochlear nucleus, it is clearly a binaural structure in that it receives indirect input from the contralateral cochlear nucleus via MNTB (e.g., Harrison and Warr, 1962; Spangler et al, 1985; Glendenning et al, 1985). The cells in MNTB are innervated by large-diameter fibers (Spangler et al, 1985) that arise from globular cells in AVCN (Harrison and Warr, 1962; van Noort, 1969; Warr, 1969, 1972, 1982; Tolbert et al, 1982) and end in calyces of Held, indicating that transmission at this relay is rapid. Finally, LSO and MSO differ in their ascending connections, as is discussed later.

The fact that there are differences in the connections to MSO and LSO has the following implications: cells in MSO probably receive the same type of input from both ears, e.g., excitatory (Clark, 1969). The interposed MNTB neuron in the pathway to LSO from the contralateral AVCN raises the possibility that the input from one ear is excitatory, whereas that from the other ear is inhibitory. Electron microscopy studies of cells in LSO show that the soma and proximal dendrites receive terminals that contain small flat synaptic vesicles, presumably inhibitory synapses; the most distal parts of the dendrites receive terminals that contain large round vesicles, presumably excitatory synapses (Cant, 1984). Because the electrophysiological evidence is reviewed elsewhere, it is sufficient to note that, in general, cells in LSO are "EI" cells, that is, are excited by auditory stimulation of the ipsilateral ear and inhibited by stimulation of the contralateral ear; in addition, these cells are sensitive to binaural intensity differences. Many cells in MSO are "EE" cells, that is, are excited by stimulation of either ear, but EI cells are also found in MSO (Goldberg and Brown, 1969). Many cells in MSO are sensitive to binaural time differences. However, it should be noted that the exact locations of cells that have different binaural response properties have not been identified by modern anatomical techniques for marking individual cells. Moreover, with a few exceptions (Galambos et al, 1959; Goldberg and Brown, 1969; Tsuchitani, 1977), the existing electrophysiological evidence was obtained without marking electrode penetrations; therefore there is considerable doubt as to the exact location of EE and EI cells within the superior olivary complex.

Models of Binaural Interaction in MSO and LSO

The bilateral innervation of the cells in MSO, discovered by Stotler (1953), is remarkably similar to the neural model of a place code for binaural time differences proposed by Jeffress (1948). Figure 5-7A illustrates Jeffress' model as seen in light of Stotler's finding. In this model each neuron in MSO is viewed as a coincidence detector that is excited when the inputs to the medial and lateral dendrites coincide (Fig. 5-7A). Each input to an MSO cell has a set delay. Any given pair of unequal delays will offset some small difference in stimulus onset time at the two ears when a sound originates nearer one ear than the other. Thus the paired delays are the basis for transforming small differences in the time of arrival of a sound at the two ears into a place code at the MSO. Jeffress proposed that differences in fiber length could provide the appropriate delays. The neural circuitry for such a system of delay lines has been described in the avian homolog of MSO, the nucleus laminaris (Young and Rubel, 1983; see Fig. 5-12).

Another possible mechanism for producing delays is suggested by the projection of fibers of the auditory nerve across slabs of cells in AVCN (see Fig. 5-2). Figure 5-7B shows schematically how an auditory nerve fiber might contact several different cells to produce at each cell a slightly different delay with respect to the onset of the stimulus. Note that the fiber in this schematic diagram is not different in principle from the fiber in Figure 5-2B. For convenience in illustrating this point, the delays are represented by fiber length, but differences in the configuration of the terminals (as in Fig. 5-2B) would be an equally plausible mechanism for producing different delays at different cells in AVCN. The total delay in the arrival of an impulse at the dendrite of an MSO cell would then be the sum of the delays along the auditory nerve fiber at AVCN plus the delay introduced by the nerve impulse traveling from AVCN to the MSO cell.

A second model of binaural interaction, first proposed by von Bekesy (1930) and further elaborated by van Bergeijk (1962) is mentioned here because its design fits some of the connections and functions of LSO. This model was designed to explain the psychophysical phenomenon of a trading relationship between binaural time differences and binaural intensity differences when the two cues are independently manipulated via earphones. Van Bergeijk applied the model to MSO because, at the time, connections from MNTB to LSO were not known. From evidence already reviewed, LSO seems a more likely candidate for this type of processing. In this model the location of sound is encoded by the level of excitation in one bilaterally innervated nucleus relative to the level of excitation in its contralateral counterpart. The level of excitation in each nucleus is a result of excitatory input from one ear and inhibitory input from the other ear. Applying the model to LSO, the ipsilateral input would be excitatory; the contralateral, inhibitory. The model requires that incoming fibers branch for some distance within LSO. It incorporates the idea of different delays from the two ears, but as a recruitment mechanism, not as a mechanism for producing a place code for binaural delays. The model postulates that in one dimension of LSO, e.g., anterior to posterior, there is recruitment of cells over time, and in the orthogonal dimension there is recruitment with increasing intensity. Considering first the time dimension, impulses arriving from the two sides would travel toward the center of the nucleus and produce a wave of excitation or inhibition; when the excitatory and inhibitory waves meet they would cancel one another. Thus, if the sound leads in time in one ear, there would be a greater excited area in the LSO ipsilateral to the leading ear due to the greater travel time

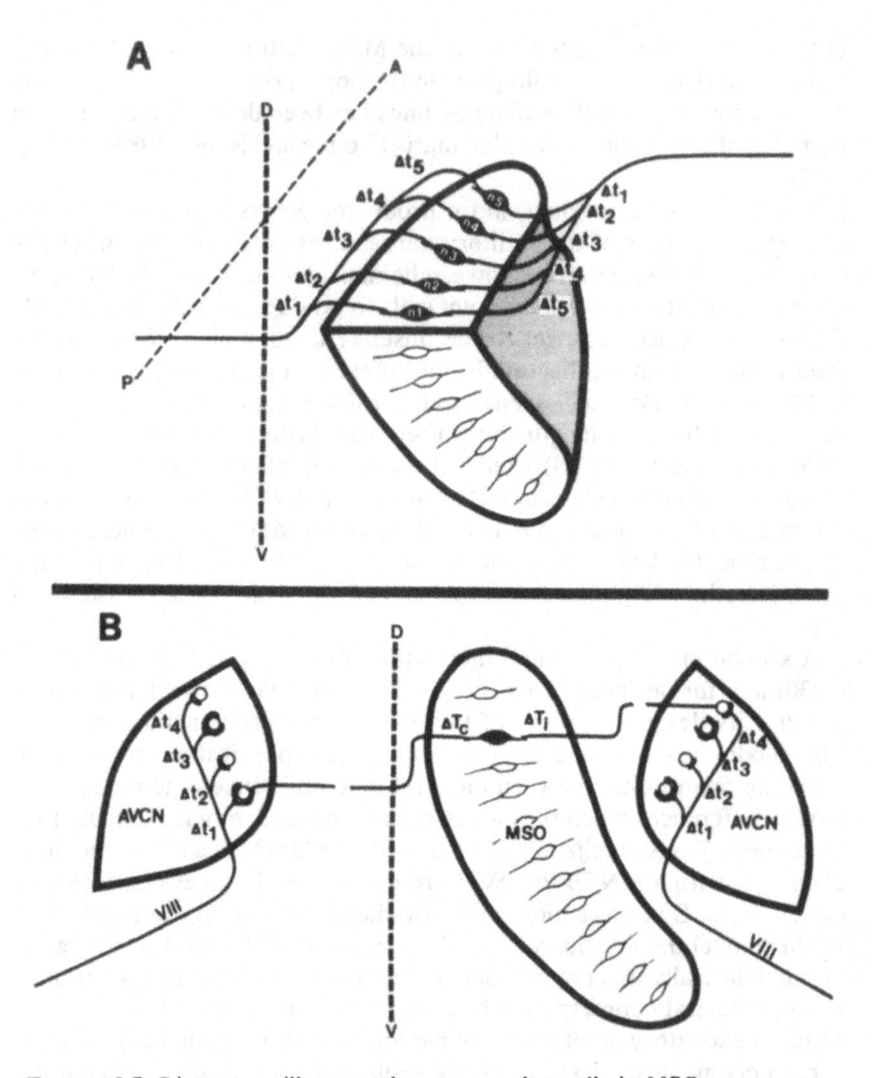

FIGURE 5.7. Diagram to illustrate the concept that cells in MSO operate as coincidence detectors so that temporal differences at the two ears are converted into an anatomical location (i.e., a "place code") in the nervous system. A: Model proposed by Jeffress (1948) as it might be arranged in MSO. A hypothetical array of neurons receives input from two different "delay lines". The neural array is shown in a schematic horizontal section through MSO so that the innervated cells are all located within an isofrequency plane. Each neuron receives bilateral input. Considering first the input from the ipsilateral side (right), the incoming fibers are arranged so that each successive neuron in the array receives a slightly different delay (Δt) relative to the onset of the stimulus. Because the input from the contralateral side (left) has a similiar set of delays arranged in the reverse direction, some neurons in the array will receive simultaneous input for a given position of a sound source in space. For example, if a sound directly in front reaches both ears at the same time and yields Δt_3 at both the left and the right, then neuron n_3

of the excitatory input prior to encountering inhibitory input from the opposite side. Along the intensity dimension, the ratio of excited area to inhibited area on a given side would similarly be proportional to the intensity of the sound. Thus, the ratio of excited to inhibited area would depend on both relative intensity and relative onset of the stimulus at the two ears. Laterality would be encoded by comparing the amount of excited area in one LSO relative to the excited area in the contralateral LSO. The neural basis for the comparison mechanism presumably depends on ascending connections to the inferior colliculus. However, there is little in what is known about these connections to suggest how such a comparison might work.

To progress further with models of the sort just described it is necessary to have a more detailed picture of the connections to MSO and LSO, especially the connections along the anterior-posterior plane. The fact that dendrites of cells in both MSO and LSO are distributed for a considerable distance along this plane adds an interesting complication to the simple picture shown in Figure 5-7A. What is the consequence of this extensive dendritic branching for the performance of the models? The answer depends on knowledge of how individual fibers from AVCN contact individual cells in MSO or LSO.

Origin of Pathways to the Superior Olivary Complex

Osen (1969b) proposed the important idea that the pathways to LSO and MSO have separate origins in AVCN, specifically that the pathway to LSO originates from small spherical cells whereas the pathway to MSO originates from large spherical cells. These two types of spherical cells have a different distribution in AVCN, large spherical cells being located more anteriorly than small spherical cells (Osen, 1969a), although the two

---

will fire. If the sound is moved to the right, then the sound reaches the right ear before the left and a longer delay will be required at the right input to MSO than at the left in order to compensate for the delay of the sound. In this case it might be neuron n_1 that would receive simultaneous inputs. **B**: Possible mechanisms by which delays might be introduced before the level of MSO, which is now illustrated in a schematic frontal section. An auditory nerve fiber (VIII) is shown in each AVCN to branch and contact an array of cells along an isofrequency plane in AVCN. A different delay would be introduced, depending on which of these cells then contacted the dendrite of the cell in MSO. In the example shown here, the delay at the medial dendrite would be $\Delta t_1 + \Delta T_c$ (the delay in the contralateral cochlear nucleus, Δt_1, plus the delay of the fiber from the cochlear nucleus, ΔT_c); the delay at the lateral dendrite would be $\Delta t_1 + \Delta t_2 + \Delta t_3 + \Delta T_i$ (the sum of the delays at the ipsilateral cochlear nucleus plus the delay in the ipsilateral fiber from the cochlear nucleus, ΔT_i). If ΔT_c and ΔT_i were equal, these delays would provide simultaneous input from a sound originating from somewhere to the right of the midline.

types of cells overlap somewhat (Brawer et al, 1974). Recent studies of retrograde transport support the general principle of Osen's proposal, with some modifications. The MSO does receive input from restricted parts of anterior AVCN, from the large spherical cell area but also from a subdivision of the small spherical cell area (Cant and Casseday, 1986). The LSO receives input from these areas (Warr, 1982; Glendenning et al., 1985) and also from more posterior parts of AVCN (Warr, 1982; Rouiller and Ryugo, 1984; Cant and Casseday, 1986). These different populations of cells may provide different transformations of the stimulus, potentially providing LSO with a type of input different from that to MSO (Kiang et al, 1973; Rhode et al, 1983; Rouiller and Ryugo, 1984). Thus, division of the binaural system into parallel pathways begins at its origin and is completed as fibers from AVCN diverge into separate pathways to LSO and MSO. A separate pathway to MNTB can be identified on the basis of cell type of origin because its input comes from globular cells in AVCN (Warr, 1982). A later section addresses the question of whether the pathways from MSO and LSO remain separate at higher levels.

Comparative Observations Concerning the Superior Olivary Complex

The structure of the superior olivary complex in mammals varies considerably from one species to another. It has been proposed that the presence, absence, or relative development of its component nuclei may provide clues to their function (Harrison and Irving, 1966; Masterton et al, 1975). On the one hand, LSO is present and appears to be similar in structure in all mammalian species examined. As would be expected from its connections to LSO, MNTB is also present in all species of mammals. On the other hand, the development of MSO is quite variable. In some species, such as in *Galago* (Fig. 5-8), it is large and more prominent than LSO;

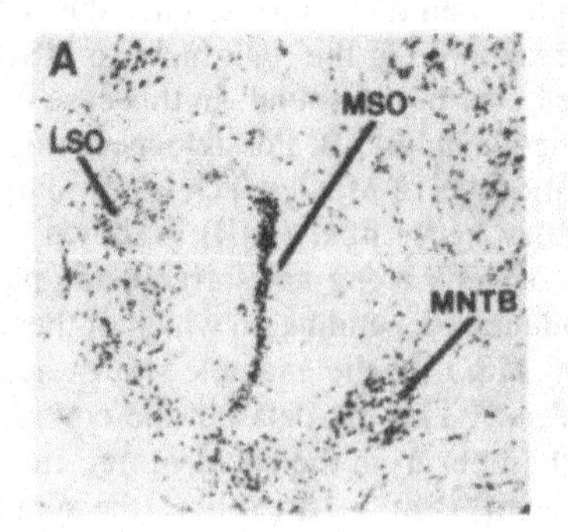

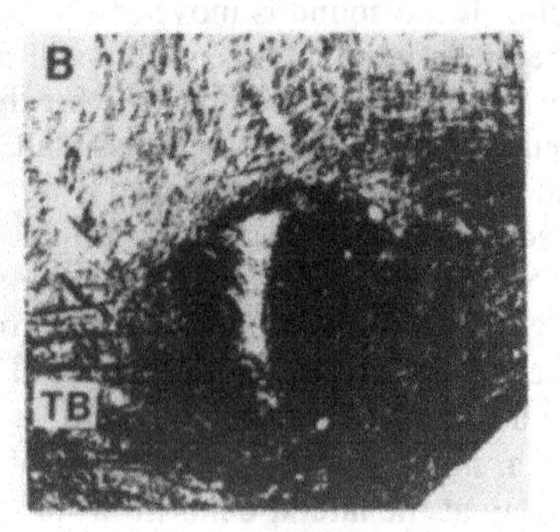

FIGURE 5.8. Photomicrographs demonstrating the appearance of MSO in the frontal plane in a primate, *Galago senegalensis*: **A:** Nissl stain of the left superior olive. **B:** a fiber stain of the right superior olive.

in other species MSO is entirely absent. Two theories have been proposed to account for such species variations. Harrison and Irving (1966) suggested that the development of MSO is related to the development of visual pathways, so that MSO tends to be large in diurnal animals and small or even absent in some nocturnal mammals. Masterton and his colleagues (Masterton and Diamond, 1967; Masterton et al, 1975) suggested that the size of MSO is related to the use of binaural cues. Animals with large heads have a large range of binaural time cues available to them and thus should have a large MSO. Animals with small heads have a small range of binaural time cues available to them and apparently rely mostly on binaural intensity cues. Small animals also have a greater potential capacity to use binaural intensity cues because their range of hearing usually extends to higher frequencies than does the range of large mammals (Masterton, et al, 1969).

The study of echolocating bats has resulted in evidence that at first appears inconsistent with either of these theories. Some of these small nocturnal mammals have a well-developed MSO such as that seen in *Pteronotus parnellii* (Fig. 5-9) and in *Rhinolophus ferrumequinum* (Schweizer, 1981). These bats rely on echolocation for survival and have poorly developed visual systems. Although their range of hearing extends above 100 kHz, the low-frequency end of the range may not extend to the frequencies at which phase information could be coded into binaural time cues. Can timing information be found elsewhere in the bat's auditory environment, such as in the echoes of high-frequency sonar pulses? Both *Pteronotus* and *Rhinolophus* use long constant frequency echolocation cries. The echo of a long constant-frequency pulse can be modulated by

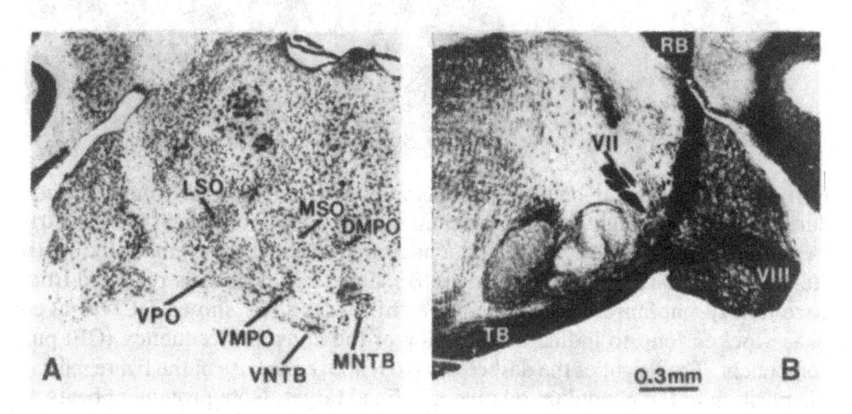

FIGURE 5.9. Photomicrographs demonstrating the appearance of the superior olivary complex in the frontal plane in the echolocating bat *Pteronotus parnellii*, one of the few species of bats with a large MSO. **A**: cell bodies (Nissl stain) of the left side of the brain stem. **B**: A fiber stain showing the right side of a section adjacent to that shown in **A**. Note the dense axonal plexus around the cell bodies of MSO (from Zook and Casseday, 1982a).

the wing beats of a moth that the bat is pursuing (Goldman and Henson, 1977). Alternatively, if the echo overlaps with the outgoing pulse while the bat is flying, beat frequencies can occur (Fig. 5-10). Either of these modulations in the envelope of the stimulus contains timing information that could be used as binaural time cues for localizing targets. This possibility is suggested by the ability of human observers to detect binaural time differences on the basis of ongoing cues of amplitude variation of high-frequency stimuli (McFadden and Pasanen, 1976). Finally, an MSO has been identified only in species of bats that emit constant-frequency pulses. Most bats emit short frequency-modulated pulses, and in such bats an MSO has not been identified. The superior olivary complex appears to vary greatly among species of bats, and the possibility that this variation is related to differences in echolocation pulse design merits further study.

Studies of the avian auditory system reveal that part of the binaural pathway is remarkably similar to that of mammals. Nucleus laminaris in the bird is homologous to MSO in mammals (Boord, 1968; Young and Rubel, 1983). The bird does not have an obvious homolog to LSO. Figure 5-11 shows the nucleus laminaris in a chicken. The pattern of projections

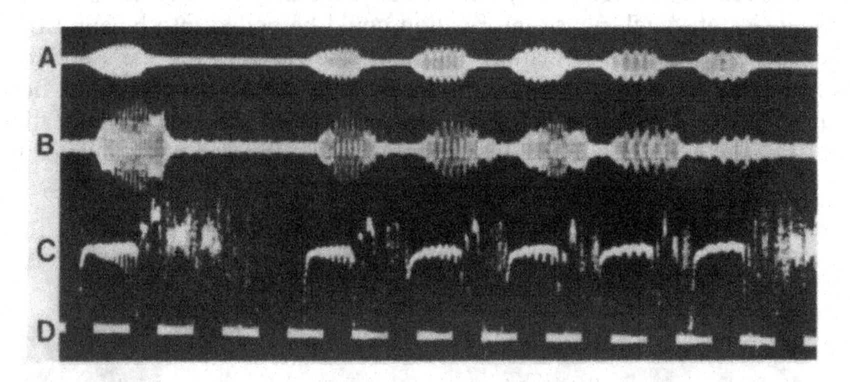

FIGURE 5.10. Cochlear microphonic (CM) potentials recorded from *Pteronotus* during simulated "flight" on a pendulum swinging toward a target. The top trace (A) shows emitted pulse detected by a microphone under the animal's mouth (see Henson et al, 1982). The second trace (B) shows CM potentials recorded from a chronically implanted electrode. The third trace (C) shows DC voltage of phase-locked loop to indicate frequency of the constant-frequency (CF) pulse component. The height of the dashed trace (D) above the base of the figure indicates the position of the pendulum relative to a fixed target. Note prominent beats registered in CM and microphone records. The beat frequency represents the frequency difference between the emitted pulse and the CF components of the echo. The frequency of the beat becomes lower in relation to each successive pulse because as the pendulum slows down, from left to right in the tracings, the difference between vocalization frequency and echo frequency decreases (unpublished records, O.W. Henson and J.B. Kobler).

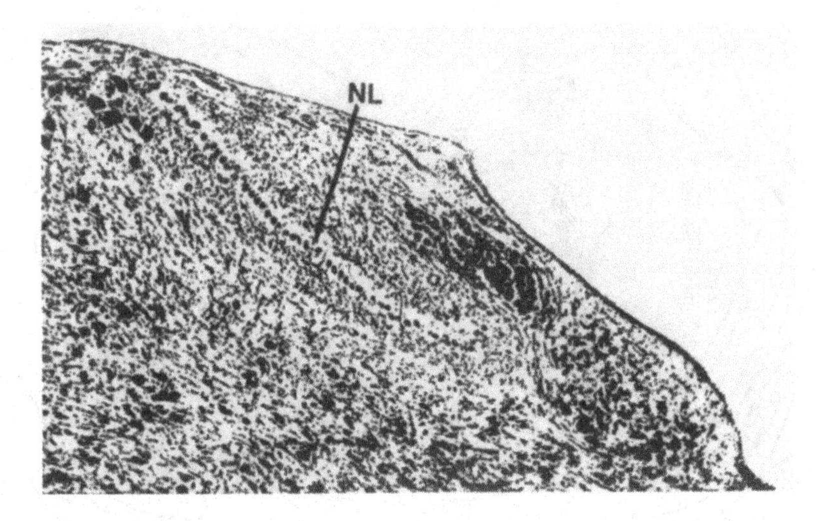

FIGURE 5.11. Photomicrograph demonstrating the appearance of nucleus laminaris in a Nissl-stained frontal section through the brain stem of the chicken. Nucleus laminaris is the avian homolog of MSO (photomicrograph courtesy of E.W. Rubel; see also Rubel and Parks, 1975; Rubel et al, 1984).

of single axons to nucleus laminaris (Fig. 5-12) supports the idea that appropriate delays may be determined by fiber length, as suggested by Jeffress (1948). Further evidence in birds suggests that binaural delays may be organized orthogonally to the tonotopic arrangement of nucleus laminaris (Young and Rubel, 1983).

Ascending Pathways From MSO and LSO

Earlier it was established that MSO and LSO are separate divisions of the binaural system. Now the question arises of whether the separation continues in the efferent pathways from LSO and MSO. The answer is complicated by observations that projections from LSO and MSO diverge and converge at their targets. LSO and MSO project to two separate structures, the dorsal nucleus of the lateral lemniscus and the central nucleus of the inferior colliculus (Glendenning and Masterton, 1983); that is, both LSO and MSO diverge into two pathways. However, because LSO and MSO each contributes to both of these pathways, there is convergence at each target. Moreover, both LSOs provide input to both inferior colliculi, whereas the pathway from MSO is ipsilateral, a point that can be illustrated by observing the distribution of labeled cells in MSO and LSO after HRP injections in the inferior colliculus. The same pattern is seen in all eutherian species in which MSO is present (Beyerl, 1978; Roth et al, 1978; Adams, 1979; Jones, 1979; Schweizer, 1981; Zook and

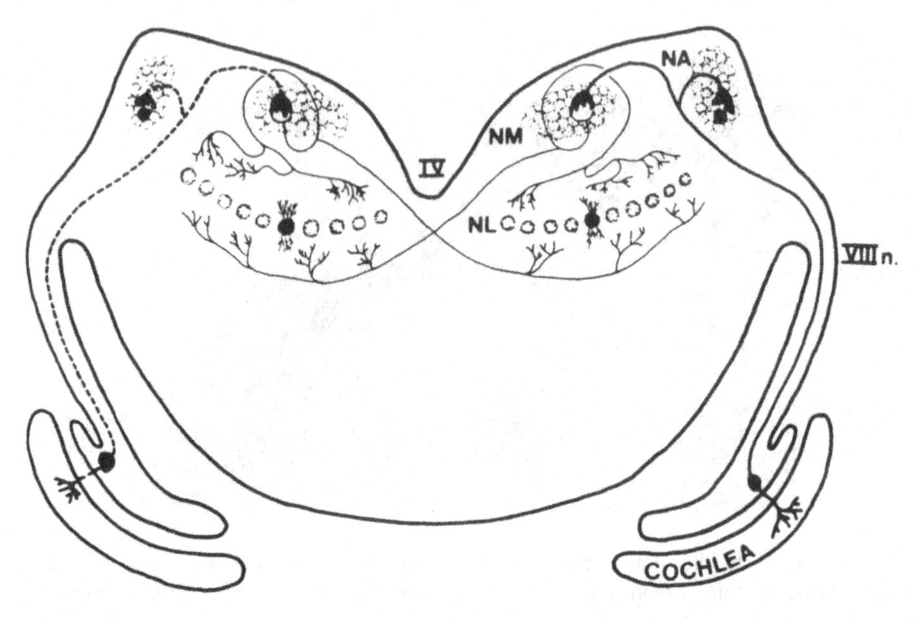

FIGURE 5.12. Brain stem auditory pathways in the chicken. Fibers from the eighth cranial nerve terminate in nucleus magnocellularis (NM) which is homologous to AVCN in mammals, and in nucleus angularis (NA), which is homologous to DCN. Nucleus laminaris (NL) consists of a tonotopically organized array of bipolar cells. Fibers from nucleus magnocellularis bifurcate to project to the dorsal dendrites of nucleus laminaris ipsilaterally and to the ventral dendrites contralaterally. A single fiber contacts cells across the entire width of the nucleus. Note that the circuitous route of the ipsilateral axon lengthens this pathway. The bird appears to have no structure homologous to LSO. (Drawing courtesy of E.W. Rubel. Permission granted from E. Rubel, Hearing Science: Recent Advances, College Hill Press, San Diego, CA. 1984.)

Casseday, 1982b) and is useful evidence for evaluating whether or not MSO exists in a given species (Fig. 5-13). However, in the North American opossum many MSO cells project bilaterally (Willard and Martin, 1984). The bilateral projections from LSO arise mainly from the medial or high-frequency part, whereas mostly ipsilateral projections arise from the lateral or low-frequency part (Glendenning and Masterton, 1983). When it is recalled that low frequencies have a greater representation in MSO than in LSO, it becomes evident that the laterality of projections from both nuclei is related to tonotopic organization. Areas of high frequencies project bilaterally from LSO whereas areas of low frequency (MSO or LSO) project ipsilaterally. Experiments in which two retrograde tracers were used, one in each inferior colliculus, demonstrate that at least some of the bilateral

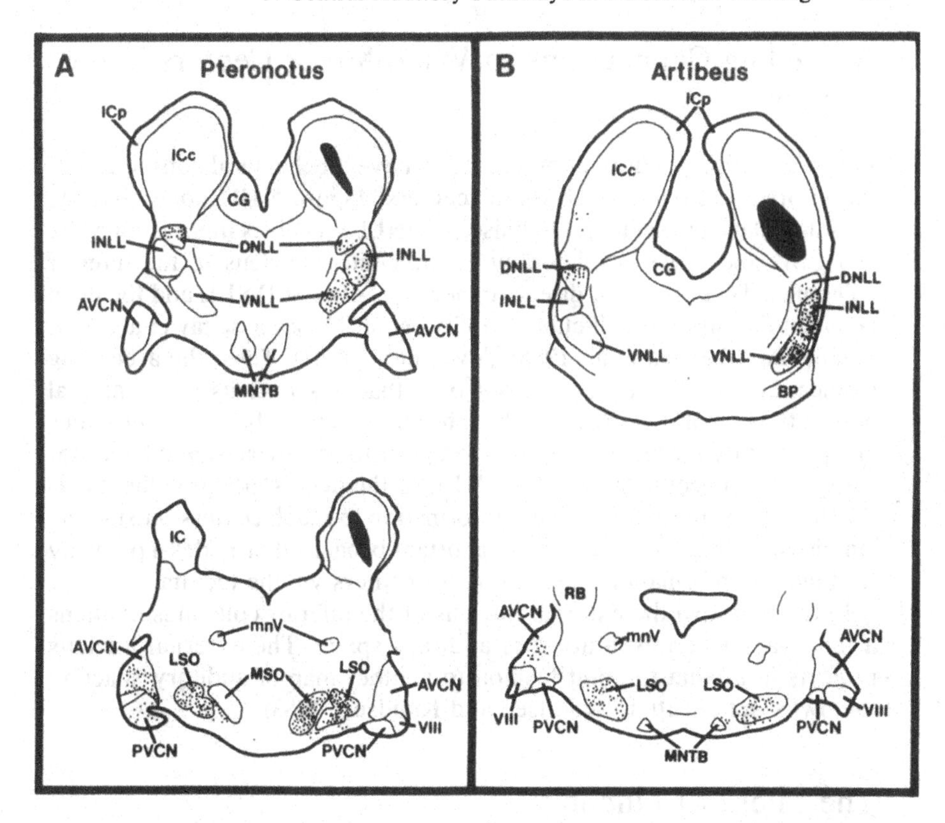

FIGURE 5.13. Pattern of labeled cells seen after HRP injections in the inferior colliculus in two different species. **A**: echolocating bat with an MSO (*Pteronotus parnellii*). **B**: Echolocating bat that lacks an MSO (*Artibeus jamaicensis*). Both A and B show frontal sections through the brain stem with injection sites indicated in black and labeled cells as dots. Note that in both species labeled cells are seen bilaterally in LSO; in the MSO of *Pteronotus*, labeled cells are seen only ipsilateral to the injection. In the lateral lemniscus of both species labeled cells are found bilaterally in the dorsal nucleus but only ipsilaterally in the intermediate and ventral nuclei. This pattern is the same as that seen in the cat. (Figure modified from Zook and Casseday, 1980.)

projections arise from cells that send collaterals to both inferior colliculi (Glendenning and Masterton, 1983; Willard and Martin (1984). Because the inferior colliculus is the target of a number of pathways in addition to those from the LSO and MSO, the question of the organization of these pathways within the inferior colliculus is addressed later. First we shall consider briefly a possible route for mediating behavior in response to the direction of sound.

Ascending Connections to Visuo-Motor Centers in the Tectum

For information about auditory space to have a behavioral consequence, the information must reach motor centers. Figure 5-1B shows that the dorsal nucleus of the lateral lemniscus (DNLL) receives input from MSO and LSO and projects bilaterally to the central nucleus of the inferior colliculus. There is also a direct connection between DNLL and the deep layers of the superior colliculus (Grafova et al, 1978; Casseday et al, 1979; Kudo, 1981; Tanaka et al, 1985; Covey et al, 1987). Thus, the ascending connections to DNLL suggest not only that this nucleus is an integral part of the binaural pathway to the inferior colliculus, but that it provides an opportunity for bilateral integration prior to and bypassing the inferior colliculus. The connection from DNLL to the deep superior colliculus is the first opportunity for binaural information to reach centers outside the lemniscal auditory system and is important because it provides a pathway for binaural information to reach motor centers via the tectum.

In the barn owl the external nucleus of the inferior colliculus contains a topographic representation of auditory space. The external nucleus projects in a point-to-point fashion to another map of auditory space in the owl's optic tectum (Knudsen and Knudsen, 1983).

The Inferior Colliculus

Structure of the Inferior Colliculus

Only a few features of the cytoarchitecture of the inferior colliculus relevant to the main purpose of this chapter will be summarized. First, the largest division of the inferior colliculus is the central nucleus. This structure is the main target of ascending fibers from AVCN and the superior olivary complex. The central nucleus does not contain the clear anatomical subdivisions seen at lower levels such as between LSO and MSO. Thus, it is necessary to look for some other cytoarchitectonic aspect of organization with which the ascending pathways may be related. An important clue may be found in the appearance of the central nucleus after Golgi impregnation; the dendrites of disc-shaped cells are oriented so as to form laminae parallel to the ascending fibers from the lateral lemniscus (Morest and Oliver, 1984; Oliver and Morest, 1984). These dendritic laminae, as seen in the frontal plane, are tilted slightly from ventrolateral to dorsomedial in the central part of the nucleus. In the lateral part the dendritic laminae are oblique to the frontal plane. Projections that may be related to these laminae will be described after reviewing the pathways that converge at the central nucleus of the inferior colliculus.

The central nucleus is capped dorsally and laterally by several subdivisions that can be further distinguished using Golgi techniques (Morest and Oliver, 1984; Oliver and Morest, 1984). Certain of these distinctions, for example, the line between the central nucleus and the dorsal cortex, are difficult to see in Nissl-stained material. Because this is the only stain that can be used with most techniques for tracing connections, the regions outside the central nucleus are here referred to collectively as pericentral areas. These pericentral areas, which receive descending input from auditory cortex (Diamond et al, 1969; Casseday et al, 1976; Willard and Martin, 1983), are only mentioned in passing because it is unclear what their connections might be with ascending pathways related to directional hearing.

Convergence of Pathways at the Inferior Colliculus

A number of pathways, including those from LSO and MSO, converge at the central nucleus of the inferior colliculus. The targets of LSO, MSO, and AVCN in the inferior colliculus of the tree shrew and the bat are shown in Figure 5-14. Two points are illustrated. First, the targets of LSO and MSO lie within the target of AVCN. Second, there may be species differences within this general plan of organization. In the tree shrew, the targets of MSO and LSO overlap only partially with each other, and together their targets occupy only part of the target area of AVCN. In the bat the targets of all these nuclei overlap almost completely with one another. The difference in the relative size of the MSO-LSO target may reflect the relative importance of directional hearing for these two species. The organization in the inferior colliculus of the cat may be somewhere between that of bat and tree shrew. In the central part of the central nucleus of the cat, MSO and LSO targets overlap, but the medial part is largely the LSO target whereas the lateral part is largely the MSO target (Glendenning et al, 1981; Henkel and Spangler, 1983; Aitkin and Schuck, 1985). Experiments on the opossum show that the contralateral projections from

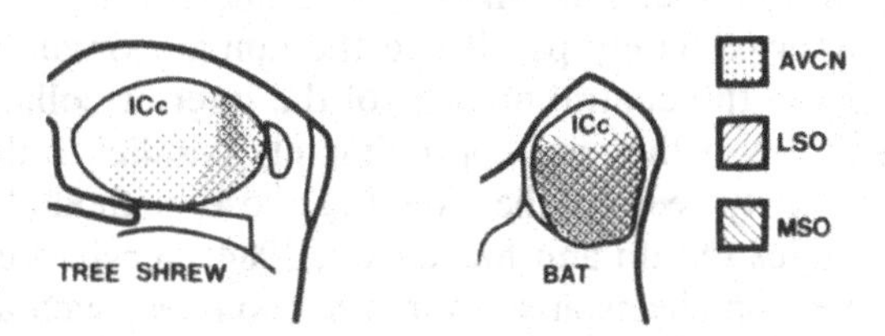

FIGURE 5.14. Comparism of projection targets of several brain stem nuclei in the inferior colliculus of two different species. The tree shrew (left) is an example of a species in which there is considerable separation of targets; at the other extreme is the bat in which the targets of all three structures overlap almost completely and extend throughout most of the central nucleus of the inferior colliculus.

LSO occupy a target in the central nucleus that is within the target of AVCN (Willard and Martin, 1983). The ipsilateral projections in these experiments are difficult to evaluate because of contributions from the dorsomedial periolivary nucleus, which projects diffusely to the ipsilateral inferior colliculus (Casseday and Covey, 1983).

Note that the targets of the indirect pathways from AVCN via MSO and LSO are contained within the target of the direct pathway from AVCN. Thus, direct "monaural" pathways project to the same area as indirect "binaural" pathways. This point is important for a later discussion of monaural pathways. Finally, it should be added that in the tree shrew the target of the dorsal nucleus of the lateral lemniscus is virtually the same as the MSO target (Covey and Casseday, unpublished observation), so that the dorsolateral part of the central nucleus is the site of convergence of two "binaural" pathways and one "monaural" pathway. The response properties of most binaural cells in the inferior colliculus differ in several ways from the response properties of cells in LSO or MSO. Such differences are no doubt a result of the convergence of several pathways as mentioned above.

Projection Sheets Within the Inferior Colliculus

With current techniques it is possible to examine the relations of ascending projections to the inferior colliculus in light of the laminar organization of the central nucleus. Recent studies of anterograde transport to the inferior colliculus from the superior olive reveal projections that form thin sheets in the central nucleus (Casseday and Covey, 1983; Henkel and Spangler, 1983). Figure 5-15 shows an example of a projection sheet after an injection of WGA-HRP in LSO of the cat. The following features of these projection sheets provide clues to the basic organization of terminals from LSO and MSO within the central nucleus:

(1) They extend throughout most of the central nucleus.
(2) They are about 100 μm wide in cross-section and are separated from one another by about 100 μm (Fig. 5-15 and 5-16).
(3) They are approximately parallel to the laminar organization of disc-shaped cells in the central nucleus of the inferior colliculus.
(4) In the central part they are approximately parallel to the contours of metabolic activity seen by the 2-deoxyglucose method after stimulation with pure tones (Nudo and Masterton, 1984; Servière et al, 1984).
(5) Similar projection sheets arise from other sources, such as subdivisions of the cochlear nuclei (Oliver, 1984a; Zook and Casseday, 1985) and lateral lemniscus (Kudo, 1981; Zook and Casseday, 1985).

Taken together these findings suggest the hypothesis that each sheet of ascending projections and the cells with which the sheet is associated form a unit of organization in the inferior colliculus.

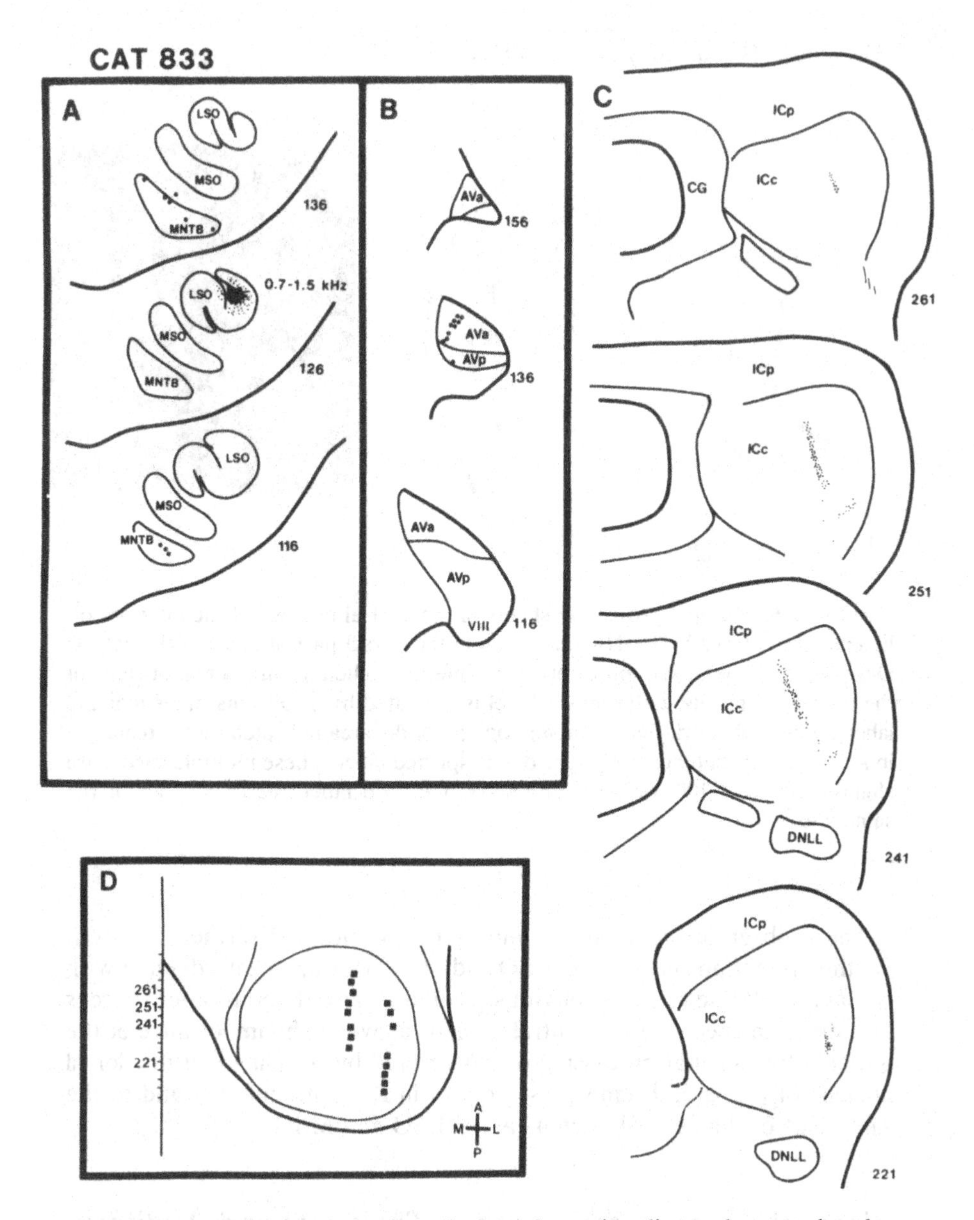

FIGURE 5.15. Projection sheets from the lateral superior olive to the central nucleus of the inferior colliculus and their relationship with input to the superior olive from AVCN in the cat. **A**: Location of a very small injection of WGA-HRP in LSO as seen in frontal sections. **B**: Band of labeled cells seen in frontal sections through AVCN. **C**: Two sheets of transported label in frontal sections through the central nucleus of the inferior colliculus arranged in a rostral to caudal direction from top to bottom. The main sheet of label is about 100 μm wide and is tilted slightly from dorsomedial to ventrolateral. A much smaller sheet of label tilted in the reverse orientation is seen ventrolateral to the main one. **D**: Horizontal reconstruction of the inferior colliculus to illustrate that these sheets extend throughout most of the rostrocaudal dimension of the central nucleus. The midpoint of each sheet in the frontal sections is shown as a square projected onto the horizontal reconstruction. Section numbers are indicated on the bar at the midline.

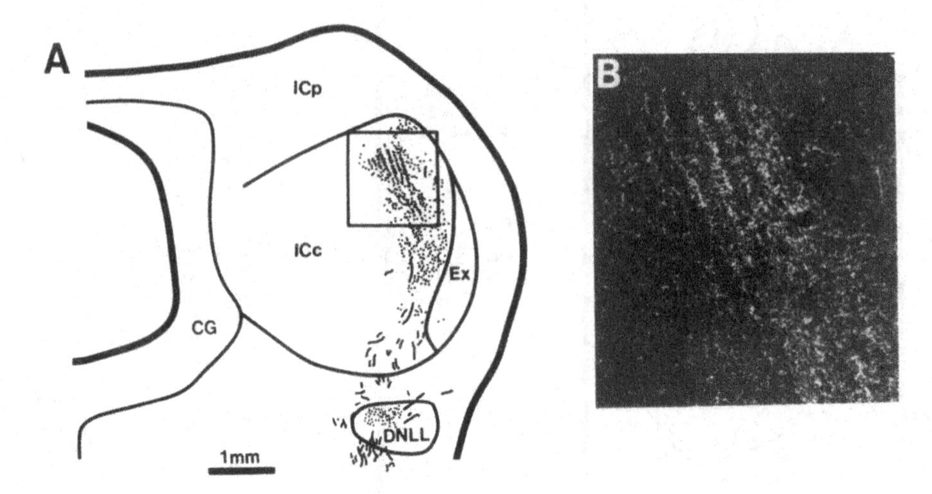

FIGURE 5.16. Multiple projection sheets in the central nucleus of the inferior colliculus after a large WGA-HRP injection in the dorsal part of MSO of the cat. **A**: Drawing of a frontal section through the inferior colliculus and dorsal nucleus of the lateral lemniscus; anterograde label is indicated by small dots; lines indicate labeled fibers. **B**: Dark-field photomicrograph of the area indicated by the rectangle in **A**. The bright patches are grains of transported label. These multiple sheets are slightly less than 100 μm wide and separated by a label-free area of about the same width.

These observations raise an important question for further research. Do the projection sheets from LSO and MSO overlap or interdigitate with one another? The question obviously could be asked about other sources of projection sheets to the central nucleus as well. The importance of the question for binaural processing is underscored by recalling that the dorsal nucleus of the lateral lemniscus projects in the same manner and to the same part of the central nucleus as do LSO and MSO.

Binaural Representation at the Thalamus and Auditory Cortex

Pathways From the Inferior Colliculus to the Medial Geniculate Body

The central nucleus of the inferior colliculus has already been identified as the target of most ascending auditory pathways, binaural and monaural. For this reason examination of the projections of the central nucleus to the thalamus is crucial to an attempt to discover the link between multiple pathways to the inferior colliculus and multiple pathways from the medial geniculate body. The central nucleus is the main source of projections to the ventral division of the medial geniculate body (Moore and Goldberg, 1963). The sources of these projections are disc-shaped and stellate cells (Oliver, 1984b). The pericentral areas project to regions other than the ventral division (Casseday et al, 1976; Oliver and Hall, 1978; Calford and Aitkin, 1983).

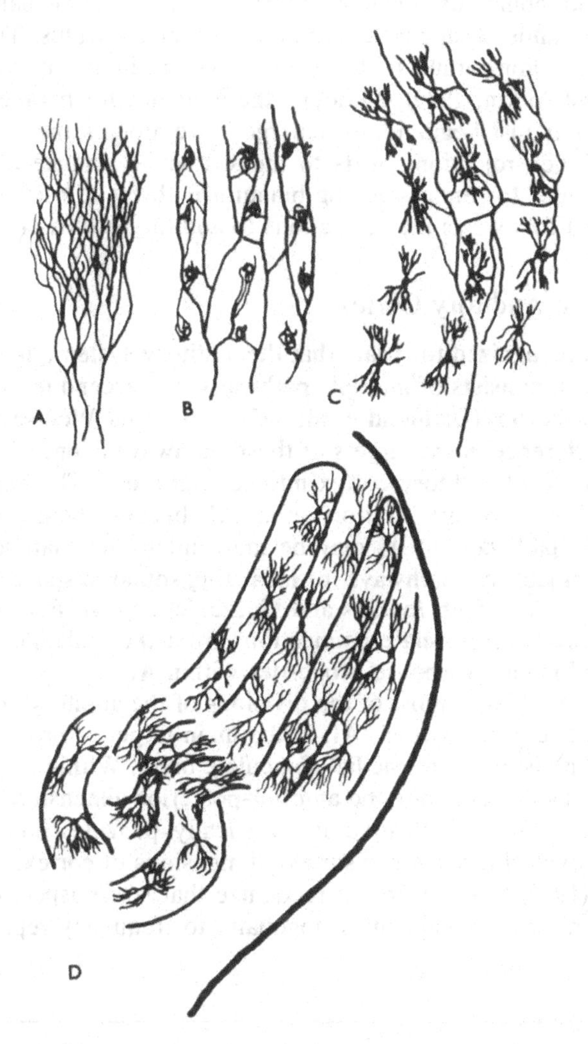

FIGURE 5.17. Orientation of cells in frontal sections through the medial geniculate body of the cat. The solid dark lines in (**D**) indicate the orientation of dendritic laminae as seen in Golgi reacted preparations. **A**, overlapping branches of afferent fibers in the lateral division; **B**, details of the terminals of these fibers; **C**, single afferent fiber branching along three dendritic laminae; **D**, dendritic laminae in the ventral division (Morest, 1965). Reproduced with permission from Journal of Anatomy, London, Cambridge University Press.

The laminar organization (Morest, 1965) of the ventral divisions of the medial geniculate body (Fig. 5-17) may be a key element in the organization of binaural pathways to the thalamus. From the central nucleus of the inferior colliculus the projections form sheets that parallel the laminae of the medial geniculate body (Andersen et al, 1980). Further, the projection sheets in the medial geniculate body may be very closely related to those in the inferior colliculus: small injections of HRP in the medial geniculate body show bands of labeled cells in the inferior colliculus. These bands have an orientation parallel to the dendritic laminae in the inferior colliculus (Calford and Aitkin, 1983). Although the evidence for parallel sheets of connections is just beginning to emerge, it reinforces the idea that the relation of the projection sheets to the structural laminae may contain important clues for understanding binaural pathways. Further evidence on this point follows as the projections to auditory cortex are examined.

Pathways to Auditory Cortex

It has been recognized for years that the auditory system, like other sensory systems, consists of multiple pathways that ascend in parallel from thalamus to cortex (Diamond et al, 1958; Rose and Woolsey, 1958). In fact, the difference in the targets of these pathways is one of the criteria for subdivision of auditory cortex into several areas. This section deals mainly with the primary auditory area, AI, because behavioral studies suggest that pathways to AI may be more important than some of the other thalamocortical pathways for localizing sound in space (Neff et al, 1956; Strominger, 1959; Jenkins and Merzenich, 1984). Further, there is recent evidence to correlate connections from the thalamus with areas differing in binaural response properties within AI.

Figure 5-18A shows most of the divisions of the auditory cortical field in the cat. The primary area, AI, is the main target of projections from the ventral division of the medial geniculate body. Within AI, frequency sensitivity is oriented along the anterior-posterior dimension so that isofrequency contours, which are drawn vertically, provide a convenient reference for exploring the other surface dimensions of cortex.

Tunturi (1952) was the first to recognize that some aspect of binaural organization may be mapped orthogonally to frequency representation.

▷

FIGURE 5.18. A: Main divisions within the auditory cortex of the cat showing the organization of AI with respect to frequency and binaural properties. In AI isofrequency contours run approximately in a dorsal to ventral direction, with frequency increasing from caudal to rostral. Orthogonal to this organization are groups of cells segregated to EE and EI bands. The most dorsal EE band has been reported to have some properties that distinguish it from the other EE bands and has been termed dorsal zone (DZ). (Modified from Merzenich et al, 1979. Reproduced with

A

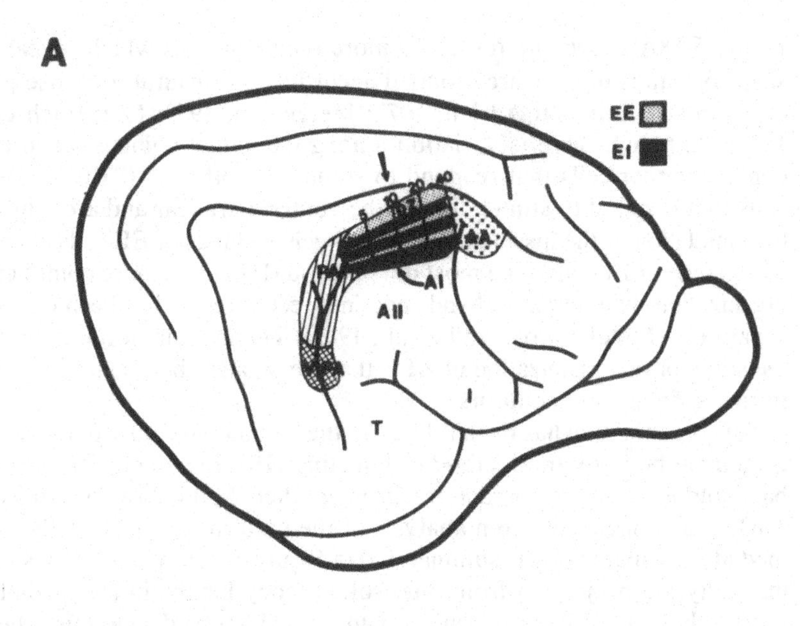

B

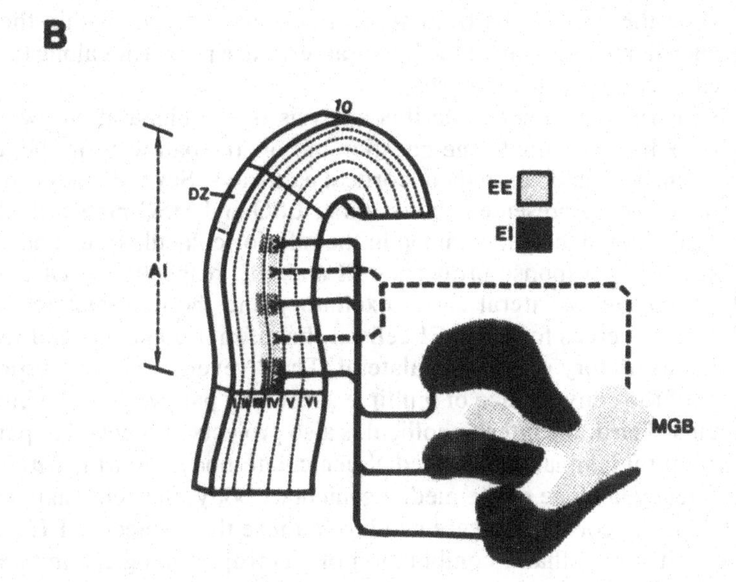

permission from Michael Merzenich, Exp. Brain Res. Springer-Verlag, 1979.) **B**: Connections from one lamina in the medial geniculate body to cortical layers III and IV in one isofrequency contour in AI of the cat. The lamina is viewed as if projected on edge and flattened. The binaural strips in AI appear to have a corresponding organization within the laminae of the medial geniculate body. (Modified from a drawing by Middlebrooks and Zook, 1983. Reproduced with permission from The Journal of Neuroscience, Society for Neuroscience, 1983.)

Figure 5-18A shows the results of more recent studies which reveal that strips or bands of cells are grouped according to binaural response properties in AI (Imig and Adrian, 1977; Merzenich, 1979; Merzenich et al, 1979). Although there is variation among individual animals, each band contains either cells that respond to stimuli at both ears ("EE" cells) or cells that respond to stimulation of the contralateral ear and are inhibited by stimulation of the ipsilateral ear ("EI" cells). Bands of EE cells alternate with bands of EI cells. The most dorsal band (DZ) has a more complicated organization than the other bands and in a recent report has been classified separately (Middlebrooks and Zook, 1983). The most important point for the study of the organization of AI is that the binaural bands are orthogonal to the isofrequency contours.

This organization has been related to the laminar structure of the medial geniculate body by the method of injecting HRP in one cortical binaural band and a second retrograde tracer in another (Middlebrooks and Zook, 1983). The conclusion from analysis of the pattern of labeled cells in the medial geniculate body is summarized in Figure 5-18B, which shows schematically the projection from one isofrequency lamina in the medial geniculate body to one isofrequency contour in AI. According to this scheme, the lamina as a whole determines the frequency representation at the cortex, while the projections from separate clusters of cells within the isofrequency lamina determine the binaural response properties along the isofrequency contours at the cortex.

It is instructive to conclude this analysis of the binaural pathways of the cortex by returning to the question of how the pathways in the lower brain stem are related to thalamocortical pathways. Several clues are seen from the evidence presented above. First, LSO and MSO project in sheets that parallel the dendritic laminae in the inferior colliculus. Second, LSO and MSO have response properties EI and EE, respectively. Of course, at the LSO the ipsilateral ear is excitatory and the contralateral ear is inhibitory, whereas for most EI cells at the inferior colliculus and rostral to it, the excitatory ear is contralateral. This "reversal" is not surprising in view of the convergence of multiple ascending pathways at the inferior colliculus. Third, the inferior colliculus also projects in sheets that parallel the structural laminae in the medial geniculate body. Fourth, within the isofrequency laminae of the medial geniculate body, the cells that project to EI bands at cortex are segregated from those that project to EE bands. The conclusion is that a combination of electrophysiological and axonal tracing techniques might reveal whether the binaural arrangement within a lamina of the medial geniculate body is related to the laminar arrangement of disc-shaped cells in the inferior colliculus, or whether the EI-EE arrangement of cells in the medial geniculate body is a consequence of a transformation that takes place in the inferior colliculus via convergence or divergence of sheets of projections from LSO and MSO.

Monaural Pathways

The hypothesis that some pathways are monaural in function arises from a behavioral experiment in which cats were trained to localize sound after one cochlea had been destroyed (Neff and Casseday, 1977; see also Chapter 9). Although binaural interaction was abolished in these animals, auditory information could obviously ascend on either side of the brain at levels above the superior olivary complex (Fig. 5-1). After the animals had learned to localize sound, auditory cortex was ablated from one hemisphere only. A schematic view of the anatomical consequences of ablation of the cortex ipsilateral or contralateral to the intact cochlea is shown in Figure 5-19. The behavioral result was that an animal's ability to localize sound was only affected after ablation of auditory cortex contralateral to the intact cochlea, and in these animals the loss was devastating. The animals could not relearn to localize sound, and only one of three animals was able to learn to approach a continuously pulsing sound. The animals in which the cortex was ablated on the side ipsilateral to the intact cochlea showed no loss in their ability to localize sound.

The importance of this result is that pathways from one ear to the contralateral cortex are necessary for localizing sound on the basis of monaural

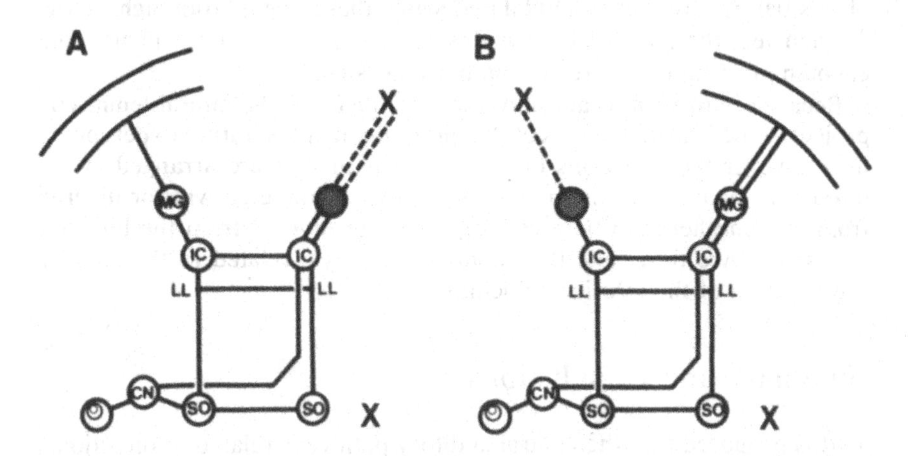

FIGURE 5.19. Two combinations of lesions with very different behavioral effects on the cat's ability to localize sound in space. **A**: Ablation of cochlea and auditory cortex are on the same side; consequently, the pathways from the remaining cochlea to the contralateral auditory cortex are nonfunctional. An animal with this combination of lesions is unable to localize even though there are pathways to the cortex ipsilateral to the intact cochlea. **B**: Ablations of cochlea and auditory cortex are on opposite sides leaving intact the pathway from the cochlea to the contralateral auditory cortex but destroying the ipsilateral pathway. An animal's ability to localize sound is unimpaired after this combination of lesions.

cues alone. Although it is not possible to say that these same pathways serve a monaural function for localization in a binaural animal, the fact that a number of pathways cross the brain stem and bypass the binaural centers (Fig. 5-1) supports the idea that some pathways transmit monaural information only; this information might then be used in sound localization.

It is likely that the intermediate and ventral nuclei of the lateral lemniscus are major components in monaural pathways. Except for a small bilateral component, the connections of these nuclei (Fig. 5-1A) seem to serve mainly for transmitting information to one inferior colliculus from the contralateral cochlear nucleus (Warr, 1966, 1969; Roth et al, 1978; Glendenning et al, 1981; Zook and Casseday, 1982b). It is significant that these nuclei are hypertrophied and highly differentiated in echolocating bats (Schweitzer, 1981; Zook and Casseday, 1982a). These animals rely not only on binaural cues for orientation, but also are adept at a type of localization that may not depend on binaural processing, detecting the distance of objects in space from the temporal or spectral characteristics of echoes from their biosonar calls (Simmons et al, 1974; Simmons, 1979). Recent experiments on the big brown bat (Covey and Casseday, 1986) show that in one nucleus of the lateral lemniscus, frequency organization is compressed in such a way that it is likely that this nucleus extracts narrow frequency bands from complex signals. Because the biosonar signal of this bat is a frequency-modulated signal that sweeps from high to low frequencies, the idea arises that this nucleus provides a mechanism for encoding temporal features of the biosonar signal.

Because both ventral and intermediate nuclei of the lateral lemniscus project to the central nucleus of the inferior colliculus, future experiments must answer the questions of how these pathways are arranged in the inferior colliculus. Do the monaural pathways converge with or diverge from one another? Do they converge with or diverge from the binaural pathways? Are they, like the binaural pathways, related to the laminar organization of the inferior colliculus?

Summary and Conclusions

1. It is proposed that the central auditory pathways related to directional hearing are composed of two types of systems, binaural and monaural pathways.
2. The binaural pathways arise from bilateral projections from the anteroventral cochlear nucleus to the medial superior olives and lateral superior olives in part via the medial nucleus of the trapezoid body.
3. The binaural pathways ascend to the inferior colliculus directly from the superior olivary complex and indirectly via the dorsal nucleus of the lateral lemniscus.
4. Binaural pathways in the thalamus are arranged in EE and EI zones within isofrequency laminae of the medial geniculate body, and these

EE and EI zones are organized as bands orthogonal to the isofrequency contours in AI of the auditory cortical field.

5. Monaural pathways may also play a role in directional hearing. Likely candidates for monaural pathways are those that originate in the cochlear nucleus and project directly to the contralateral inferior colliculus, bypassing the superior olivary complex, or project to the contralateral inferior colliculus via the ventral and intermediate nuclei of the lateral lemniscus.
6. An important question is how the pathways in the lower brain stem are linked to the pathways in the forebrain. The answer will probably arise from examination of the arrangement of projection sheets in the inferior colliculus and medial geniculate body.

Acknowledgments. This work was supported by National Institutes of Health grant NS 21748 and National Science Foundation grants BNS 82-17357 and BNS 85-20441.

References

Adams, J.C. (1979). Ascending projections to the inferior colliculus, J. Comp. Neurol. 183, 519–538.

Aitkin, L.M., Phillips, S.C. (1984). The interconnections of the inferior colliculi through their commissure. J. Comp. Neurol. 228, 210–216.

Aitkin, L.M., Schuck, D. (1985). Low frequency neurons in the lateral central nucleus of the cat inferior colliculus receive their input predominantly from the medial superior olive. Hear. Res. 17, 87–93.

Andersen, R.A., Roth, G.L., Aitkin, L.M. and Merzenich, M.M. (1980). The efferent projections of the central nucleus of the inferior colliculus in the cat. J. Comp. Neurol. 194, 649–662.

Beyerl, B.D. (1978). Afferent projections to the central nucleus of the inferior colliculus in the rat. Brain Res. 145, 209–223.

Boord, R.L. (1968). Ascending projections of the primary cochlear nuclei and nucleus laminaris in the pigeon. J. Comp. Neurol. 133, 523–542.

Brawer, J.R., Morest, D.K. (1975). Relations between auditory nerve endings and cell types in the cat's anteroventral cochlear nucleus seen with the Golgi method and Nomarski optics. J. Comp. Neurol. 160, 491–506.

Brawer, J.R. Morest, D.K., Kane, E.C. (1974). The neuronal architecture of the cochlear nucleus of the cat. J. Comp. Neurol. 155, 251–300.

Cajal, S., Ramon y (1909). Le Systeme nerveux de l'homme et des vertebres. Instituto Ramon y Cajal, Madrid.

Calford, M.B., Aitkin, L.M. (1983). Ascending projections to the medial geniculate body of the cat: Evidence for multiple parallel auditory pathways through thalamus. J. Neurosci. 3, 2365–2380.

Cant, N.B. (1984). The fine structure of the lateral superior olivary nucleus of the cat. J. Comp. Neurol. 227, 63–77.

Cant, N.B., Casseday, J.H. (1986). Projections from the anteroventral cochlear nucleus to the lateral and medial superior olivary nuclei. J. Comp. Neurol. 247, 457–476.

Casseday, J.H., Covey, E. (1983). Laminar projections to the inferior colliculus as seen from injections of wheat germ agglutinin-horseradish peroxidase in the superior olivary complex of the cat. Neurosci. Abstr. p. 766.

Casseday, J.H., Diamond, I.T., Harting, J.K. (1976). Auditory pathways to the cortex, *Tupaia glis*. J. Comp. Neurol. 166, 303–340.

Casseday, J.H., Jones, D.R., Diamond, I.T. (1979). Projections from cortex to tectum in the tree shrew, *Tupaia glis*. J. Comp. Neurol. 185, 253–292.

Casseday, J.H., Neff, W.D. (1975). Auditory localization: Role of auditory pathways in brain stem of the cat. J. Neurophysiol. 38, 842–858.

Clark, G. (1969). The ultrastructure of nerve endings in the medial superior olive of the cat. Brain Res. 14, 293–305.

Covey, E., Casseday, J.H. (1986). Connectional basis for frequency representation in the nuclei of the lateral lemniscus of the bat, *Eptesicus fuscus*. J. Neurosci. 6, 2926–2940.

Covey, E., Hall, W.C., Kobler, J.M. (1987). Subcortical connections of the superior colliculus in the mustache bat, *Pteronotus parnellii*. J. Comp. Neurol, in press.

Diamond, I.T., Chow, K.L., Neff, W.D. (1958). Degeneration of caudal medial geniculate body following cortical lesions ventral to auditory area II in cat. J. Comp. Neurol. 109, 349–362.

Diamond, I.T., Jones, E.G., Powell, T.P.S. (1969). The projection of the auditory cortex upon the diencephalon and brain stem in the cat. Brain Res. 15, 305–340.

Fekete, D.M., Rouiller, E.M., Liberman, M.C., Ryugo, D.K. (1984). The central projections of intracellularly labeled auditory nerve fibers in cats. J. Comp. Neurol. 229, 432–450.

Fernandez, C., Karapas, F. (1967). The course and termination of the striae of Monakow and Held in the cat. J. Comp. Neurol. 131, 371–386.

Galambos, R., Schwartzkopff, J., Rupert, A. (1959). Microelectrode study of superior olivary nuclei. Am. J. Physiol. 197, 527–536.

Glendenning, K.K., Brunso-Bechtold, J.K., Thompson, G.C. Masterton, R.B. (1981). Ascending auditory afferents to the nuclei of the lateral lemniscus. J. Comp. Neurol. 197, 673–703.

Glendenning, K.K., Masterton, R.B. (1983). Acoustic chiasm: Efferent projections of the lateral superior olive. J. Neurosci. 3, 1521–1537.

Glendenning, K.K., Hutson, K.A., Nudo, R.J., Masterton, R.B. (1985). Acoustic chiasm II: Anatomical basis of binaurality in lateral superior olive of cat. J. Comp. Neurol. 232, 261–285.

Goldberg, J.M., Brown, P.B. (1968). Functional organization of the dog superior olivary complex: an anatomical and electrophysiological study. J. Neurophysiol. 31, 639–656.

Goldberg, J.M., Brown, P.B. (1969). Response of binaural neurons of dog superior olivary complex to dichotic tonal stimuli: Some physiological mechanisms of sound localization. J. Neurophysiol. 32, 613–636.

Goldberg, J.M., Moore, R.Y. (1967). Ascending projections of the lateral lemniscus in the cat and monkey. J. Comp. Neurol. 129, 143–156.

Goldman, L.J., Henson, O.W., Jr. (1977). Prey recognition and selection by the constant frequency bat, *Pteronotus p. parnellii*. Behav. Ecol. Sociobiol. 2, 411–419.

Grafova, I., Ottersen, O.P., Rinvik, E. (1978). Mesencephalic and diencephalic afferents to the superior colliculus and periaqueductal gray substance demonstrated by retrograde axonal transport of horseradish peroxidase in the cat. Brain Res. 146, 205–220.

Harrison, J.M., Irving, R. (1966). Visual and nonvisual auditory systems in mammals. Science 154, 738–743.

Harrison, J.M., Warr, W.B. (1962). A study of the cochlear nuclei and ascending auditory pathways of the medulla. J. Comp. Neurol. 119, 341–379.

Held, H. (1893). Die zentrale Gehörleitung. Arch. Anat. Physiol. 17, 201–248.

Henkel, C.K., Spangler, K.M. (1983). Organization of the efferent projections of the medial superior olivary nucleus in the cat as revealed by HRP and autoradiographic tracing methods. J. Comp. Neurol. 221, 416–428.

Henson, O.W. Jr., Pollak, G.D., Kobler, J.B., Henson, M.M., Goldman, L.J. (1982). Cochlear microphonic potentials elicited by biosonar signals in flying bats, *Pteronotus p. parnellii.* Hearing Res. 7, 127–147.

Imig, T.J., Adrian, H.O. (1977). Binaural columns in the primary field (AI) of cat auditory cortex. Brain Res. 138, 241–257.

Jeffress, L.A. (1948). A place theory of sound localization. J. Comp. Physiol. Psychol. 41, 35–39.

Jenkins, W.M., Masterton, R.B. (1982). Sound localization: Effects of unilateral lesions in central auditory system. J. Neurophysiol. 47, 987–1016.

Jenkins, W.M., Merzenich, M.M. (1984). Role of cat primary auditory cortex for sound-localization behavior. J. Neurophysiol. 52, 819–847.

Jones, D.R. (1979). Auditory pathways in the brainstem of the tree shrew (*Tupaia glis*). Doctoral thesis, Duke University.

Jones, D.R., Casseday, J.H. (1979). Projections of auditory nerve in the cat as seen by anterograde transport methods. Neuroscience 4, 1299–1313.

Kiang, N.Y.-S., Morest, D.K., Godfrey, D.A., Guinan, J.J., Jr., Kane, E.C. (1973). Stimulus coding at caudal levels of the cat's auditory nervous system: I. Response characteristic of single units. In: Basic Mechanisms in Hearing, Moller, A.R. (ed.). pp. 455–478. New York: Academic Press.

Knudsen, E.I., Knudsen, P.F. (1983). Space-mapped auditory projections from the inferior colliculus to the optic tectum in the barn owl (*Tyto alba*). J. Comp. Neurol. 218, 187–196.

Kudo, M. (1981). Projections of the nuclei of the lateral lemniscus in the cat: An autoradiographic study. Brain Res. 221, 57–69.

Lorente de No, R. (1933). Anatomy of the eighth nerve. III. General plan of structure of the primary cochlear nuclei. Laryngoscope 43, 327–350.

Masterton, R.B., Diamond, I.T. (1967). Medial superior olive and sound localization. Science 155, 1696–1697.

Masterton, R.B., Heffner, H., Ravizza, R. (1969). Evolution of human hearing. J. Acoust. Soc. Am. 45, 966–985.

Masterton, R.B., Jane, J.A., Diamond, I.T. (1967). Role of brainstem auditory structures in sound localization. I. Trapezoid body, superior olive and lateral lemniscus. J. Neurophysiol. 30, 341–360.

Masterton, B., Thompson, G.C., Bechtold, J.K., RoBards, J. (1975). Neuroanatomical basis of binaural phase-difference analysis for sound localization: A comparative study. J. Comp. Physiol. Psychol. 89, 379–386.

McFadden, D., Pasanen, E.G. (1976). Lateralization at high frequencies based on interaural time differences. J. Acoust. Soc. Am. 59, 634–639.

Merzenich, M.M. (1979). Some recent observations on the functional organization of the central auditory nervous system. In: Brain Mechanisms of Sensation, Katsuki, Y., Norgren, R., Sato, M. (eds.). pp. 3–19. New York: John Wiley & Sons.

Merzenich, M.M., Andersen, R.A., Middlebrooks, J.C. (1979). Functional and topographic organization of the auditory cortex. Exp. Brain Res. (Suppl. 2), 61–75.

Middlebrooks, J.C., Zook, J.M. (1983). Intrinsic organization of the cat's medial geniculate body identified by projections to binaural response-specific bands in the primary auditory cortex. J. Neurosci. 3, 203–224.

Moore, C.N., Casseday, J.H., Neff, W.D. (1974). Sound localization: The role of the commissural pathways of the auditory system of the cat. Brain Res. 82, 13–26.

Moore, R.Y., Goldberg, J.M. (1963). Ascending projections of the inferior colliculus in the cat. J. Comp. Neurol. 121, 109–136.

Morest, D.K. (1965). The laminar structure of the medial geniculate body of the cat. J. Anat. (Lond.) 99, 143–160.

Morest, D.K. (1968). The collateral system of the medial nucleus of the trapezoid body of the cat, its neuronal architecture and relation to the olivo-cochlear bundle. Brain Res. 9, 288–311.

Morest, D.K., Oliver, D.L. (1984). The neuronal architecture of the inferior colliculus in the cat: Defining the functional anatomy of the auditory midbrain. J. Comp. Neurol. 222, 209–236.

Neff, W.D., Casseday, J.H. (1977). Effects of unilateral ablation of auditory cortex on monaural cat's ability to localize sound. J. Neurophysiol. 40, 44–52.

Neff, W.D., Fisher, J.F., Diamond, I.T., Yela, M. (1956). Role of auditory cortex in discrimination requiring localization of sound in space. J. Neurophysiol. 19, 500–512.

Nudo, R.J., Masterton, R.B. (1984). 2-deoxyglucose studies of stimulus coding in the brainstem auditory system of the cat. In: Contributions to Sensory Physiology, Vol. 8. Neff, W.D. (ed.). pp. 79–97. New York: Academic Press.

Oliver, D.L. (1984a). Dorsal cochlear nucleus projections to the inferior colliculus in the cat: A light and electron microscopic study. J. Comp. Neurol. 224, 155–172.

Oliver, D.L. (1984b). Neuron types in the central nucleus of the inferior colliculus that project to the medial geniculate body. Neuroscience 11, 409–424.

Oliver, D.L., Hall, W.C. (1978). The medial geniculate body of the tree shrew, *Tupaia glis*. II. Connections with the neocortex. J. Comp. Neurol. 182, 459–494.

Oliver, D.L., Morest, D.K. (1984). The central nucleus of the inferior colliculus in the cat. J. Comp. Neurol. 222, 237–264.

Osen, K.K. (1969a). Cytoarchitecture of the cochlear nuclei in the cat. J. Comp. Neurol. 136, 453–484.

Osen, K.K. (1969b). The intrinsic organization of the cochlear nuclei in the cat. Acta Oto-laryngol. 67, 352–359.

Pfeiffer, R.R. (1966). Anteroventral cochlear nucleus: Wave forms of extracellularly recorded spike potentials. Science 154, 667–668.

Rhode, W.S., Oertel, D., Smith, P.H. (1983). Physiological response properties of cells labeled intracellularly with horseradish peroxidase in cat ventral cochlear nucleus. J. Comp. Neurol. 213, 448–463.

Rose, J.E., Woolsey, C.N. (1958). Cortical connections and functional organization of the thalamic auditory system of the cat. In: Biological and Biochemical Bases of Behavior. Harlow, H., Woolsey, C.N. (eds.). pp. 127–150. Madison: University Wisconsin Press.

Roth, G.L., Aitkin, L.M., Andersen, R.A., Merzenich, M.M. (1978). Some features of the spatial organization of the central nucleus of the inferior colliculus of the cat. J. Comp. Neurol. 182, 661–680.

Rouiller, E.M., Ryugo, D.K. (1984). Intracellular marking of physiologically characterized cells in the ventral cochlear nucleus of the cat. J. Comp. Neurol. 225, 167–186.

Rubel, E.W., Parks, T.N. (1975). Organization and development of brain stem auditory nuclei of the chicken: tonotopic organization of n. magnocellularis and n. laminaris. J. Comp. Neurol. 164, 411–434.

Rubel, E., Barn, D., Deitch, J., Durham, D. (1984). Auditory system development. In: Hearing Science, Berlin, C. (ed.). College Hill Press.

Scheibel, M.E., Scheibel, A.B. (1974). Neuropil organization in the superior olive of the cat. Exp. Neurol. 43, 339–348.

Schwartz, I.R. (1977). Dendritic arrangements in the cat medial superior olive. Neuroscience 2, 81–101.

Schwartz, I.R. (1984). Axonal organization in the cat medial superior olivary nucleus. In: Contributions to Sensory Physiology, Vol. 8. Neff, W.D. (ed.). pp. 99–129. New York: Academic Press.

Schweizer, H. (1981). The connections of the inferior colliculus and the organization of the brainstem auditory system in the greater horseshoe bat (*Rhinolophus ferrumequinum*). J. Comp. Neurol. 201, 25–49.

Servière, J., Webster, W.R., Calford, M.B. (1984). Isofrequency labelling revealed by a combined [14C]-2-deoxyglucose, electrophysiological, and horseradish peroxidase study of the inferior colliculus of the cat. J. Comp. Neurol. 228, 463–477.

Simmons, J.A., Lavender, W.A., Lavender, B.A., Doroshow, C.A., Kiefer, S.W., Livingston, R., Scallet, A.C., Crowley, D.E. (1974). Target structure and echo spectral discrimination by echolocating bats. Science 186, 1130–1132.

Simmons, J.A. (1979). Perception of echo phase information in bat sonar. Science 204, 1336–1338.

Spangler, K.M., Warr, W.B., Henkel, C.K. (1985). The projections of principal cells of the medial nucleus of the trapezoid body in the cat. J. Comp. Neurol. 238, 249–262.

Stotler, W.A. (1953). An experimental study of the cells and connections of the superior olivary complex of the cat. J. Comp. Neurol. 98, 401–432.

Strominger, N.L. (1969). Subdivisions of auditory cortex and their role in localization of sound in space. Exp. Neurol. 24, 348–362.

Strominger, N.L., Strominger, A.I. (1971). Ascending brain stem projections of the anteroventral cochlear nucleus in the rhesus monkey. J. Comp. Neurol. 143, 217–242.

Takahashi, T., Moiseff, A., Konishi, M. (1984). Time and intensity cues are processed independently in the auditory system of the owl. J. Neurosci. 4, 1781–1786.

Tanaka, K., Otani, K., Tokunaga, A., Sugita, S. (1985). The organization of neurons in the nucleus of the lateral lemniscus projecting to the superior and inferior colliculi in the rat. Brain Res. 341, 252–260.

Tolbert, L.P., Morest, D.K., Yurgelun-Todd, D. (1982). The neuronal architecture of the anteroventral cochlear nucleus of the cat in the region of the cochlear nerve root: Horseradish peroxidase labeling of identified cell types. Neuroscience 7, 3031–3052.

Tsuchitani, C. (1977). Functional organization of lateral cell groups of cat superior olivary complex. J. Neurophysiol. 40, 296–318.

Tunturi, A.R. (1952). A difference in the representation of auditory signals for the left and right ears in the isofrequency contours of the right middle ectosylvian auditory cortex of the dog. Am. J. Physiol. 168, 712–727.

van Bergeijk, W.A. (1962). Variations on a theme of Bekesy: A model of binaural interaction. J. Acoust. Soc. Am. 34, 1431–1437.

van Noort, J. (1969). The Structure and Connections of the Inferior Colliculus. An Investigation of the Lower Auditory System. Leiden: van Gorcum & Co.

von Bekesy, G. (1930). Zur Theorie des Hörens. Über das Richtungshören bei einer Zeitdifferenz oder Lautstärkenungleichheit der beiderseitigen Schalleinwirkungen. Physik Z. 31, 824–835, 857–868.

Warr, W.B. (1966). Fiber degeneration following lesions in the anterior ventral cochlear nucleus of the cat. Exp. Neurol. 14, 453–474.

Warr, W.B. (1969). Fiber degeneration following lesions in the posteroventral cochlear nucleus of the cat. Exp. Neurol. 23, 140–155.

Warr, W.B. (1972). Fiber degeneration following lesions in the multipolar and globular cell areas in the ventral cochlear nucleus of the cat. Brain Res. 40, 247–270.

Warr, W.B. (1982). Parallel ascending pathways from the cochlear nucleus: Neuroanatomical evidence of functional specialization. In: Contributions to Sensory Physiology, Neff, W.D. (ed.). New York: Academic Press.

Willard, F.H., Martin, G.F. (1983). The auditory brainstem nuclei and some of their projections to the inferior colliculus in the North American opossum. Neuroscience 10, 1203–1232.

Willard, F.H., Martin, G.F. (1984). Collateral innervation of the inferior colliculus in the North American opossum: A study using fluorescent markers in a double-labeling paradigm. Brain Res. 303, 171–182.

Young, S.R., Rubel, E.W. (1983). Frequency-specific projections of individual neurons in chick brainstem auditory nuclei. J. Neurosci. 3, 1373–1378.

Zook, J.M., Casseday, J.H. (1980). Identification of auditory centers in lower brain stem of two species of echolocating bats: evidence from injection of horseradish peroxidase into inferior colliculus. In: Proceedings of the Fifth International Bat Research Conference Wilson, D.E., Gardner, A.L. (Eds.). Lubbock: Texas Tech Press.

Zook, J.M., Casseday, J.H. (1982a). Cytoarchitecture of auditory system in lower brainstem of the mustache bat, *Pteronotus parnellii*. J. Comp. Neurol. 207, 1–13.

Zook, J.M., Casseday, J.H. (1982b). Origin of ascending projections to inferior colliculus in the mustache bat, *Pteronotus parnellii*. J. Comp. Neurol. 207, 14–28.

Zook, J.M., Casseday, J.H. (1985). Projections from the cochlear nuclei in the mustache bat, *Pteronotus parnellii*. J. Comp. Neurol. 237, 307–324.

Zook, J.M., Casseday, J.H. (1987). Convergence of ascending pathways at the inferior colliculus of the mustache bat, *Pteronotus parnellii*. J. Comp. Neurol., in press.

Abbreviations Used in Figures

AA	=	Anterior auditory field
Ava	=	Anterior division of the anteroventral cochlear nucleus
AVCN	=	Anteroventral cochlear nucleus
AVp	=	Posterior division of the anteroventral cochlear nucleus
BP	=	Brachium pontis
Ce	=	Cerebellum
CG	=	Central gray
CN	=	Cochlear nucleus
DCN	=	Dorsal cochlear nucleus
DMPO	=	Dorsomedial periolivary nucleus
DNLL	=	Dorsal nucleus of the lateral lemniscus
DZ	=	Dorsal zone of AI
EE	=	Excitatory excitatory
EI	=	Excitatory inhibitory
Ex	=	External nucleus of the inferior colliculus
I	=	Insular cortex
IC	=	Inferior colliculus
ICc	=	Central nucleus of the inferior colliculus
ICp	=	Pericentral area of the inferior colliculus
INLL	=	Intermediate nucleus of the lateral lemniscus
IV	=	Fourth cranial (trochlear) nerve
LL	=	Lateral lemniscus
LSO	=	Lateral superior olive
MG	=	Medial geniculate body
MNTB	=	Medial nucleus of the trapezoid body
mnV	=	Motor nucleus of the fifth cranial (trigeminal) nerve
MSO	=	Medial superior olive
NA	=	Nucleus angularis
NL	=	Nucleus laminaris
NM	=	Nucleus magnocellularis
PA	=	Posterior auditory field
PVCN	=	Posteroventral cochlear nucleus
RB	=	Restiform body
SO	=	Superior olive
T	=	Temporal cortex
TB	=	Trapezoid body
Vest	=	Vestibular nerve
VIII	=	Eighth cranial nerve
VMPO	=	Ventromedial periolivary nucleus
VNLL	=	Ventral nucleus of the lateral lemniscus
VNTB	=	Ventral nucleus of the trapezoid body
VPA	=	Ventroposterior auditory field
VPO	=	Ventral periolivary nucleus

6
Physiological Studies of Directional Hearing

SHIGEYUKI KUWADA AND TOM C.T. YIN

Introduction

An important function of the auditory system is to locate the position of a sound source. Whereas functions such as pitch and intensity perception can be performed by one ear alone, the localization of sound depends heavily on comparisons between the acoustic inputs from the two ears. Binaural cues used for sound localization include interaural differences in level, time of arrival, and frequency spectrum of the sound. Interaural level differences are more commonly, though incorrectly (see Kuhn, 1977, and Behar, 1984), called interaural intensity differences (IIDs). Most investigations of binaural hearing have focused on the role of differences in interaural intensity and time in accordance with the duplex theory first postulated by Lord Rayleigh (1907). This theory states that IIDs resulting from the sound shadow cast by the head is the primary cue for localizing high-frequency tones, whereas interaural time differences (ITDs) between the arrival of the sound at the two ears are responsible for the localization of low-frequency tones. The basis for this distinction is best understood in terms of the relation between head size and the wavelength of the sound. For high-frequency signals, the wavelength is small compared with the size of the head, or interaural distance, and the head becomes an obstacle to the propagating sound wave. For laterally placed sounds the head attenuates the signal intensity at the distal ear, thereby creating an IID. At lower frequencies, where the wavelength is larger than the interaural distance, the head is acoustically transparent, and no IID is created. At these frequencies, ITDs are the most important cue.

Acoustical, psychophysical, and anatomical experiments using pure-tone stimuli provide strong evidence for the duplex theory (see Chapters 1–3 and 5). This review focuses on the physiological evidence that central auditory neurons process IIDs and ITDs. It begins with a description of the response properties of central auditory cells to sounds presented through externally placed speakers (free-field stimulation) or via sealed earphones (dichotic stimulation), followed by a discussion of possible neural mechanisms involved in processing IID and ITD information. For other reviews on the physiology of binaural hearing, the reader is referred to Rosenzweig (1961), Erulkar (1972), Keidel and Neff (1974), Webster and Aitkin (1975), Altman (1978), Brugge and Geisler (1978), Brugge (1980), Masterton and Imig (1984), Yin and Kuwada (1984), and Yin and Chan (1986).

Free-Field Studies

A natural way to begin is by asking if there are cells in the auditory system that are selectively responsive to a position of a sound source in space. This is important since, unlike the peripheral components of the visual and somatic sensory systems that are spatially organized, the cochlea encodes frequency, not spatial location. The most convincing evidence for the existence of central auditory neurons responsive to sound locations come from studies of the barn owl, which can use auditory cues to capture its prey in total darkness (Knudsen and Konishi, 1978; Moiseff and Konishi, 1981). In these experiments the owl was placed in the center of a semicircular hoop on which a speaker could be positioned at almost any azimuthal or elevational angle, at a constant distance from the owl. Recordings from single neurons along the lateral and anterior borders of the mesencephalicus lateralis dorsalis (MLD), the avian homologue of the inferior colliculus, revealed cells that responded only to sounds emanating from small, restricted areas in space. Most of the receptive fields are ellipsoid in shape with a longer elevational axis. In some cases, an inhibitory area surrounding the excitatory receptive field was detected; stimuli positioned in such areas suppressed spontaneous activity. At higher intensities the receptive field properties showed a responsive area in front (best area) and another behind (secondary area) the owl, but both at the same azimuth. The best areas of 14 such neurons are plotted in Figure 6-1. In each case, the stimuli were noise bursts encompassing the cell's responsive frequency range.

An important feature of these "limited-field" neurons was that neighboring cells responded to sounds placed in adjacent regions of auditory space. This results in a systematic map of auditory space in the MLD, which is illustrated in the sections taken from the left MLD in the three anatomical planes (Fig. 6-1). Best areas from the mid-ipsilateral to the far contralateral horizontal angles are represented in a rostral to caudal direction; upper to lower elevations are mapped in a dorsal to ventral direction. Thus, not only are there neurons in the barn owl's MLD selectively responsive to the position of a sound in space, but these neurons are also anatomically arranged so that their receptive fields form an orderly rep-

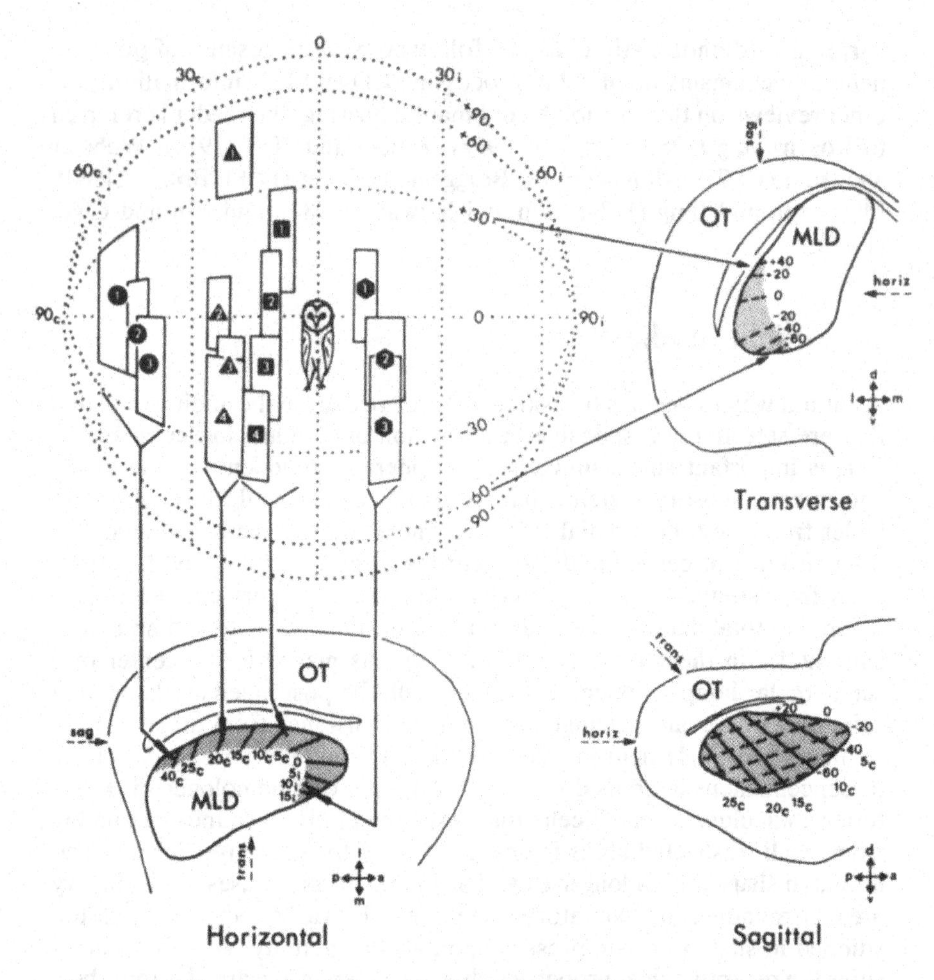

FIGURE 6.1. Topographical map of auditory space in the mesencephalicus lateralis dorsalis (MLD) of the owl based on the center of the neuron's best area. At the upper left are plots of best areas from 14 cells superimposed on an imaginary globe that diagrams the relation of the sound source to the position of the owl. The symbols depict different electrode penetrations, and the numbers represent the order in which the cells were encountered. The topographical distribution of these neurons are shown in transverse, horizontal, and sagittal sections through the MLD. The stippled areas represent the space-mapped regions. Solid lines reflect isoazimuthal contours (horizontal and sagittal sections), whereas dotted lines represent isoelevation contours (transverse and sagittal sections). Dashed arrows depict the planes of the other two sections, and the solid crossed arrows define the section orientation: a, anterior; p, posterior; d, dorsal; v, ventral; m, medial; and l, lateral. The optic tectum (OT) is also shown on each section. (From Knudsen and Konishi, 1978A, 1978B.)

resentation of auditory space. A similar map of auditory space that is a result of direct anatomical projections from the MLD also exists in the optic tectum of the owl (Knudsen, 1982). These tectal neurons respond to both acoustic and visual stimuli, and the spatial positions of the receptive fields of these two modalities are in register.

In mammals the situation is not as clear. A map of auditory space has been found in the deep and intermediate layers of the superior colliculus of the guinea pig (Palmer and King, 1982; King and Palmer, 1983) and cat (Middlebrooks and Knudsen, 1984). This topographical representation of auditory space resembles the retinotopic map of visual space present in the superficial layers of the superior colliculus: cells with contralateral visual and auditory fields near the frontal midline are located rostrally and those with far contralateral fields are found caudally. Preliminary results in superior colliculus recordings in awake monkeys suggest a similar organization for visual and auditory receptive fields, with the additional finding that the position of the auditory receptive field for some cells is dependent on eye position (Jay and Sparks, 1984). For example, with the eyes fixated straight ahead, the optimal response of a superior colliculus neuron may be to a sound source 10 degrees to the right of midline. When the monkey fixates his eyes to some other position, the optimal neural response is still to a sound source 10 degrees to the right of the fixation point; even though the head and ears have remained stationary. Thus it seems that the discrepancy between the position of a sound source and the visual axis is mapped onto the superior colliculus. The coordination of the motor map of saccadic eye movements and auditory space would enable eye movements to be accurately programmed to fixate an invisible sound source.

In other areas of the mammalian auditory system, a map of auditory space has yet to be found. Studies of the cat auditory cortex (Middlebrooks and Pettigrew, 1981) and central nucleus of the inferior colliculus (Semple et al, 1983) demonstrated cells with two types of receptive fields: hemifield and axial. Hemifield neurons have receptive fields that cover all or most of the contralateral sound field whereas axial neurons have smaller restricted receptive fields that are aligned with the acoustic axis of the pinna and shift with movements of the pinna (Middlebrooks and Pettigrew, 1981; Semple et al, 1983). Both cell types are tuned to higher frequencies (>3 kHz) and are thought to reflect an IID sensitivity, but neither type are topographically organized. Recent studies by Moore et al (1984) showed low-frequency cells in the central nucleus of the inferior colliculus with restricted receptive fields, and such cells may correspond to the ITD-sensitive cells found using dichotic stimuli (see below). Some suggestion for a topographical representation of stimulus azimuth along the rostrocaudal dimension has been found from the free-field studies of Aitkin et al (1985). No doubt, this will be an active area of research in the next few years.

Dichotic Stimulation: Sensitivity to IIDs

Free-field studies have been useful in elucidating the functional role of binaural neurons. However, to understand the relevant stimulus cues and underlying mechanisms, it is necessary to use dichotic stimuli where the signals to each ear can be independently varied. Most electrophysiological studies have investigated the effects of varying IIDs and ITDs. Consistent with the duplex theory, to pure-tone stimuli, low-frequency cells are sensitive to ITDs, whereas high-frequency neurons are sensitive only to IIDs.

In cat, the maximal IID created by the shadowing effect of the head and amplification by the proximal pinna approaches 40 dB for stimuli positioned along the acoustical axis of either ear (Phillips et al, 1982; Semple et al, 1983). Sensitivity to IIDs in this range has been demonstrated in the superior olivary complex (Boudreau and Tsuchitani, 1968; Goldberg and Brown, 1968, 1969; Caird and Klinke, 1983), dorsal nucleus of the lateral lemniscus (Brugge et al, 1970), inferior colliculus (Rose et al, 1966; Geisler et al, 1969; Stillman, 1972), superior colliculus (Schechter et al, 1981; Wise and Irvine, 1983; Hirsch et al, 1985; Yin et al, 1985), medial geniculate body (Aitkin and Webster, 1972), and auditory cortex (Brugge et al, 1969; Brugge and Merzenich, 1973; Imig and Adrian, 1977; Phillips and Irvine, 1981). Figure 6-2 shows the sensitivity of a neuron in the dorsal nucleus of the lateral lemniscus to IIDs (Brugge et al, 1970). In Figure 6-2A, each curve illustrates the responses to changes in IID generated by holding the intensity to the contralateral ear constant, while the intensity to the ipsilateral ear was varied. There was a sharp decrease in discharge rate as the ipsilateral intensity was increased, regardless of the contralateral intensity level. In Figure 6-2B, the responses are normalized and plotted as a function of IID. The similar slopes indicate that the sensitivity of the cell to changes in IID is independent of absolute intensity levels. The range of dichotically created IIDs to which such neurons are sensitive are similar to those generated in the free field.

The type of IID sensitivity just described is first seen in high-frequency neurons of the lateral superior olive, where they are excited by ipsilateral

FIGURE 6.2. Effects of varying IIDs on the discharge rate of a neuron in the cat dorsal nucleus of the lateral lemniscus. **A:** The intensity of the signal to the ipsilateral ear was systematically increased from 10 to 90 dB sound pressure levels (SPL) while holding the contralateral stimulus intensity constant at 7 SPLs. Numbers beside each curve are the constant SPLs at the contralateral ear. **B:** The data from A are replotted as a percentage of the number of spikes with respect to contralateral stimulation alone as a function of interaural intensity differences. In all cases the stimulus to both ears was a 6.4-kHz tone burst, 200 ms in duration, presented 20 times. (From Brugge et al, 1970. Reproduced with permission from J.F. Brugge, 1970 Journal of Neurophysil. 33:441-558.)

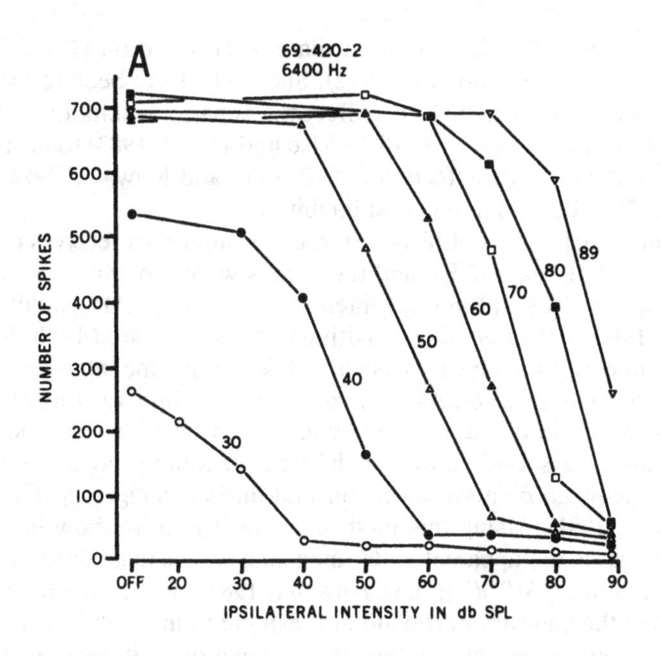
A
69-420-2
6400 Hz
NUMBER OF SPIKES
IPSILATERAL INTENSITY IN db SPL
0
100
200
300
400
500
600
700
OFF
20
30
40
50
60
70
80
90
30
40
50
60
70
80
89

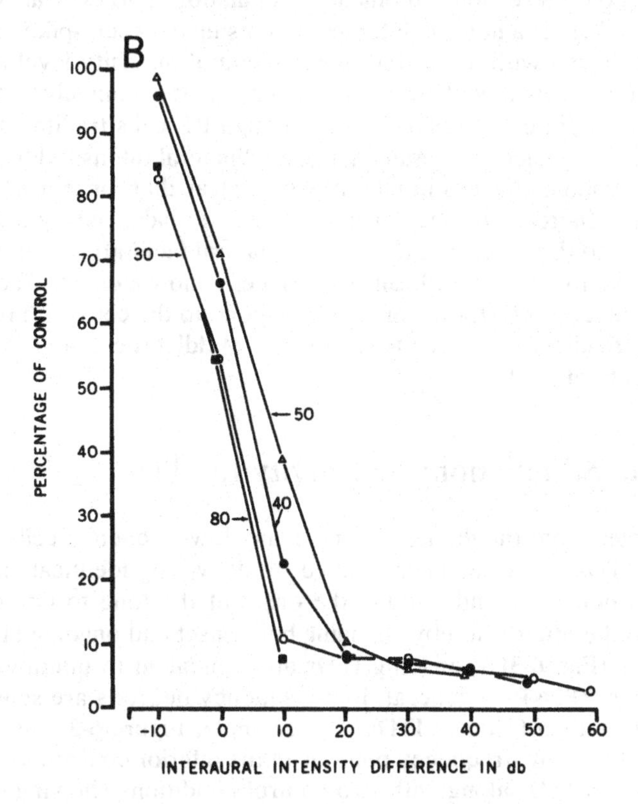
B
PERCENTAGE OF CONTROL
INTERAURAL INTENSITY DIFFERENCE IN db
0
10
20
30
40
50
60
70
80
90
100
-10
0
10
20
30
40
50
60
30
50
40
80

stimulation and inhibited by contralateral stimulation (Boudreau and Tsuchitani, 1968; Guinan et al, 1972). Such cells have been termed "EI" (Goldberg and Brown, 1968) to reflect the separate excitatory-inhibitory influences from each ear, "EO/I" (Wise and Irvine, 1983) to describe the monaural and binaural effects, or "BI" (Yin and Kuwada, 1984; Hirsch et al, 1985) to designate binaural inhibition.

The other major class of high-frequency binaural cells receives excitatory inputs from both ears (EE), and these cells were thought to be sensitive to overall changes in binaural intensity level, not IIDs (Goldberg and Brown, 1968). Usually, IID sensitivity has been tested by holding the stimulus intensity at one ear constant, while varying the intensity delivered to the other ear (Fig. 6-2A). This procedure, while varying IIDs, also changes the mean binaural intensity level. A procedure that more accurately mimics a sound source at different azimuthal positions involves keeping the mean binaural level constant during changes in IID (Phillips and Irvine, 1981). Using this method, many EE cells show binaural facilitation that is not apparent under monaural-stimulating conditions, hence the designation "BF" (Yin and Kuwada, 1984). For example, some EE neurons in the auditory cortex do not respond to monaural stimulation of either ear, but do respond to binaural stimulation (Kitzes et al, 1980). In such cells, there is a non-monotonic relationship between spike count and IID when tested with a constant mean binaural intensity level (see Fig. 6-13). Similar cells have been found in the superior colliculus (Wise and Irvine, 1983; Yin et al, 1985). Thus, although EE cells traditionally have been thought to encode overall changes in binaural intensity level and EI cells as encoding changes in IID, it is now clear that for many EE cells, an intensity increase in one ear is not perfectly balanced by a decrease in intensity to the other ear; they show binaural facilitation and therefore are sensitive to IIDs as well. Since these cells show a distinct facilitatory peak to changes in IID, they might correspond to the cells with restricted receptive fields seen in free-field studies (Middlebrooks and Pettigrew, 1981; Moore et al, 1984).

Dichotic Stimulation: Sensitivity to ITDs

As predicted from the duplex theory, many low-frequency cells are sensitive to ITDs. This has been studied by delivering identical sinusoidal stimuli to both ears and varying the onset of the tone to one ear with respect to the other, thereby changing both onset and ongoing ITDs. For pure tones (Fig. 6-3D), ongoing ITDs are equivalent to interaural phase differences (IPDs). In general, low-frequency neurons are sensitive to IPDs, rather than to onset ITDs. For example, Figure 6-3 illustrates responses of two low-frequency neurons in the inferior colliculus of the cat to changes in ITD, along with two control conditions showing that IPD

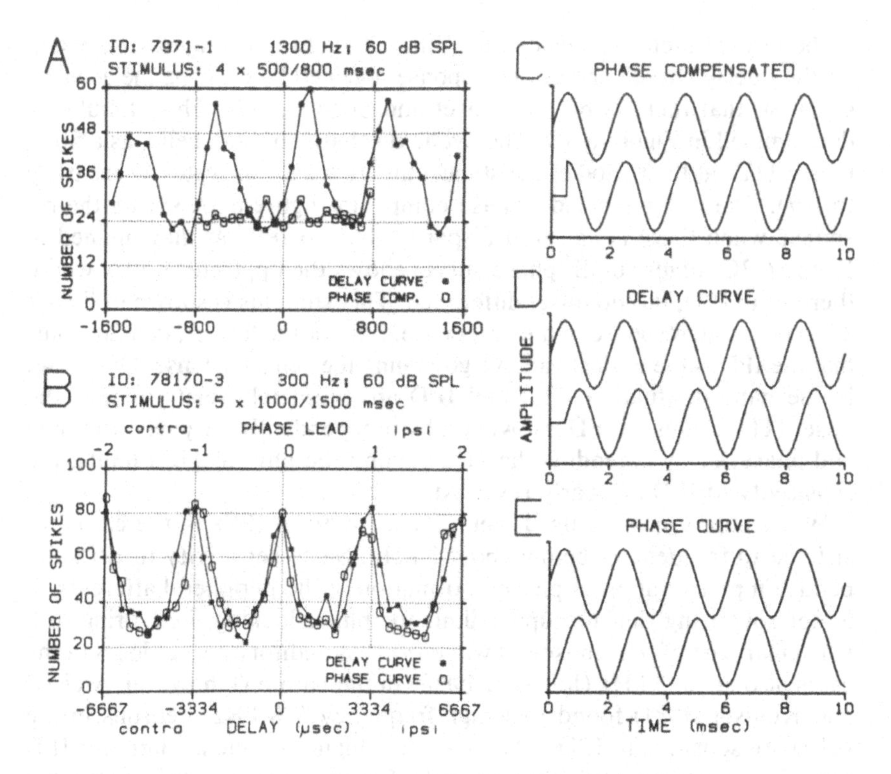

FIGURE 6.3. Responses of two inferior colliculus neurons to changes in interaural delay and interaural phase along with a schematic of the three types of binaural stimuli used. In A the delay curve (★) shows the response to a 1,300-Hz tone burst with a delay in the contralateral ear (-) with respect to the ipsilateral ear (+). In D is a schematic representation of this delay stimulus. The upper and lower traces represent the signals to each ear. Here, there are both onset and ongoing interaural time differences. The phase curve (0) illustrates the response of the same neuron to a stimulus in which the interaural phase was set equal to zero at each delay. The stimulus is shown in C. In this case, there are no ongoing interaural time differences, but onset time differences are still present. In B is a comparison of the response of another inferior colliculus neuron to changes in interaural delay (★) to the response when interaural phase was varied (0) while holding the interaural delay at zero. This latter stimulus is illustrated in E; only ongoing time differences are present. The actual acoustic stimuli were ramped with a rise/fall time of 3.9 ms. (From Kuwada and Yin, 1983. Reproduced with permission from Kuwada and Yin, 1983 and Journal of Neurophysiology.)

is the critical factor (Kuwada and Yin, 1983). In both Figures 6-3A and 3B the "delay curve" shows the response obtained by delaying the acoustic signal so that there is both an onset and ongoing ITD. This stimulus is diagrammed in Figure 6-3D. The cyclic discharge of both cells, as a function of ITD, at the period of the stimulating frequency indicates a sensitivity to IPD. The curve labeled "phase comp" in Figure 6-3A shows the response when there is an onset disparity, but no IPD, as diagrammed in Figure 6-3C. Figure 6-3B (phase curve) shows the opposite control where there is an IPD, but no onset difference. This stimulus is shown in Figure 3E. The cyclic response in the latter case, but not the former, demonstrates that the IPD is the critical variable governing the cell's response. Of course, in the natural situation, ITD and IPD are intimately related, since the onset ITD creates an IPD. However, by independently varying onset time and phase, or by extending the ITD beyond the physiological limits, the sensitivity to IPD is clearly revealed.

When pure tones are used, neural sensitivity to IPDs in the cat is restricted to frequencies below about 3 kHz. Since sensitivity to IPDs depends on preservation of phase information in the peripheral afferents, it is not surprising that the upper limit for phase-locking, i.e., firing to a particular part of the sinusoidal waveform, of auditory nerve fibers in the cat is also about 3 kHz (Johnson, 1980). In the barn owl, however, Moiseff and Konishi (1981) found that high-frequency 6–8 kHz) neurons in the MLD are sensitive to IPDs; this is a much higher frequency limit for IPD sensitivity as compared with mammals. Recent experiments indicate that neurons in the barn owl's nucleus magnocellularis, the avian homologue of the anteroventral cochlear nucleus, are able to phase-lock at these high frequencies (Sullivan and Konishi, 1984). Thus, the cat and the owl seem to use similar mechanisms to encode ITDs, but the owl does so at higher frequencies because its auditory nerve fibers can preserve phase information at those high frequencies.

Although the ITD stimulus tests the sensitivity of cells at discrete phase differences, it does not examine the response to continuous phase changes, like those resulting from a tonal source moving along the azimuth. Such dynamic phase changes can be generated by a binaural beat stimulus. A diagram of a binaural beat stimulus and the assumed movement of the sound source about the head is shown in Figure 6-4. In this example the frequency delivered to the contralateral ear is slightly higher by the beat frequency (f_b) than that delivered to the ipsilateral ear. When the tones are turned on, the contralateral signal begins to lead. When the interaural phase $\phi = 0.5$, the tones are 180 degrees out of phase, after which the ipsilateral signal leads until they are back in phase at $\phi = 1.0$. The presumed movement created by such a stimulus is shown in Figure 6-4B. Since the tones are delivered dichotically, the perceived movement of the sound source will be within the head (see Hafter, this volume), but for purposes of illustration, this movement has been projected outside the

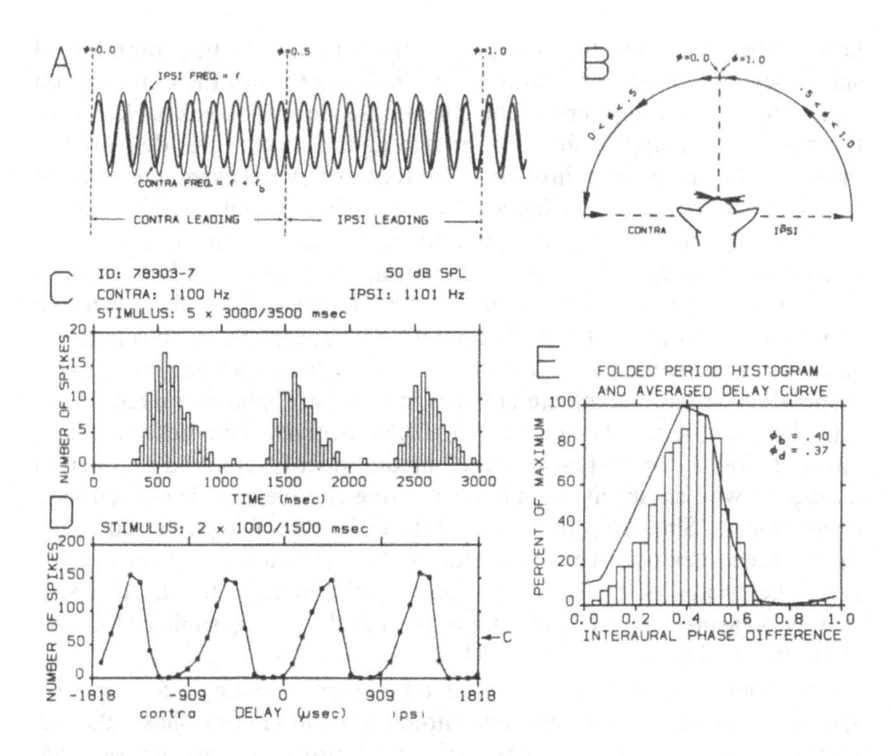

FIGURE 6.4. **A:** Binaural beat stimulus. To illustrate the continuously changing shift in interaural phase, we have superimposed the frequencies to each ear. In this example the contralateral frequency is greater than the ipsilateral frequency by f_b, resulting in an initial contralateral phase lead changing to an ipsilateral lead at $\Phi = 0.5$. A complete cycle of interaural phase change from $\Phi = 0.0$ to 1.0 will occur at a rate of f_b Hz. **B:** Presumed movement created by a binaural beat stimulus in A. For illustrative purposes, the movement is projected outside the head, and the arrows indicate the direction of movement. At $\Phi = 0.0$, or equivalently 1.0, the two frequencies are in phase. This simulates a sound source directly in front of the animal. As the contralateral signal begins to lead, the source moves toward the contralateral ear until it is fully lateralized; then, after a variable period of time which is a function of the stimulating frequency (dotted line), it jumps to the ipsilateral ear and moves back toward the midline. The speed of the simulated movement is directly related to the magnitude of f_b, and changing directions is accomplished by reversing the frequencies to the two ears. **C:** Response of an inferior colliculus neuron to a binaural beat stimulus with f_b = -1 Hz. **D:** Response of the same neuron to changes in interaural delay using similar stimulus parameters as in C. **E:** Comparison between the binaural beat and interaural delay response. For the binaural beat condition, the response was folded on f_b = 1 Hz (period histogram) while the delay curve was folded on the period of the stimulating frequency (solid line). Averaged responses to both stimuli have been normalized. (From Yin and Kuwada, 1984.)

head. When $\phi = 0.0$, the sound is on the midline. As the contralateral signal begins to lead, the sound moves laterally toward the contralateral ear. Near $\phi = 0.5$, it jumps to the ipsilateral side and moves back toward the midline as ϕ approaches 1.0. Switching the frequencies delivered to the two ears changes the direction of movement, whereas increases in the speed of movement are achieved by increasing the difference between the two frequencies. A comparison of the responses of an inferior colliculus neuron to an ITD and a binaural beat stimulus is shown in Figure 6-4C–E. To generate the binaural beat, a 1,100-Hz tone was delivered to the contralateral ear and a 1,101-Hz tone to the ipsilateral ear. During the 3-second duration of the tone, the neuron responded with three bursts, corresponding to three complete changes in interaural phase created by the 1-Hz-beat frequency. To compare the IPD sensitivity with these two stimulus conditions, the response to the binaural beat is replotted as a period histogram with an abscissa equivalent to one complete cycle of interaural phase change. Similarly, the interaural delay function is replotted to reflect the average response over one period of the stimulating frequency (1,100 Hz). The comparisons, shown in Figure 6-4E, indicate tht the IPD sensitivity measured by these two types of stimuli is quite similar (Kuwada et al, 1979; Yin and Kuwada, 1983a).

The binaural beat stimulus has the advantage of testing a neuron's response to changes in the rate and direction of interaural phase change. Figure 6-5 shows the responses of three inferior colliculus neurons that exhibit different dynamic phase sensitivities. The data are presented as period histograms and are organized such that the two columns represent opposite directions of phase change, and the rows from top to bottom reflect decreasing rates of phase change. The responses in Figure 6-5A illustrate the most common finding: the cell responds with approximately the same number of spikes, independent of changes in rate or direction. Figure 6-5B shows responses from a neuron sensitive to slow rates of interaural phase change. The responses are greater at low-beat frequencies and less at high-beat frequencies. Cells that respond more vigorously at higher rates of phase change have also been observed (not shown). Figure 6-5C illustrates a direction-sensitive cell that responds only when the frequency of the signal to the ipsilateral ear is higher, which corresponds to movements of the sound source toward the ipsilateral ear.

Directional movement of a sound source has also been simulated by using dichotic click trains in which the ITDs between clicks were systematically increased and decreased. Using such a stimulus, Altman (1978) found directionally sensitive neurons in the inferior colliculus, medial geniculate, and auditory cortex. However, in the superior olivary complex he did not find cells that showed this sensitivity. Using free-field stimuli, direction- and motion-sensitive cells have been found in the cat superior colliculus (Gordon, 1973) and auditory cortex (Sovijari and Hyvarinen, 1974). Thus, some auditory cells are not only capable of responding to a

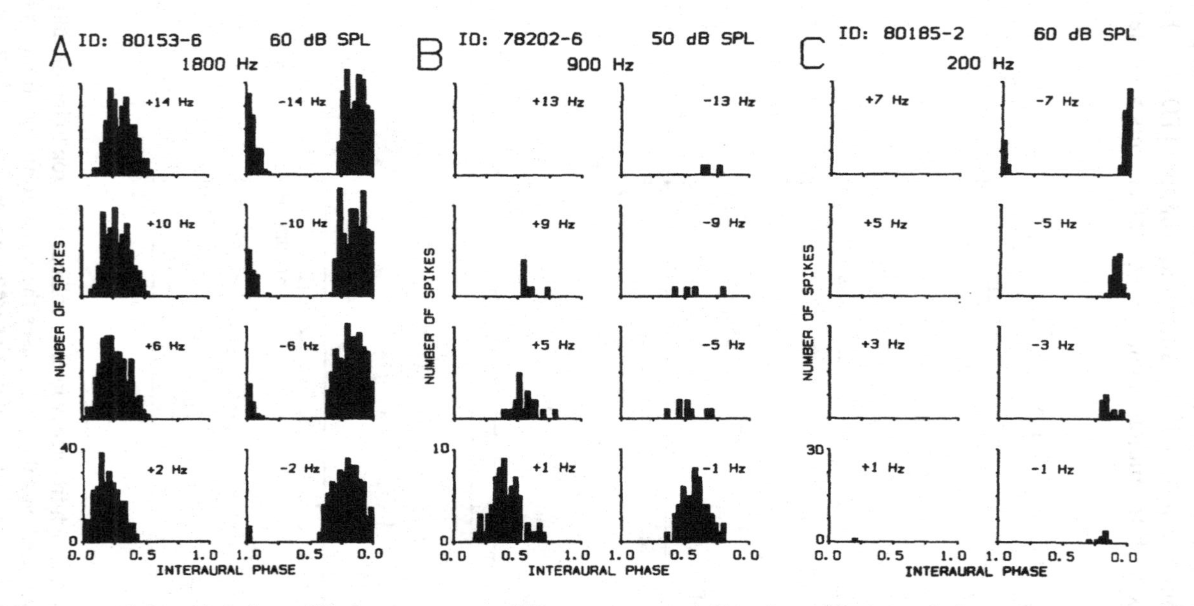

FIGURE 6.5. Responses of three inferior colliculus neurons to different rates and direction of interaural phase change created by binaural beat stimuli. Responses are shown in the form of period histograms constructed by grouping the responses that occur in bins equivalent to 1/32 of the period of the beat frequency (f_b). In each set of histograms, (A, B, and C), the left and right columns reflect the responses to the two directions of interaural phase change, and the rows from top to bottom to reflect decreasing rate of interaural phase change. **A:** Neuron that responds equally to changes in rate and direction of interaural phase change. **B:** Rate-sensitive cell that prefers low rates of interaural phase change in either direction. **C:** Direction-sensitive neuron that prefers interaural phase changes in a particular direction. (From Yin and Kuwada, 1984.)

particular sound location, but also may be sensitive to the speed and direction of a moving sound source.

Although IPD is an effective cue for localization of low-frequency tones and IID is important at high frequencies, a departure from the duplex theory is evident from psychophysical studies showing that ITDs can be detected in complex high-frequency signals (see Hafter, this volume).

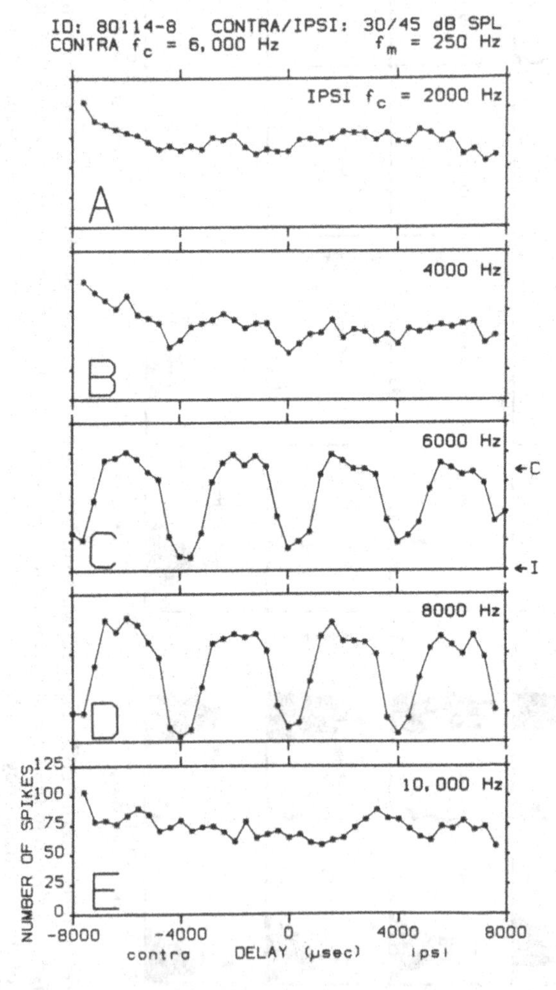

FIGURE 6.6. Responses of a high-frequency inferior colliculus neuron to interaurally delayed amplitude-modulated signals. In all cases the signal to the contralateral ear was the same, whereas the ipsilateral carrier frequency was varied between 2,000 and 1,000 Hz. The modulation envelope was trapezoidal in shape with a rise/fall time of 488 μs and the modulation frequency was 250 Hz. Arrows represent firing levels to monaural stimulation of the contralateral and ipsilateral ear. (From Yin and Kuwada, 1984.)

Presumably, temporal information is extracted from the low-frequency envelopes of these complex high-frequency signals. Physiological support for this hypothesis has been found by Yin et al (1984) using amplitude-modulated tones. Figure 6-6 shows the results from one cell in which the interaural delay curves were obtained from tones that were amplitude modulated at 250 Hz with a trapezoidal envelope. The carrier frequency to the contralateral ear was kept at 6kHz. Each panel illustrates the effects of varying ITD at different ipsilateral carrier frequencies. When the ipsilateral carrier frequency is within the neuron's frequency range, a clear sensitivity to the IPD of the modulating waveform is evident (Fig. 6-6C,D). However, when the ipsilateral carrier frequency is outside this range, no such sensitivity is evident (Figs. 6-6A,B,E). Just as the IPD sensitivity of low-frequency neurons depends on the preservation of phase information in the peripheral afferents, IPD sensitivity of high-frequency neurons must depend on a similar preservation of phase information to the envelope frequency. In this regard, high-frequency cochlear nucleus neurons can phase-lock to the envelope of amplitude-modulated signals with high-frequency carriers (Moller, 1974). Thus, it appears that high-frequency neurons can be sensitive to IPDs in a manner similar to their low-frequency counterparts. Since most stimuli in the natural environment are complex, this phase sensitivity to the envelope waveform may have functional significance.

Central Processing of IID and ITD Sensitivity

Physiological, anatomical, and behavioral experiments suggest that the processing of IID and ITD cues occur in different auditory nuclei. In the primary site of binaural interaction, the superior olivary complex, it has been shown that high-frequency cells in the lateral superior olive are sensitive to IIDs, whereas low-frequency cells in the medial superior olive are sensitive to ITDs (Boudreau and Tsuchitani, 1968; Goldberg and Brown, 1969; Caird and Klinke, 1983). Consistent with the expectations of the duplex theory, the lateral superior olive has a disproportionately large high-frequency representation, whereas low-frequency neurons predominate in the medial superior olive (Boudreau and Tsuchitani, 1968; Guinan et al, 1972). An animal's ability to localize low- and high-frequency sounds also seems to be reflected in the relative size of these two nuclei. Animals with small heads generally have a larger lateral superior olive and are unable to localize low-frequency tones, whereas those with large heads have a larger medial superior olive and are able to localize low-frequency tones (Irving and Harrison, 1967; Masterton et al, 1975). It might be supposed that the difference between these nuclei primarily reflects a difference in the portions of the frequency spectrum represented in each structure, rather than a functional difference in the processing of IIDs and

ITDs. Evidence from the barn owl suggests that this is not the case. Cells in the posterior division of nucleus ventralis lemnisci lateralis, the avian homologue of the lateral superior olive, respond to IIDs, whereas cells in nucleus laminaris, homologue of the medial superior olive, respond exclusively to ITDs (Moiseff and Konishi, 1983). Konishi and his colleagues demonstrated that in the MLD, the borders of the receptive field in the azimuthal and elevational directions depend on ITDs and IIDs, respectively. IIDs in the elevational dimension occur because the barn owl's ears are situated asymmetrically in the facial ruff: one ear is directed upward, the other downward (Knudsen and Konishi, 1979).

Neuronal Mechanisms of ITD Sensitivity

Thus far, this review has been concerned with physiological studies of directional hearing, with an emphasis on the effects of changing ITD, IPD, and IID. This section considers the neural mechanisms by which neurons combine information from the two ears and extract these binaural cues. At the outset it should be emphasized that much of the evidence supporting these models is inferential. This review is limited to models of binaural interaction applied to neural responses, rather than models directed toward human psychoacoustical data (for a review of psychoacoustical models, see Colburn and Durlach, 1978).

Coincidence Mechanisms

A model first formalized by Jeffress (1948) provides the rudiments for many of the contemporary models of binaural interaction, and it is illustrated in Figure 6-7. Excitatory inputs from each ear converge upon binaural cells. Because of the differences in conduction delays, the input from each ear will reach the binaural cells at slightly different times. The key assumption is that each binaural cell is maximally excited when the two inputs arrive simultaneously, and such cells are often referred to as coincidence detectors. The mathematical operation of cross-correlation will yield equivalent results.

A primary site of binaural convergence is the superior olivary complex. For low-frequency tones, the inputs to the medial superior olive are phase-locked to the stimulating waveform. Goldberg and Brown (1969) showed that cells in this nucleus exhibit the properties expected of coincidence detectors. Figures 6-8A and B show period histograms obtained with monaural stimulation of the contralateral and ipsilateral ear, respectively, with a 443.7-Hz tone. The effects of varying interaural delay on discharge rate and vector strength are illustrated in Figure 6-8C, while the period histograms show the phase-locked responses at four different delays (Figs.

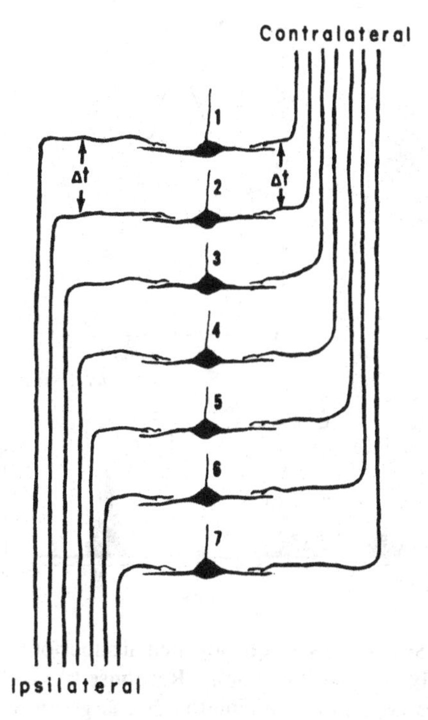

FIGURE 6.7. Diagram of Jeffress's place theory of binaural interaction. Neurons 1–7 receive separate ipsilateral and contralateral inputs. The conduction time to reach these cells is represented by the length of these inputs. Since the neurons act as coincidence detectors, neuron 4 would respond optimally when both inputs are activated simultaneously, whereas neuron 1 requires a contralateral delay of $6 \times \Delta t$ for maximum excitation. (From Goldberg and Brown, 1969,) adapted from Jeffress, 1948. Reproduced with permission from Goldberg and Brown 1969, Journal of Neurophysiology.)

6-8D–G). When the contralateral signal was delayed by 1,600 μs, the discharge rate and the degree of phase-locking were maximal (Fig. 6-8F). Since this delay of 1,600 μs corresponds to the difference between the peak phase angle of the monaural responses, this cell is maximally responsive when the inputs from the two ears arrive in close temporal coincidence. Similar evidence for a coincidence mechanism in the medial superior olive has also been reported by Crow et al (1978), Chan and Yin (1984), and Yin and Chan (1986), in the dorsal nucleus of the lateral lemniscus by Brugge et al (1970), and in the inferior colliculus by Kuwada et al (1984). Physiological evidence for a cross-correlation mechanism has been provided recently by studying the responses of inferior colliculus neurons to ITDs of noises correlated to different degrees (Yin and Chan, 1986).

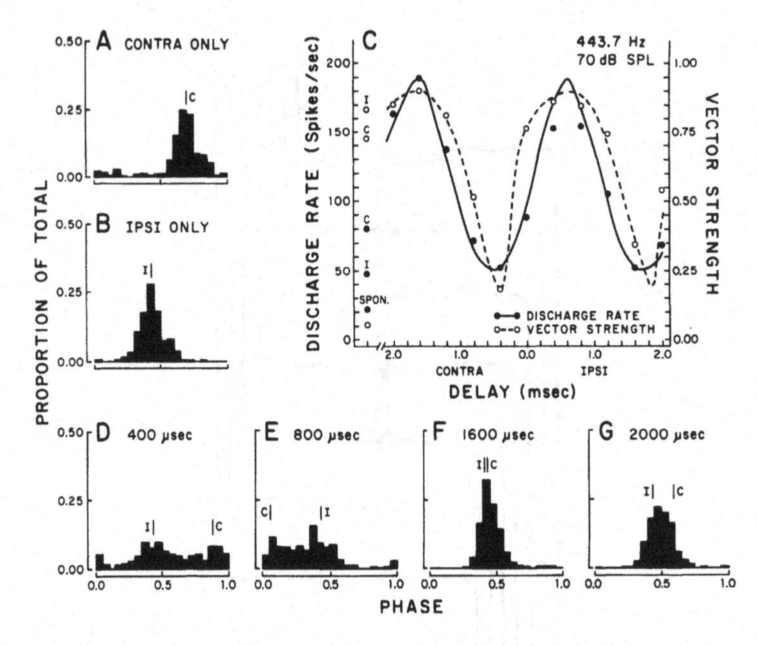

FIGURE 6.8. Responses of a low-frequency medial superior olivary neuron to monaural and interaurally delayed tone bursts. Response (A, B, D–G) are in the form of period histograms created by binning the discharge on the period of the stimulating frequency (443.7 Hz). **A, B:** Phase-locked responses to contralateral (C) and ipsilateral (I) stimulation alone. Vertical bars depict the mean phase angle. **C:** Discharge rate and vector strength (degree of synchrony) as a function of interaural delay. Mean phase angle and vector strength were calculated by the vector-averaging method. Vector strength of 1.0 indicates that all spikes fall into one bin of the period histogram; a value 0 indicates an equal distribution across all bins. **D–G:** Period histograms derived from delays of the contralateral stimulus at the values indicated. Bars represent the estimated position of the mean phase angles based on the monaural responses (**A, B**). (From Goldberg and Brown, 1969. Reproduced with permission from Goldberg and Brown, (1969 and Journal of Neurophysiol.)

Characteristic Delay

An important issue, first raised by Rose et al (1966), is the effect of changing frequency on the responses of neurons sensitive to IPD. When they plotted the responses to interaural delays at several different frequencies on the same time axis for a given neuron, a few cells responded with either maximal or minimal discharges at a particular, or characteristic, delay, which was defined as the delay at which the periodic discharge curves for different frequencies reach the same relative amplitude. In their examples, the common relative amplitude occurred either at the point of maximal or minimal responses, but they emphasized that there was no reason to expect the relative amplitude to be restricted to these peaks or troughs. The principle of characteristic delay indicates that neurons are capable of signaling IPDs independent of the stimulus frequency, and this hypothesis has support from work done in various auditory nuclei (Brugge et al, 1969; Geisler et al, 1969; Brugge et al, 1970; Moushegian et al, 1971; Stillman, 1971; Aitkin and Webster, 1972; Brugge and Merzenich, 1973; Benson and Teas, 1976). A complication to the concept of characteristic delay is that ITDs can vary as a function of stimulus frequency. In cat, ITDs as large as 435 μs at frequencies below 1 kHz have been measured; a value significantly greater than that predicted from interaural distance alone (Roth et al, 1980; also see Kuhn, this volume).

Since in previous studies characteristic delay was qualitatively assessed and the sample size was small, Yin and Kuwada (1983b) made a quantitative assessment of characteristic delay for many neurons in the cat's inferior colliculus. Figure 6-9 illustrates their analytical procedure using simulated triangular delay curves. Three different characteristic delay responses are simulated. For each type, two cycles of the delay curves at six different frequencies are plotted. Figure 6-9A shows a response in which all of the interaural delay curves peak at a delay of +300 μs. On the right is a plot of interaural phase *v* frequency. This curve is linear with a slope of 300 μs and a phase intercept of 0.0. If the responses all have a common minimum point, as in Fig 9B, then the phase *v* frequency plot is again linear and has a slope corresponding to the characteristic delay (−100 μs). However, in this case the phase intercept is 0.5; reflecting the fact that the point of common relative amplitude is half a cycle from the peak. If the phase *v* frequency plot is linear, but does not have a phase intercept of either 0.0 or 0.5, as in Figure 6-9C, then the interaural delay curves have the same relative amplitude at an ITD corresponding to the slope of the phase *v* frequency plot. Thus, characteristic delay is indicated by a linear relationship between interaural phase and frequency. The slope of the line specifies the value of the characteristic delay, and the phase intercept indicates whether the characteristic delay occurs at the peak (0.0), trough (0.5), or somewhere in between.

Figure 6-10 illustrates this analysis applied to the neural responses of four inferior colliculus neurons. On the left are the responses to interaural delay stimuli to several frequencies, superimposed on a common time axis. In the middle are composite curves derived by averaging all of the responses on the left, and on the right are the phase *v* frequency plots. In Figure 6-10A the cell shows a characteristic delay at the peak, which is consistent with the linear phase *v* frequency plot with a phase intercept near 0.0. In Figure 6-10B the characteristic delay occurs at the minimal point, and, here, the phase intercept is near 0.5. The most common finding in the inferior colliculus is illustrated in Figure 6-10C, where it is not obvious from examination of the superimposed delay curves whether the

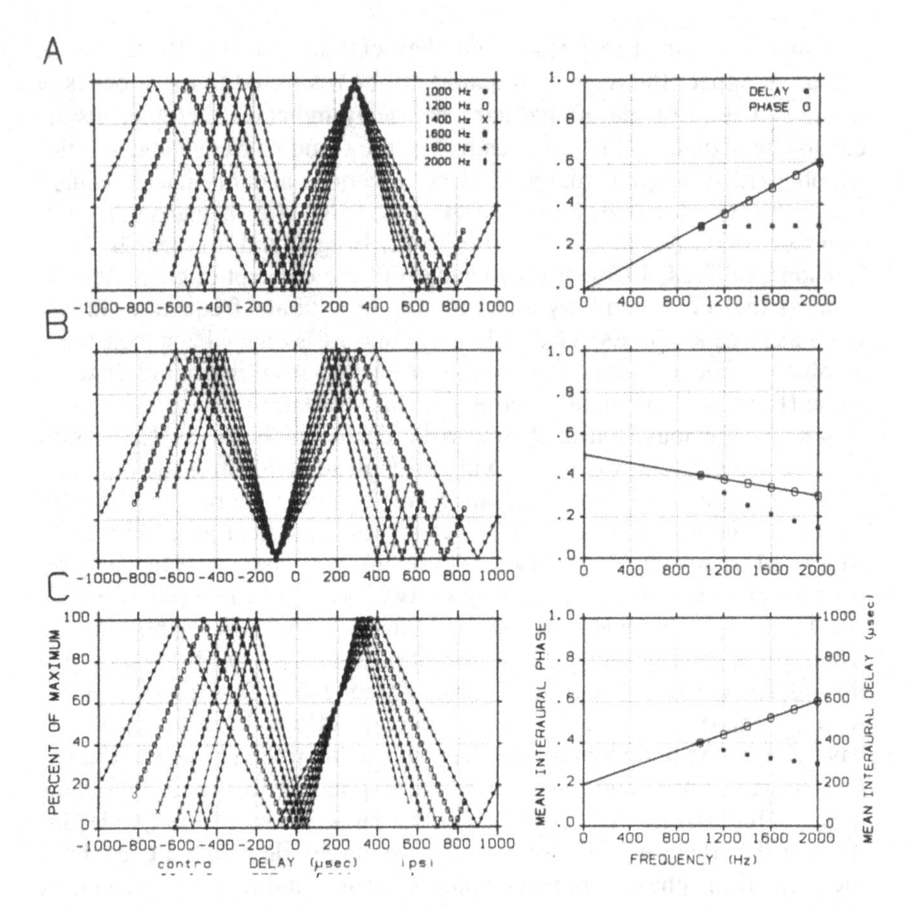

FIGURE 6.9. Simulation of three characteristic delay responses. On the left are the superimposed interaural delay curves simulated as triangular waveforms for frequencies between 1 to 2 kHz. Each interaural delay curve is plotted twice, once on either side of zero delay. On the right are plots of mean interaural phase *v* stimulating frequency. Each phase value is also plotted as an equivalent interaural delay by multiplying by the period of the stimulating frequency. **A:** Simulated response that has a characteristic delay of + 300 μs at the maximum point on the interaural delay curves. The phase *v* frequency plot is linear and has a phase intercept of 0.0 with a slope of +300 μs. **B:** Simulated response that has a characteristic delay of -100 μs at the minimum point on the delay curves. The phase *v* frequency curve is linear and has an intercept of 0.5 with a slope of -100μs. **C:** Simulated response that has a characteristic delay at +200μs at the 60% relative amplitude point. The phase *v* frequency plot is linear and has an intercept of 0.2 with a slope of +200 μ. (From Yin and Kuwada, 1983b.)

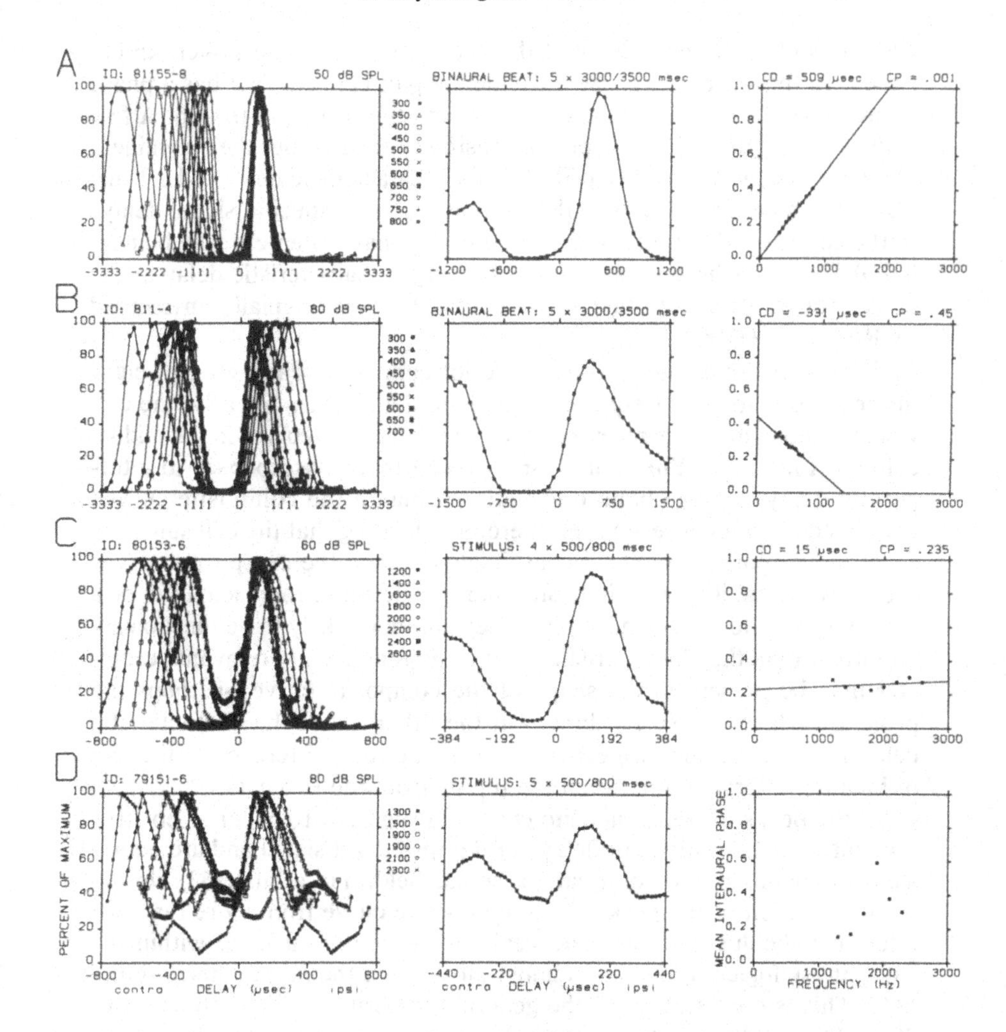

FIGURE 6.10. Responses of four low-frequency inferior colliculus neurons and the application of the characteristic delay analysis outlined in Figure 6-9. On the left are illustrated the normalized delay curves to different frequencies superimposed on a common time axis for four neurons (A–D). Two cycles of the averaged interaural delay at each frequency are plotted. The middle column shows the composite curves derived by averaging the delay curves to the left. On the right are plots of mean interaural phase as a function of stimulating frequency with a least squares regression line fitted to the data points. The slope and phase intercept are specified as characteristic delay (CD) and characteristic phase (CP), respectively. **A:** Cell that shows a CD at the peak. **B:** Another cell with a CD at the trough. **C:** Cell with a CD to a contralateral delayed signal of 15μs, but the point of common relative amplitude is neither at the peak or trough (CP = 0.235). **D:** Cell that does not show a CD since the phase v frequency plot is not linear. (From Yin et al, 1983.)

cell has a characteristic delay at the peak, trough, or somewhere in between. In this case, the phase v frequency plot is linear, but has a phase intercept of 0.25, which is halfway between the peak and trough. A few cells, as in Figure 6-10D, display a sensitivity to IPD, but their nonlinear phase v frequency plots disqualify them as characteristic delay cells. Thus, cells may have characteristic delays that occur at specific signal delays to the contralateral or ipsilateral ear, and the phase intercept may be near 0.0, 0.5, or anywhere in between, reflecting a characteristic delay at the peak, trough, or some other relative amplitude as originally envisioned by Rose et al (1966).

If, for a wide-band noise signal, inferior colliculus cells were to weight the responses to each frequency equally and sum them linearly, the cell would respond in the manner shown by the composite curves in the middle column (Fig. 6-10). Yin et al (1986) showed that the responses to interaurally delayed, wide-band noise stimuli match the composite curves computed from tonal responses, thereby indicating that the cell sums the spectral components of the signal linearly. Since the composite curves show less variability with intensity and maybe more representative of a cell's response to wide-band stimuli, they are more likely to be functionally important than the characteristic delay. The functionally relevant parameter may be present in the shape of the composite curve—perhaps the peak, trough, or its steep slope. On the other hand, the characteristic delay reflects the neural mechanism; it is the time difference at the site of binaural interaction between the inputs from the two ears.

All the peaks of the composite curves in Figure 6-10 occur at positive time differences, which are delays of the ipsilateral signal and are equivalent to sounds in the contralateral sound field. Yin et al (1983) plotted the distribution of the peaks of the composite curve from more than 600 inferior colliculus neurons. The peaks were tightly bunched within the cat's physiological range and almost all were in the contralateral sound field. This is consistent with the general observation that at all levels of the auditory system rostral to the decussating fibers of the superior olivary complex, the primary influence is from sounds presented on the contralateral side of the head (Masterton and Imig, 1984).

A Computer Model

The Jeffress coincidence model provides a framework for explaining interaural phase and time sensitivity, but it cannot incorporate some of the physiological results such as the modulation of the discharge above and below the responses to monaural stimulation (Fig. 6-4D), or speed and direction sensitivity (Fig. 6-5B,C). Sujaku et al (1981) developed a computer model that is able to simulate many of these response properties. The model is similar to Jeffress' with the addition of two presynaptic inhibitory collaterals from each side. These collaterals allow the input from one side to influence the input from the other side. The inputs to the model are Poisson point processes in which the rate parameter is modulated by a half-wave rectified sinusoid to simulate the phase-locking of eighth nerve fibers. Each input can be delayed by a fixed time ($D_1 - D_4$) and shifted by a fixed phase (ϕ_1 and ϕ_2). At each of the four synaptic sites, specific rules govern the presynaptic transmitter release and resulting postsynaptic response. Figure 6-11 compares the responses of an inferior colliculus neuron to interaural delay (Fig. 6-11A) and binaural beat stimuli (Fig. 6-11B) with the responses of the computer model (Figs. 6-11C,D). It is evident that the model can closely simulate the IPD sensitivity displayed in the neural responses to both stimuli.

Figure 6-12 shows that the model responds like characteristic delay cells in the inferior colliculus. In these examples the three cases differ only in the values of time delays, D_1 and D_2, and phase shifts, ϕ_1 and ϕ_2. The parameters were chosen to illustrate characteristic delays at the peak, trough, and at some other relative amplitude. In all cases the values of $\phi_1 - \phi_2$ determine the phase intercept, $D_1 - D_2$ set the characteristic delay, and the value of the phase intercept determines the relative amplitude at which the characteristic delay occurs. There is a striking similarity between the neural responses (Fig. 6-10) and those generated by the model (Fig. 6-12). With a suitable choice of parameters, this model

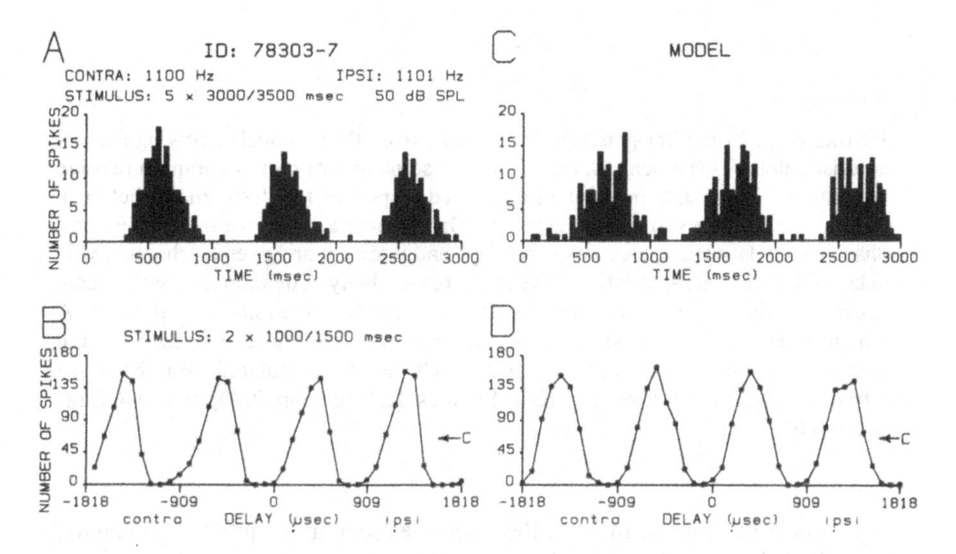

FIGURE 6.11. Comparison of neural and computer-simulated model responses to binaural beat and interaural delay stimuli. **A:** Responses of an inferior colliculus neuron to a binaural beat stimulus. **B:** Responses of the same neuron to changes in interaural delay. **C:** Responses of the model to the same binaural beat stimulus used in A. **D:** Responses of the same model to the same interaural delay stimulus used in B.

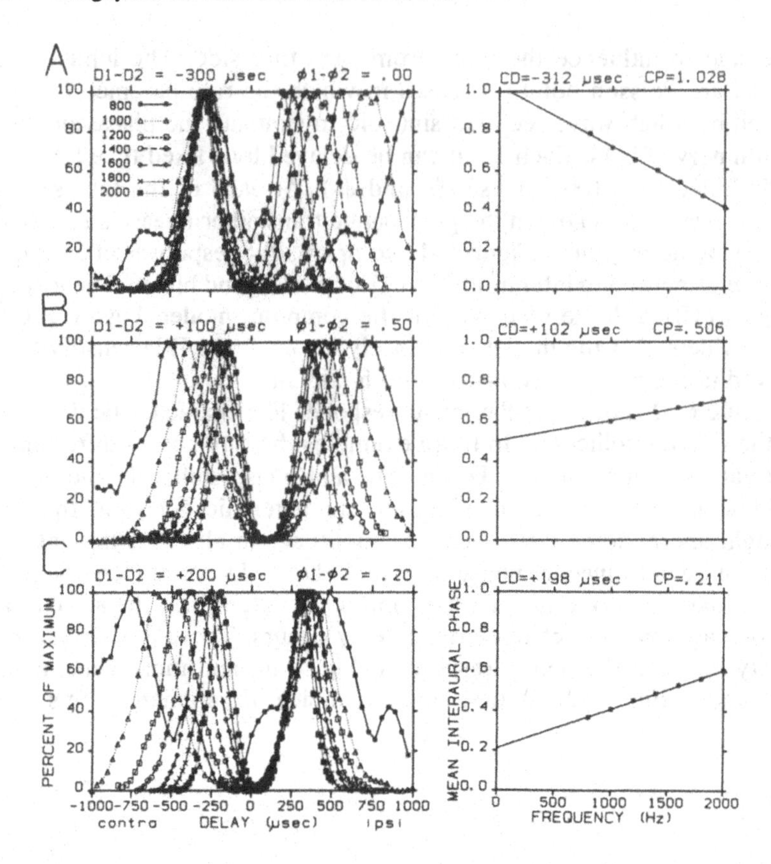

FIGURE 6.12. Computer-generated responses from three models showing a characteristic delay at the peak **A**, trough **B**, and some other point of common relative amplitude **C**. For each model the normalized responses to 7 frequencies between 800 to 2,000 Hz are shown (left column). On the right are plots of mean interaural phase *v* stimulating frequency and the slope and intercept are given as characteristic delay (CD) and characteristic phase (CP), respectively. The models are the same except for the values of D_1 and D_2, and Φ_1 and Φ_2, which are as indicated. A binaural beat (f_b = 1 Hz) stimulus was used to generate the responses. Note the high correspondence between the values of CD and CP calculated from the phase *v* frequency plots and their theoretical values derived from D_1-D_2 and Φ_1-Φ_1, respectively.

is also able to simulate most of the responses seen in the inferior colliculus, including the directional- and speed-sensitive response properties previously described (Fig. 6-5). This modified coincidence model is meant to be a general schema of binaural interaction and suggests that these excitatory and inhibitory elements may be necessary parts of the neuronal circuitry involved in detecting IPD.

Neuronal Mechanisms of IID Sensitivity

What about the neural mechanisms involved in detecting IIDs? Since response latency changes as a function of stimulus intensity, Jeffress (1948) proposed that the neural mechanisms for sensitivity to IIDs could also involve temporal coincidence—the so-called latency hypothesis. Evidence supporting such a model was obtained by Yin et al (1985) in the cat superior colliculus. Figure 6-13 shows the responses of a high-frequency superior colliculus neuron to changes in IID while the mean binaural intensity was kept constant. The discharge was a non-monotonic function of IID. This cell did not respond to monaural stimulation of either ear, but did to binaural stimulation with the proper IID. The IID function broadens with increasing average intensity level, and in this case the curves are symmetrical with respect to zero IID. Such cells in the superior colliculus responded only at the stimulus onset, and the latency of that response decreased with increasing intensity level. Since the response latency contains information about stimulus amplitude, the cellular mechanisms of IID detection could involve an analysis of the delay between the arrival times of the monaural inputs. To test this experimentally, Yin et al (1985) compared the IID and ITD sensitivity of these cells and found that the IID functions at zero delay (Fig. 6-13A) were similar in shape to the ITD functions for isointensity stimuli (Fig. 6-13B). The ITD curves are also non-monotonic and symmetrical about zero ITD. ITDs in the millisecond range, far outside those useful for sound localization, were required to modulate the responses. These large ITDs are, however, in the range of neural latency changes that occur with these variations in stimulus inten-

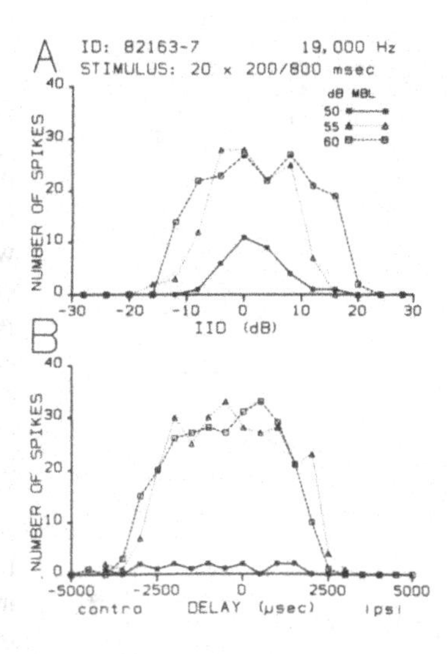

FIGURE 6.13. Responses of a superior colliculus neuron to changes in IID and ITD. **A:** Responses to changes in IID at three mean binaural levels (MBL). The cell is most sensitive at zero IID at all MBLs tested. **B:** Responses of the same neuron to changes in ITD at three SPLs where the intensity to each ear was the same. The cell is most sensitive at zero ITD at all SPLs. (Adapted from Yin et al, 1985.)

sity. In contrast to the ongoing ITD sensitivity of low-frequency neurons, high-frequency IID-sensitive neurons use onset ITD cues.

The above results are consistent with the latency model of IID sensitivity that also uses temporal coincidence (Yin et al, 1985). In this case, the model (Fig. 6-14) relies on the dependence of response latency on sound intensity to generate changes in the time required for excitation from the two sides to reach the binaural cell. It assumes that afferents from each ear are excitatory (BF) and each input creates a transient wave of subthreshold excitation in the binaural cell. The curves indicate diagrammatically the excitatory postsynaptic potentials resulting from stimulation of the contralateral and ipsilateral ear. The wide bar indicates the degree of temporal coincidence and, hence, the response of the binaural cell. Illustrated are the model and responses expected if, for isointensity stimulation, the contralateral and ipsilateral latencies are approximately equal. Thus, maximal coincidence occurs at zero IID. Changes in IID to either side result in the excitatory potentials occurring at different times, causing the degree of temporal coincidence to decrease, and hence the discharge rate to fall more or less symmetrically on either side. It is easy to see that with isointensity stimuli, the response will also decrease when either stimulus is delayed in time. If for isointensity stimuli the response latency to ipsilateral stimulation is shorter than to contralateral stimulation, maximal coincidence will not occur at zero IID, but rather at some IID where the contralateral intensity is greater. Likewise, with changes in ITD, temporal coincidence will not occur until the contralateral stimulus is given a time advantage by delaying the ipsilateral signal. For these cells IID and ITD functions are asymmetrical about zero (not shown). Also obtainable are

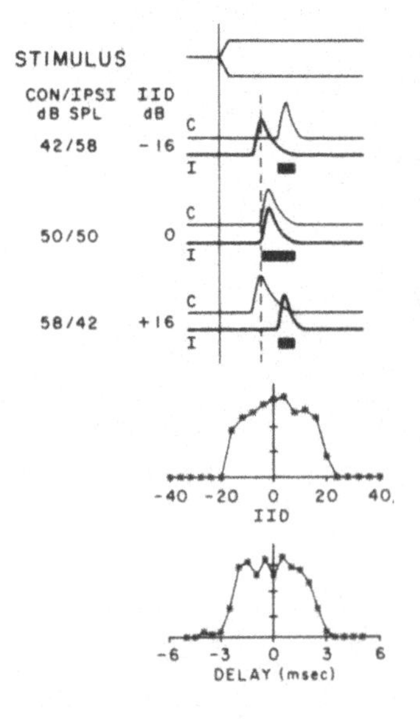

FIGURE 6.14. A model of IID sensitivity in which both inputs are excitatory and have the same latency at zero IID. At the top is a schematic of the stimulus envelope. In the next three diagrams are shown the hypothesized excitatory inputs when the intensity to the ipsilateral ear is higher, equal to, and lower than the contralateral signal. At zero IID, the excitatory latencies are the same and thus arrive at the binaural cell in coincidence, resulting in a maximal response. When the IIDs are not zero, the inputs do not arrive in coincidence, resulting in a less than optimal response. The wide bars indicate the degree of coincidence. At the bottom are shown the IID and ITD functions for a superior colliculus neuron with maximal sensitivity at zero IID and ITD. (Adapted from Yin et al, 1985.)

analogous results that simulate the other major class of cells sensitive to IID—those that are excited by stimulation of one ear and inhibited by stimulation of the other ear (BI). For these cells it is assumed that the contralateral response is suprathreshold and the ipsilateral response is inhibitory.

We have discussed common neural mechanism underlying IPD and IID sensitivity, namely, one that relies on temporal coincidence detection at the level of the single cell. The model of IPD sensitivity is applicable to cells that use the temporal information present in the phase-locked responses of the inputs to the binaural cell. Since phase-locking has been shown to the low-frequency envelopes of complex high-frequency signals (Moller, 1974; Rees and Moller, 1983), this model may also be appropriate for IPD sensitivity in high-frequency neurons (Fig. 6-6). The model of IID sensitivity relies on the dependence of response latency to stimulus intensity, and in this case the onset ITD is the critical variable. The IID model is incomplete, since it does not account for the expected changes in discharge rate that must occur as a result of changes in overall stimulus intensity. However, it does account for many of the properties of high-frequency cells sensitive to IID when tested with long ITDs (Kitzes et al, 1980; Yin et al, 1985).

Future Research

Although it is well established that binaural cells can use IID and IPD cues, little is known about how this information is transformed at higher levels of the neuraxis. For example, IPD sensitivity is seen at all levels of the auditory system, starting at the superior olivary complex. The most extensive data come from inferior colliculus neurons, whereas information regarding other auditory areas is limited, making comparisons difficult. Preliminary examination of the responses of medial superior olive and inferior colliculus neurons suggests that inhibitory influences may be more prevalent in the inferior colliculus, since the discharge rates to changes in IPD for inferior colliculus neurons often are below the response levels to monaural stimulation. This inhibition sharpens the peak and slopes of the cyclic discharge and functionally may serve to increase localization acuity. Sensitivity to IID also appears to undergo a transformation. Below the level of the inferior colliculus, IID functions are monotonic or sigmoidal in shape (Fig. 6-2), whereas non-monotonic IID functions (Fig. 6-13) are seen in the space-mapped region of the MLD, superior colliculus, and auditory cortex. These non-monotonic cells would be expected to generate the restricted receptive fields seen under free-field conditions. Further studies are needed to verify these observations.

To comprehend the neural mechanisms involved in processing IPD and IID, it is necessary to study the functional relationships between different auditory nuclei. Although the anatomical pathways have been studied in some detail, little is known about the functional circuity. Thus, future studies should involve direct stimulation and recordings from interconnected nuclear groups, as well as intracellular recordings to directly observe the interplay between excitatory and inhibitory synaptic events.

Traditionally, free-field and dichotic stimulation studies have been conducted separately, and, with one exception (Moiseff and Konishi, 1981), the same cell has not been studied under both types of stimulus conditions. Thus, it can only be inferred that the results of dichotic studies are directly applicable to those in the free field. To directly compare the information from these two approaches, experiments combining both techniques are necessary.

As a final note, the vast majority of physiological studies of binaural hearing have been conducted using anesthetized animals. These anesthetics are known to have powerful influences on synaptic transmission. Techniques for chronic recordings with minimal discomfort to the animal are available, and the use of such methods in auditory research will be a necessary direction in the near future. Furthermore, recordings from brain stem auditory nuclei have shown that neural activity to the same auditory stimulus changes as a function of behavioral state (Ryan et al, 1984). Such findings question the traditional view that the major auditory nuclei are entirely sensory in function. Clearly, the anesthetised animal is not suitable for examining these influences.

Acknowledgments. We appreciate the participation of T.J. Buunen, J.C.K. Chan, J.A. Hirsch, R.K. Kochhar, Y. Sujaku, J. Syka, and R.E. Wickesberg in our experiments. We are especially grateful to Dr. Jerzy E. Rose, who suggested that we search for the "gold" in the inferior colliculus.

The preparation of this manuscript was supported by the National Institutes of Health grants NS18027 (to S. Kuwada), EY02606 (to T.C.T. Yin), and NS12732 (to J.E. Hind).

References

Aitkin, L.M., Pettigrew, J.D., Calford, M.B., Phillips, S.C., Wise, L.Z. (1985). Representation of stimulus azimuth by low-frequency neurons in the inferior colliculus of the cat. J. Neurophysiol. 53, 43–59.

Aitkin, L.M., Webster, W.R. (1972). Medial geniculate body of the cat: organization and responses to tonal stimuli of neurons in the ventral division. J. Neurophysiol. 35, 365–380.

Altman, J.A. (1978). Sound Localization. Neurophysiological Mechanisms. Tondorf, J. (ed.). pp. 1–188. Chicago: Beltone Institute for Hearing Research.

Behar, A. (1984). Intensity and sound pressure level. J. Acoust. Soc. Am. 76, 632.

Benson, D.A., Teas, D.C. (1976). Single unit study of binaural interaction in the auditory cortex of the chinchilla. Brain Res. 103, 313–338.

Boudreau, J.C., Tsuchitani, C. (1968). Binaural interaction in the cat superior olive S segment. J. Neurophysiol. 31, 442–454.

Brugge, J.F. (1980). The neurophysiology of the central auditory and vestibular systems. In: Otolaryngology. Paparella, M.M., Shumrick, D.A., Meyerhoff, W.L., Side, A.B., (eds.). pp. 253–296. Philadelphia: W.B. Saunders.

Brugge, J.F., Anderson, D.J., Aitkin, L.M. (1970). Response of neurons in the dorsal nucleus of the lateral lemniscus of the cat to binaural stimuli. J. Neurophysiol. 33, 441–458.

Brugge, J.F., Dubrovsky, L.M., Aitkin, L.M., Anderson, D.J. (1969). Sensitivity of single neurons in the auditory cortex of cat to binaural tonal stimulation: Effects of varying interaural time and intensity. J. Neurophysiol. 32, 1005–1024.

Brugge, J.F., Geisler, C.D. (1978). Auditory mechanisms of the lower brainstem. Ann. Rev. Neurosci. 1, 363–394.

Brugge, J.F., Merzenich, M.M. (1973). Responses of neurons in the auditory cortex of the macaque monkey to monaural and binaural stimulation. J. Neurophysiol. 36, 1138–1158.

Caird, D., Klinke, R. (1983). Processing of binaural stimuli by cat superior olivary complex neurons. Exp. Brain Res. 52, 385–399.

Chan, J.C.K., Yin, T.C.T. (1984). Interaural time sensitivity in the medial superior olive of the cat: comparisons with the inferior colliculus. Soc. Neurosci. Abstr. 10. 844.

Colburn, H.S., Durlach, N.I. (1978). Models of binaural interaction. In: Handbook of Perception. Vol. IV. Carterette, E.C., Friedman, M.P. (eds.). pp. 467–518. New York: Academic Press.

Crow, G., Rupert, A.L., Moushegian, G. (1978). Phase-locking in monaural and binaural medullary neurons: implications for binaural phenomena. J. Acoust. Soc. Am. 64, 493–501.

Erulkar, S.D. (1972). Comparative aspects of spatial localization of sound. Physiol. Rev. 52, 237–360.

Geisler, C.D., Rhode, W.S., Hazelton, D.W. (1969). Response of inferior colliculus neurons in the cat to binaural acoustic stimuli having wide-band spectra. J. Neurophysiol. 32, 960–974.

Goldberg, J.M., Brown, P.B. (1968). Functional organization of the dog superior olivary complex: an anatomical and electrophysiological study. J. Neurophysiol. 31, 639–656.

Goldberg, J.M., Brown, P.B. (1969). Response of binaural neurons of dog superior olivary complex to dichotic tonal stimuli: some physiological mechanisms of sound localization. J. Neurophysiol. 32, 613–636.

Gordon, B.G. (1973). Receptive fields in deep layers of cat superior colliculus. J. Neurophysiol. 36, 157–178.

Guinan, J.J., Norris, B.E., Guinan, S.S. (1972). Single auditory units in the superior olivary complex II: location of unit categories and tonotopic organization. Int. J. Neuroscience 4, 147–66.

Hirsch, J.A., Chan, J.C.K., Yin, T.C.T. (1985). Responses of neurons in the cat's superior colliculus to acoustic stimuli. I. Monaural and binaural response properties. J. Neurophysiol. 53, 726–745.

Imig, T.J., Adrian, H.O. (1977). Binaural columns in the primary field (AI) of the cat auditory cortex. Brain Res. 138, 241–257.
Irving, R., Harrison, J.M. (1967). The superior olivary complex and audition: A comparative study. J. Comp. Neurol. 130, 77–86.
Jay, M.F., Sparks, D.L. (1984). Auditory receptive fields in the primate superior colliculus shift with changes in eye position. Nature 309, 345–347.
Jeffress, L.A. (1948). A place theory of sound localization. J. Comp. Physiol. Psych. 41, 35–39.
Johnson, D.H. (1980). The relationship between spike rate and synchrony in responses of auditory-nerve fibres to single tones. J. Acoust. Soc. Am. 68, 1115–1122.
Keidel, W.D., Neff, W.D. (1974). Handbook of Sensory Physiology. Auditory System: Anatomy and Physiology. New York: Springer-Verlag.
King, A.J., Palmer, A.R. (1983). Cells responsive to free-field auditory stimuli in guinea-pig superior colliculus: Distribution and response properties. J. Physiol. (Lond.) 342, 361–381.
Kitzes, L.M., Wrege, K.S., Cassady, J.M. (1980). Patterns of responses of cortical cells to binaural stimulation. J. Comp. Neurol. 192, 455–472.
Knudsen, E.I. (1982). Auditory and visual maps of space in the optic tectum of the owl. J. Neuroscience 2, 1177–1194.
Knudsen, E.I. and Konishi, M. (1978A). A neural map of auditory space in the owl Science, 200, 795–797.
Knudsen, E.I., Konishi, M. (1978B). Space and frequency are represented separately in the auditory midbrain of the owl. J. Neurophysiol. 41, 870–884.
Knudsen, E.I., Konishi, M. (1979). Mechanisms of sound localization in the barn owl (Tyto alba). J. Comp. Physiol. 133, 13–21.
Kuhn, G.F. (1977). Model for the interaural time differences in the azimuthal plane. J. Acoust. Soc. Am. 62, 157–167.
Kuwada, S., Yin, T.C.T., (1983). Binaural interaction in low frequency neurons in inferior colliculus of the cat I: Effects of long interaural delays, intensity and repetition rate on the interaural delay function. J. Neurophysiol. 50, 981–999.
Kuwada, S., Yin, T.C.T., Syka, J., Buunen, T., Wickesberg, R.E. (1984). Binaural interaction in low frequency neurons in inferior colliculus of the cat. IV. Comparison of monaural and binaural response properties. J. Neurophysiol. 51, 1306–1325.
Kuwada, S., Yin, T.C.T., Wickesberg, R.E. (1979). Responses of the cat inferior colliculus neurons to binaural beat stimuli: Possible mechanisms for sound localization. Science 206, 585–588.
Masterton, R.B., Imig, T.J. (1984). Neural mechanisms for sound localization. Ann. Rev. Physiol. 46, 275–287.
Masterton, R.B., Thompson, G.C., Bechtold, J.K., RoBards, M.J. (1975). Neuroanatomical basis of binaural phase-difference analysis for sound localization: A comparative study. J. Comp. Physiol. Psych. 89, 379–386.
Middlebrooks, J.C., Knudsen, E.I. (1984). A neural code for auditory space in the cat's superior colliculus. J. Neurosci. 4, 2621–2634.
Middlebrooks, J.C., Pettigrew, J.D. (1981). Functional classes of neurons in the primary auditory cortex of the cat distinguished by sensitivity to sound location. J. Neurophysiol. 1, 107–120.

Moiseff, A., Konishi, M. (1981). Neuronal and behavioral sensitivity to binaural time differences in the owl. J. Neuroscience. 1, 40–48.

Moiseff, A., Konishi, M. (1983). Binaural characteristics of units in the owl's brainstem auditory pathway: precursors of restricted spatial receptive fields. J. Neurosci. 3, 2553–2562.

Moller, A.R. (1974). Responses of units in the cochlear nucleus to sinusoidally amplitude-modulated tones. Exp. Neurol. 45, 104–117.

Moore, D.R., Hutchings, M.E., Addison, P.D., Semple, M.N., Aitkin, L.M. (1984). Properties of spatial receptive fields in the central nucleus of the cat inferior colliculus. II. Stimulus intensity effects. Hear. Res. 13, 175–188.

Moushegian, G., Stillman, R.D., Rupert, A.L. (1971). Characteristic delays in the superior olive and inferior colliculus. In: Physiology of the Auditory System. Sachs, M.B. (ed.). pp. 245–254. Baltimore: National Education Consultants.

Palmer, A.R., King, A.J. (1982). The representation of auditory space in the mammalian superior colliculus. Nature 299, 248–249.

Phillips, D.P., Calford, M.B., Pettigrew, J.D., Aitkin, L.M., Semple, M.N. (1982). Directionality of sound pressure transformation at the cat's pinna. Hear. Res. 8, 13–28.

Phillips, D.P., Irvine, D.R.F. (1981). Responses of single neurons in physiologically defined area AI of the cat cerebral cortex: Sensitivity to interaural intensity differences. Hear. Res. 4, 299–307.

Rayleigh, Lord, (Strutt, J.W.) (1907). On our perception of sound direction. Phil. Mag. 13, 214–232.

Rees, A., Moller, A.R. (1983). Responses of neurons in the inferior colliculus of the rat to AM and FM tones. Hear. Res. 10, 301–330.

Rose, J.E., Gross, N.B., Geisler, C.D., Hind, J.E. (1966). Some neural mechanisms in the inferior colliculus of the cat which may be relevant to the localization of a sound source. J. Neurophysiol. 29, 288–314.

Rosenzweig, M.R. (1961). Development of research on the physiological mechanisms of auditory localization. Psychol. Bull. 58, 376–389.

Roth, G.L., Kochhar, R.K., Hind, J.E. (1980). Interaural time differences: implications regarding the neurophysiology of sound localization. J. Acoust. Soc. Am. 68, 1643–1651.

Ryan, A.F., Miller, J.M., Pfingst, B.E., Martin, G.K. (1984). Effects of reaction time performance on single-unit activity in the central auditory pathways of the rhesus macaque. J. Neurosci. 4, 298–308.

Schechter, P.B., J.A. Hirsch, T.C.T. Yin (1981). Auditory input to cells in the deep layers of the cat superior colliculus. Soc. Neurosci. Abstr. 7, 20.13.

Semple, M.N., Aitkin, L.M., Pettigrew, J.D., Calford, M.B., Phillips, D.P. (1983). Spatial receptive field in the cat inferior colliculus. Hear. Res. 10, 203–215.

Sovijari, A.R.A., Hyvarinen, J. (1974). Auditory cortical neurons in the cat sensitive to the direction of sound source movement. Brain Res. 73, 455–471.

Stillman, R.D. (1971). Characteristic delay neurons in the inferior colliculus of the kangaroo rat. Exp. Neurol. 32, 404–412.

Stillman, R.D. (1972). Responses of high frequency inferior colliculus neurons to interaural intensity differences. Exp. Neurol. 36, 118–126.

Sujaku, Y., Kuwada, S., Yin, T.C.T. (1981). Binaural interaction in the cat inferior colliculus: comparison of the physiological data with a computer simulated model. In: Neuronal Mechanisms of Hearing. Syka, J. Aitkin, L. (eds.). pp. 233–238. New York: Plenum Press.

Sullivan, W.E., M. Konishi (1984). Segregation of stimulus phase and intensity coding in the cochlear nucleus of the barn owl. J. Neurosci. 4, 1787–1799.

Webster, W.R., Aitkin, L.M. (1975). Central auditory processing. In: Handbook of Psychobiology. Gazzaniga, M.S., Blakemore, C. (eds.). pp. 325–364. New York: Academic Press.

Wise, L.Z., D.R.F. Irvine (1983). Auditory response properties of neurons in deep layers of cat superior colliculus. J. Neurophysiol. 49, 674–685.

Yin, T.C.T., Chan, J.C.K. (1986). Neural mechanisms underlying interaural time sensitivity to tones and noise. In: Functions of the Auditory System. Edelman, G.M., Gall, W.E. (eds.). New York: John Wiley & Sons. In press.

Yin, T.C.T., Chan, J.C.K., Irvine, D.R.F. (1986). Effects of interaural time delays of noise stimuli on low frequency cells in the cat's inferior colliculus. I. Responses to wideband noise. J. Neurophysiol. 55, 280–300.

Yin, T.C.T., Chan, J.C.K., Kuwada, S. (1983). Characteristic delays and their topographic distribution in the inferior colliculus of the cat. In: Mechanisms of Hearing. Webster, W.R. Aitkin, L.M. (eds.). Clayton, Victoria: Monash University Press.

Yin, T.C.T., Hirsch, J.A., Chan, J.C.K. (1985). Responses of neurons in the cat's superior colliculus to acoustic stimuli. II. A model of interaural intensity sensitivity. J. Neurophysiol. 53, 746–758.

Yin, T.C.T., Kuwada, S. (1983a). Binaural interaction in low frequency neurons in inferior colliculus of the cat. II. Effects of changing rate and direction of interaural phase. J. Neurophysiol. 50, 1000–1019.

Yin, T.C.T., Kuwada, S. (1983b). Binaural interaction in low frequency neurons in inferior colliculus of the cat. III. Effects of changing frequency. J. Neurophysiol. 50, 1020–1042.

Yin, T.C.T., Kuwada, S. (1984). Neuronal mechanisms of binaural interaction. In: Dynamic Aspects of Neocortical Function. Edelman, G.M., Cowan, W.C., Gall, W.E. (eds.). pp. 263–313. New York: John Wiley & Sons.

Yin, T.C.T., Kuwada, S., Sujaku, Y. (1984). Interaural time sensitivity of high frequency neurons in the inferior colliculus. J. Acoust. Soc. Am. 76, 1401–1410.

Part III Directional Hearing in Vertebrates: Animal Psychophysics and Special Adaptations

Most of what has been learned about the anatomy and physiology of the binaural nervous system was obtained from nonhuman mammals. On the other hand, much less has been known about the performance of intact, behaving animals in situations of directional hearing. Recently, however, considerable interest developed on how animals actually localize sound and respond to the various binaural cues.

This section surveys primarily the behavioral studies of directional hearing in numerous vertebrate species, but also includes, in some instances, supporting neural research.

The accuracy with which different species localize sound varies widely. Some animals are very poor localizers; others are very nearly as good as man, who appears to be the foremost localizer in the animal kingdom.

Although the variability in the performance of different animals reflects, to some degree, the diversity of head sizes and the consequent effects on the magnitude of interaural cues available to the animal (most species that have been tested are smaller than man), other factors such as head shape, neural processing, and specialized adaptations have allowed many species to achieve a level of sound localization performance otherwise unavailable to them. This can be seen strikingly, but not solely, in animals such as bats and dolphins whose *modus operandi* is intimately depended on hearing.

Knowledge of how successfully various animals localize sound and respond to binaural cues provides valuable insight not only into the evolution of sound localization, but also into binaural neural processing.

7

Mechanisms for Directional Hearing among Nonmammalian Vertebrates

RICHARD R. FAY AND ALBERT S. FENG

This chapter reviews data and theory on the mechanisms by which fishes, amphibians, reptiles, and birds extract information from sound about the location of its source. The emphasis is on behavioral and physiological data that reveal the coding and processing of information about a sound source's position in space.

Other chapters in this volume focusing on humans and mammals have treated in detail the kinds of information useful in directional hearing in those animals. The major sources of information are the "binaural" cues, or those deriving from differences in the stimulation of the two ears. These differences are caused by asymmetries in the shapes and orientations of the outer ears, by the sound travel time (distance) between the ears, and by the sound-refracting characteristics of the head and body. In this case, the ears can be viewed as two receivers sampling sound pressure waveforms from two points in space simultaneously and directional information gained from correlating various parameters (intensity, spectrum shape, phase, time, etc) by the auditory system. Other sources of information are termed "monaural" in the sense that directional information can be extracted from the input to only one ear. For example, the outer ears may be viewed as directional filters that transform the acoustic signal on its way to the tympanic membrane in ways that depend on the orientation of the ear relative to the sound's direction of propagation.

In general, these kinds of mechanisms also operate among the nonmammals and have been particularly well studied for some of the amphibians and birds. However, for directional hearing the nonmammals exhibit additional adaptations that apparently have arisen from their generally low-frequency bandwidth of hearing and their small effective interaural distance (small heads). Under the assumption that the two ears simply sample the sound pressure waveform at two points in space (as seems to be effective for the mammals), these two factors combine to significantly reduce the size of interaural differences in intensity, spectrum shape, and time. The following sections discuss the mechanisms by which nonmammalian auditory systems deal with these problems.

Fishes

It has been known for some time that sharks are attracted from considerable distances to low-frequency sound sources with particular temporal patterns of intensity fluctuation (Nelson and Gruber, 1963; Myrberg et al, 1976). However, although we have some recent speculation, theory, and data on the possible role of the macula neglecta in directional hearing (Corwin, 1981; Schuijf, 1981; van den Berg and Schuijf, 1983), both the

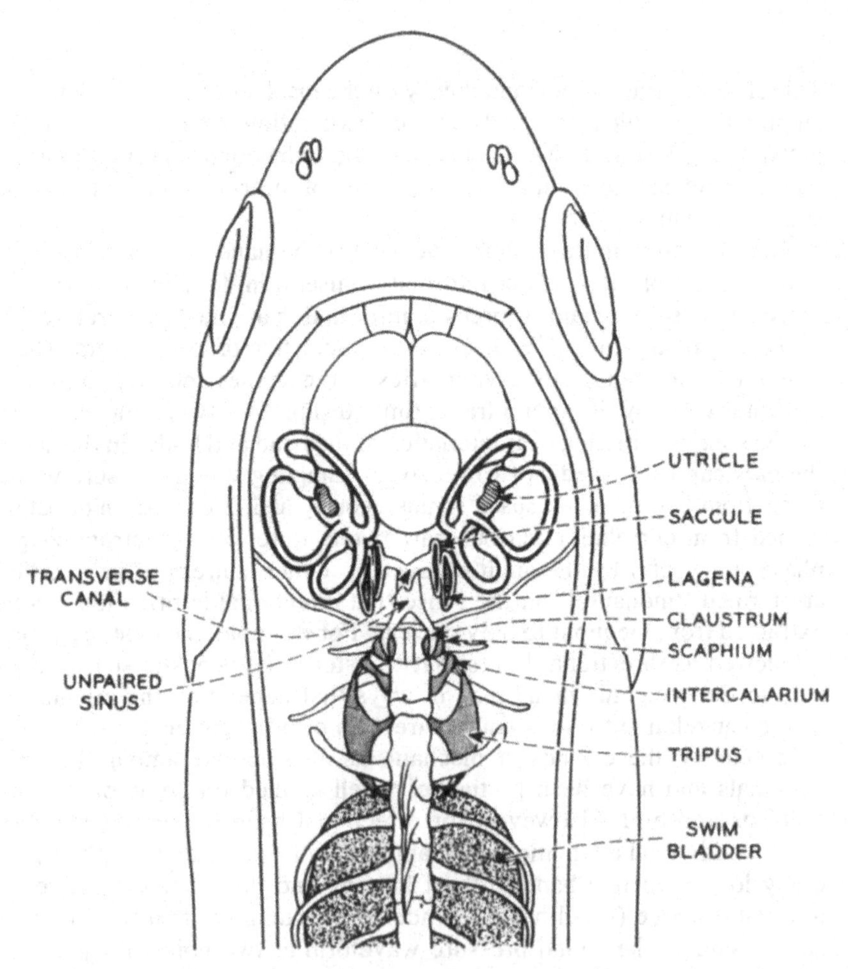

FIGURE 7.1. The ear of the ostariophysine fish *Phoxinus laevis* showing the otolith organs of the labyrinth and the Weberian ossicles (claustrum, scaphium, intercalarium, and tripus), which link the swimbladder with both ears. While nonostariophysine species lack the Weberian ossicles, analagous mechanisms exist in many species that bring the movements of the swimbladder walls to bear on the fluid systems of the two ears (von Frisch, 1938).

physical cues and the neural mechanisms that mediate attraction behavior of sharks are still unknown.

Until recently it was highly controversial whether the bony fishes possess directional hearing in any sense at all. This argument arose partly from the definition of hearing (Pumphrey, 1950), partly from the definition of underwater sound (van Bergeijk, 1967; Dijkgraaf, 1963), and partly from conflicting experimental results on this question. If hearing is defined as sound pressure reception, as was implicitly done early on (van Bergeijk, 1964; 1967), one could argue that directional hearing mediated by the ear was unlikely if not impossible, since sound pressure stimulates the ear of fishes only via the swimbladder (or other gas-filled bubble in proximity to the ear), which expands and contracts according to the sound pressure waveform. The motions of its walls are transmitted to the two ears equally and always from the same direction, regardless of the location of the sound source (Fig. 7.1). van Bergeijk and others were thus led to the conclusion that to the extent directional hearing occurs at all among the fishes, it must be mediated by the lateral line system.

Lateral Line Systems

The lateral line is a system of hair-cell receptors situated in canals (some are free-standing) over the head and flank of fishes and many aquatic amphibians (Fig. 7-2) (see reviews by Coombs et al, 1986; and Denton and Gray, 1986). The hair cells of the lateral line system are ciliated mechanoreceptors very much like those of all vertebrate auditory and vestibular systems. They are sensitive to displacement of the cilia and are direc-

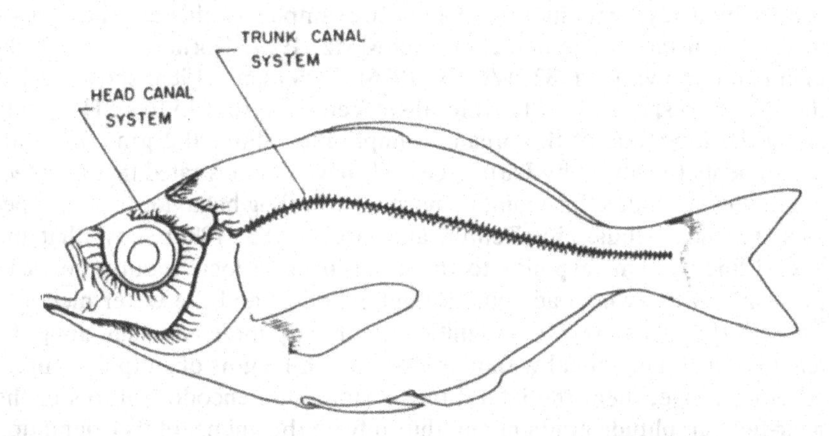

FIGURE 7.2. The extent of the lateral line system of the walleye surfperch, illustrating the extended three-dimensional array of receptors typical of this system. The trunk canal is a continuous tube covered with scales. A pore opening and a neuromast organ containing hair cells is associated with each scale along the trunk canal. (Taken from Walker, 1967.)

tionally sensitive in that displacement of the stereocilia toward the kinocilium is excitatory. The lateral line hypothesis for directional hearing was attractive for several reasons. Harris and van Bergeijk (1964) showed that the head canal organs respond to the near-field component of a dipole sound source. For the purposes of the present discussion, the near-field is defined as the region of space surrounding the sound source within which hydrodynamic particle motions (caused by the "sloshing" of the nearly incompressible water around the oscillating source) are greater in amplitude than acoustic particle motions (those normally occurring in a plane progressive sound wave for a given sound pressure level). The hydrodynamic particle motions of the near-field are further characterized by very steep and highly curved spatial gradients of amplitude within a wavelength or two from the source. The lateral line system is widely distributed over the animal's body, possibly allowing for coding the orientation of these steep spatial amplitude gradients of the near-field. This hypothesis was consistent with the available behavioral data on directional response in fishes demonstrating "localization" only for intense sources at close range (within the effective near-field). (See van Bergeijk, 1964, for a critical review of the experiments of Reinhardt, 1935; von Frisch and Dijkgraaf, 1935; and Kleerekoper and Chagnon, 1954).) Finally, van Bergeijk (1967) argued that the labyrinthine organs arose phylogenetically from the lateral line organs and that the central projections of the two systems were overlapping. It was therefore logical to think that the evolutionary history of "hearing" began with the functions of the lateral line.

Research and theory since that time have had new and different things to say about the lateral line system and its role in hearing and directional orientation. First, there is no doubt that the lateral line functions in detecting hydrodynamic motion. The best examples of this are the experiments on surface wave detection (Schwartz, 1967; Gorner, 1973; Bleckmann and Schwartz, 1982; Wilcox, 1986; Bleckmann, 1986) showing that the system responds best to water displacements from 15 to 40 Hz acting along the long axis of the organ at amplitudes above 0.2 μm. Behavior experiments (reviewed by Partridge, 1981) have demonstrated that an intact lateral line system helps maintain normal schooling behavior in some species. Physical studies by Denton and Gray (1983; 1986) show that the lateral line system responds to the differences in local motion between the fish's body (which acts like a rigid cylinder) and the water medium. These differences may vary significantly in amplitude and sign along the length of a 10-cm animal within about two fish-lengths of a dipole source. At close range, then, the lateral line system may encode features of the near-field amplitude gradient and thus inform the animal of its orientation relative to the source (Kalmijn, 1986).

Whether or not we can reasonably call this "hearing" is still controversial. An answer may lie in a comparison between the lateral line system and the more conventionally defined auditory system in fishes (otolith organs of the ear) in their responses to various mechanical inputs. Sand (1981) approached this question directly by chronically recording from the lateral line nerve during various types of mechanical stimulation. Using a dipole source at close range, Sand found a response consistent with previous data cited above, i.e., the greatest sensitivity occurring when displacements (averaging about 0.08 μm) were parallel to the canal's long axis, with a best frequency of 50 Hz. However, using acoustic standing wave stimulation with the fish in a displacement maximum, Sand found a very poor response from the lateral line, even to relatively enormous displacements (e.g., 57 μm). The conclusion here is that the lateral line responds well to near-field displacement gradients, but poorly to much larger displacements in what are essentially far-field conditions (the fish and the surrounding water mass moving together, in phase). Under these same standing wave conditions, the mass-loaded otolith organs of the ear respond with great sensitivity and in a wider bandwidth (e.g., Fay and Popper, 1975; see below).

It seems clear that van Bergeijk's view of the lateral line system as being most sensitive to near-field phenomena, and most likely mediating directional orientation to nearby sources is accurate. It may not be useful to label this function "hearing," however (Dijkgraaf, 1963). It is currently believed that the lateral line and auditory systems have parallel evolutionary and developmental histories [i.e., the labyrinth was not derived from the lateral line (Northcutt, 1980) as was previously supposed], that the central projections of the two systems show little overlap (McCormick, 1983; McCormick and Braford, 1986), that the functions of the lateral line and auditory systems are quite distinct, and that there exist mechanisms for directional hearing in the conventionally defined auditory systems of fishes. We turn now to this latter point.

Auditory Systems of Fishes

The present evidence is that fishes display aspects of directional hearing in both near- and far-fields, that the inner ear organs and not the lateral line system mediates the responses, and that the pressure waveform does, in fact, provide some useful directional information for most species. The first evidence that the organs of the ear mediated directional responses by the organism comes from the pioneering experiment by Moulton and Dixon (1967) in which the directional aspects of the tail-flip response to sound in the goldfish were shown to disappear when input to the brain from the sacculi was disrupted. Since that time, studies on the cod in free field environments (e.g., Chapman and Johnstone, 1974; Schuijf, 1975; Schuijf and Buwalda, 1975; Hawkins and Sand, 1977) have demonstrated abilities for discriminating a change in sound source location in azimuth and elevation with minimum audible angles on the order of 10 to 20 degrees, depending on signal-to-noise ratio (Fig. 7-3A). Schuijf (1975) showed that such discriminations took place in the far-field and were dependent on having both ears intact. Schuijf and Siemelink (1974) showed that directional orientations in the cod are not disrupted by surgical elimination of the lateral line system.

What is the explanation for these and other demonstrations that far-field directional hearing mediated by the ear is possible in the fishes? The auditory portions of the ears of fishes are the three otolith organs: the utricle, the saccule, and the lagena (Fig. 7-1). It is generally believed that the saccule is the primary hearing organ in most species. [In the clupeids, the utricle is considered to be the major hearing organ (Blaxter et al, 1981).] The otolith organs are composed of a patch of sensory epithelium made up of hair-cell receptors and supporting cells—the maculae, which are overlaid by a solid calcium carbonate otolith and an intervening otolith membrane. There is little doubt that stimulation occurs when a relative shearing motion is set up across the cilia, between the hair-cell bodies and the otolith.

This motion can be produced in at least two ways. The swimbladder expands and contracts in a pressure field and produces a displacement stimulus that may cause relative otolith motion (von Frisch, 1938), given an optimal pathway for the motion to be transmitted to the ear (van Bergeijk, 1967). As noted above, however, this pressure sensitivity would not be useful for extracting directional information from sound since the input to the ears would always arrive from the swimbladder and most likely stimulate both ears identically. More recently, it has been experimentally confirmed that the otolith organs respond not only to the pressure waveform via the swimbladder, but also to acoustic particle motion through an inertial stimulation principle (Fay, 1984). The otolith organs appear to act as mass-loaded accelerometers. As the animal undergoes whole-body acceleration along the axis of particle motion, an otolith's inertia causes its motion to lag behind that of surrounding tissue, and shear forces are set up across the hair-cell cilia (de Vries, 1950). In contrast to the pressure waveform, which is a scalar function, the acoustic particle acceleration waveform is a vector function with direction as well as magnitude. As outlined by Dijkgraaf (1960), it is presently believed that the axis of acoustic particle motion is coded within and between otolith organs by the relative response across arrays of hair-cell receptors oriented in various three-dimensional directions (Fay, 1984).

This hypothesis has received support from a wealth of recent studies on the patterns of hair-cell orientation within the maculae of the various otolith organs in fishes (e.g., Popper, 1977). In detail there is rich variation among species and organs in these orientation patterns; however, the general view is of a system with at least six organs (paired sacculi, lagenae, and utriculi) with nonparallel gross orientations, each containing arrays of hair cells with axes of greatest sensitivity pointing in various directions

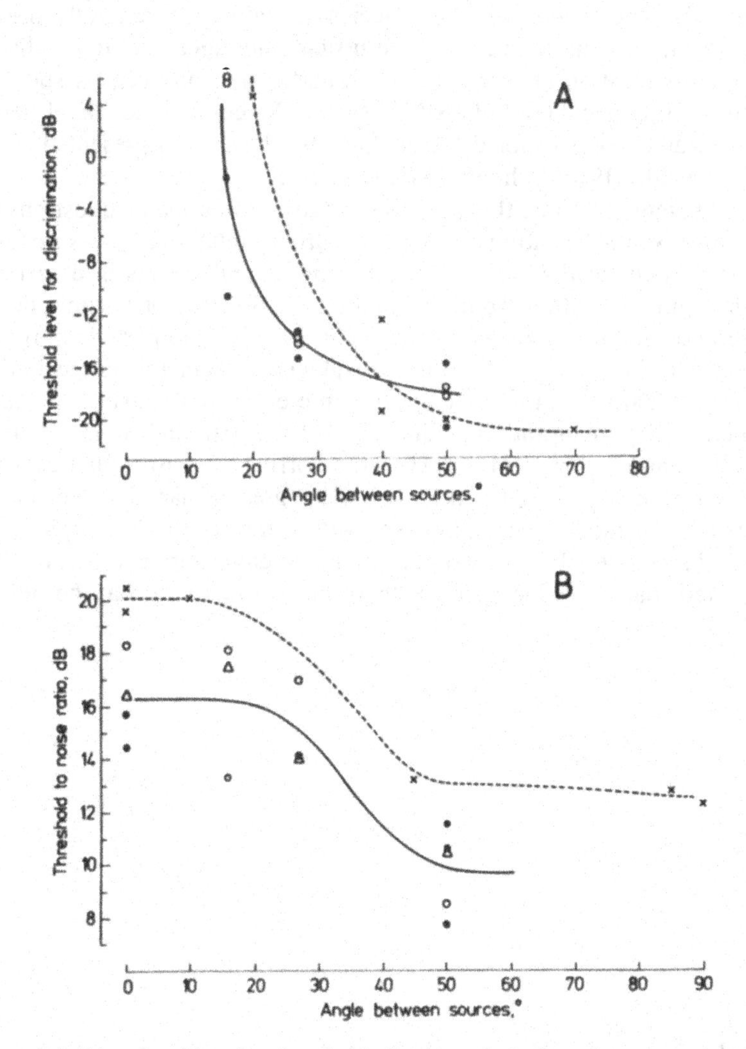

FIGURE 7.3. Directional hearing thresholds for the cod. **A:** Sound pressure threshold for the detection of a change in sound source direction as a function of angular change for a 110-Hz tone. Circles represent changes in the median vertical plane for two fish. The x and dashed line represents changes in the horizontal plane. **B:** Threshold to noise ratio for a 110-Hz tone signal detected in the presence of a broadband masking noise as a function of the angular separation between signal and noise sources. Circles and open triangles represent data from three fish with projectors in the vertical plane. The x indicates mean values with the projectors in the horizontal plane for several tone frequencies. Data from Chapman and Johnstone (1974) and Hawkins and Sand (1977). (Taken from Hawkins 1981.)

(Fig. 7-4). The sacculi of most species contain cells generally oriented fore-aft and dorsal-ventral, and the utricle and lagena contain cells with a wide distribution of orientations. In addition, many organs show curvature of the macula out of a single plane. Conceptual models of the axis determination, or "vector detector" process, have been presented (Schuijf and Buwalda, 1980; Schuijf, 1981; Popper et al, 1986).

As presented so far, this hypothesis raises three major questions concerning absolute sensitivity, motion sensitivity relative to pressure (swimbladder) input to the ear, and confusions or ambiguities in determining sound source location from information on the axis of particle motion. Several estimates of absolute behavioral sensitivity to particle movement at optimal frequencies show that displacements in the range between -160 and -200 dB with reference to 1 meter (1–100 angstroms) may be detected (Chapman and Hawkins, 1973; Chapman and Sand, 1974; Fay and Patricoski, 1980). At 200 Hz, this corresponds to a far-field sound pressure level as low as 1 dyne/cm^2, a value greater than absolute detection threshold for most species, but well within the range of biologically significant sound levels to be expected in usual environments. It is therefore expected that for some species, there are sounds that may be detected

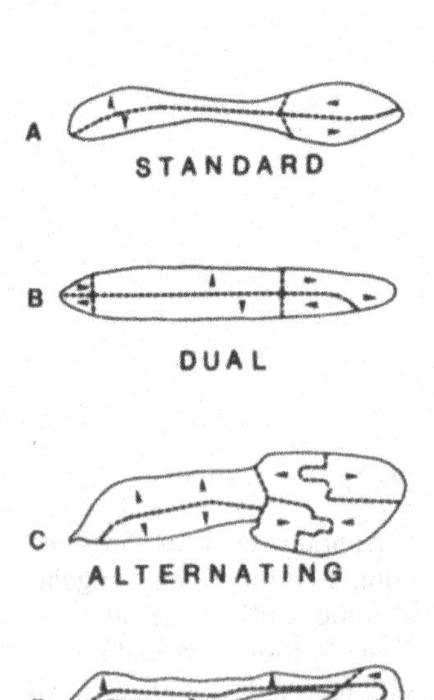

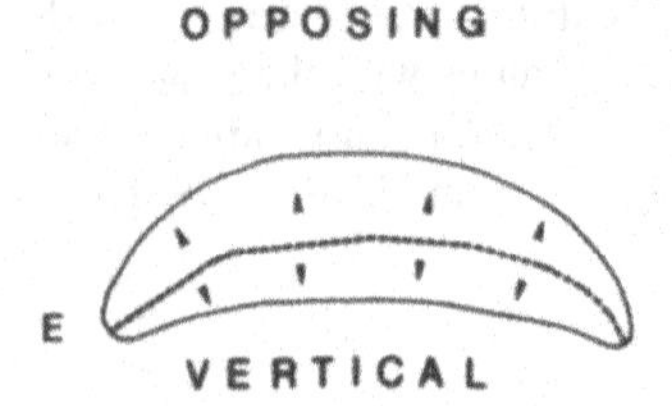

FIGURE 7.4. Patterns of directional orientation of the hair cells on the saccular maculae in fishes. The arrowheads in each region of the macula indicate the directional polarization of the hair cells (the direction from the stereocilia toward the kinocilium). Displacement along this axis is presumed to be excitatory. This figure attempts to summarize the considerable species variation in hair-cell orientation pattern. (From Popper and Coombs, 1982.)

via pressure-sensitive channels, but that are below threshold for direct particle motion "vector detection" and thus may not be localized.

The second question of relative sensitivity is particularly troublesome for those species that are specially adapted for pressure sensitivity (having efficient pathways between the swimbladder and ear) such as the goldfish. For a far-field sound well above threshold, the pressure (swimbladder mediated) input to the ear would far exceed that due to direct motion detection, and we would expect the nondirectional input to mask or otherwise interfere with the coding of the directional input (Fay, 1981; Fay et al, 1982). One possible "solution" to this problem is to focus the pressure input onto one otolith organ (e.g., the saccule) and to effectively shield the other organs (lagena and utricle) from the swimbladder. This seems to occur in the *Ostariophysi* (goldfish, catfish, minnows, etc having a highly specialized mechanical link from the swimbladder to the saccule) (Schuijf and Buwalda, 1980).

Another possibility for enhancing the relative sensitivity to particle motion involves fundamental differences between the interaural effects of the swimbladder input versus the direct motional input in the *Ostariophysi* and other sound pressure "specialists." The swimbladder's input to the two ears produces inphase responses from corresponding elements of the sacculi. However, a whole-body acceleration from the animal's side, for example, would disrupt such inphase responses and tend to produce an out-of-phase response in the two ears (since the vector points medially for one ear and laterally for the other). Any asymmetry in input from the ears to the brain would signal a particle motion component, and the asymmetry's size and polarity could contain information on its amplitude, phase, and the orientation of its axis. In this type of model, the pressure-dependent (inphase) component could be considered a carrier that could be reduced by a central subtractor mechanism (Durlach, 1972; Popper et al, 1986). In any case, more empirical work must be done before we can progress beyond this kind of speculation.

The third question recognizes the fact that a determination of the axis along which particles are moving in the far-field does not specify which end of the axis points to the source. The problem is essentially to determine the direction of wave propagation. In empirical work on the cod (Schuijf and Buwalda, 1975), it was shown that such 180° confusions do not occur in a free-field progressive wave. Further, an animal trained to orient toward the sound source in the field could be caused to orient toward or away from the source, depending on the phase relations between the particle motion and the pressure waveforms created in a sonic environment that is synthesized using standing waves. These results led to the conclusion that 180° ambiguities are "solved" through a phase comparison of the particle motion waveform with the sound pressure waveform. For example, a pressure compression peak in a sinusoid arriving from the right would lead a peak displacement to the right by a phase angle of 90° while

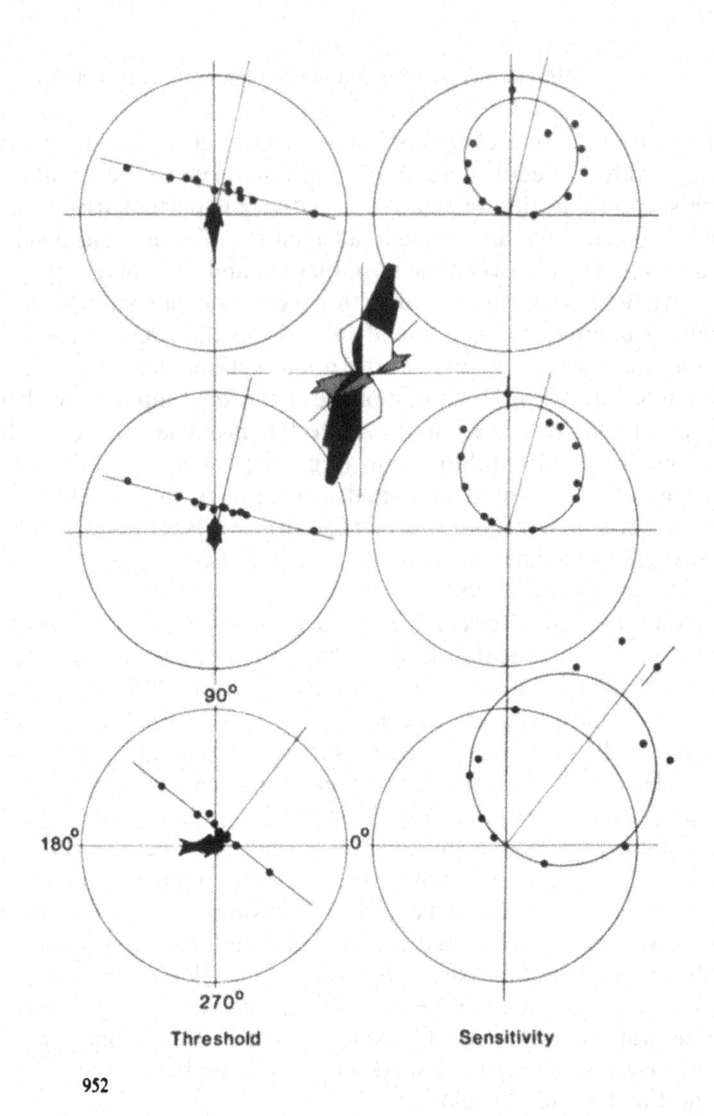

FIGURE 7.5. Polar plots of the acceleration thresholds (left) and acceleration sensitivity (right) for saccular unit 7.6; spontaneous rate = 153 ips. At the top is the X-Y plane, looking down on the fish. The middle shows the X-Z plane, looking at the fish from behind. The bottom is the Y-Z plane, looking at the fish from its right. The points plotted on the left-hand sets of coordinates show the acceleration threshold (radial distance) as a function of angle of motion (polar angle). The lines drawn through the points are least-squares best-fits. The perpendiculars to the best-fit lines define the stimulus angle having the lowest threshold for the plane shown. The distance along this line from the origin to the best-fit threshold line defines the magnitude of the lowest threshold. On the right, the data were plotted as the reciprocal of acceleration threshold, and the best-fitting circles derived from the threshold lines. The bars on the most sensitive points indicate + and − 1 dB. Also shown is a three-dimensional representation of this cell's directional sensitivity, as if the fish were viewed from the back, from the right, and from above. The black plane is saggital (Y-Z), the white is frontal (X-Z, on the plane

the reverse would be true for a signal arriving from the left. Current theories of directional hearing in fishes (Schuijf and Buwalda, 1980; Schuijf, 1981) thus require not only that the axis of motion is resolved by arrays of directionally sensitive receptor cells, but also that the motion and pressure waveforms are resolved and analyzed in the time domain. It appears that the far-field pressure waveform, which van Bergeijk (1967) believed contained no directional information, may be useful to the fish after all. Note, however, that within the near-fields of dipole sources (the best model for most biological sound sources), particle motion occurs along a curved pathway, and the direction of motion at any given point in the field may not in itself reveal the direction of the source (Kalmijn, 1986). At present, there are no adequate models of how dipole sources are located from within the near field.

Physiological Studies of Directional Hearing

Few physiological analyses of directional hearing have been carried out in fishes, since an adequate acoustic environment is difficult to achieve under conditions allowing recordings from eighth nerve units. A possible approach to this problem is to simulate far-field particle motion by producing whole-body accelerations of the animal (Fay and Olsho, 1979; Hawkins and Horner, 1981). In recent studies on the saccular, lagenar, and utricular nerves of the goldfish, Fay (1984, 1986) recorded unit responses to linear sinusoidal acceleration of the whole body along axes oriented in three dimensions. The thresholds for single units were measured for 36 axes of motion; 12 axes spaced at 15° intervals in each of the horizontal, saggital, and frontal planes of the animal. The basic finding is that in each of these planes, thresholds plotted in polar coordinates tend to fall on a straight line with a particular orientation (Fig. 7-5). This is essentially equivalent to a description of a cell's directional sensitivity as a circle (a cosine function of the input angle relative to the cell's axis of best sensitivity). These planar projections of directional sensitivity can be combined in three dimensions to reveal a "sensitivity solid," which is a sphere. The vector originating at the coordinate origin and ending at the sphere's center defines a cell's directionality and sensitivity in three dimensions.

Figure 7-6 shows distributions of these best directions for the goldfish's three otolith organs. Clearly, each organ "looks" at different classes of

of the paper), and the striped is horizontal (X-Y). The spherical coordinates of the cell's best direction vector are 76° azimuth, 42° elevation, and best sensitivity is −149 dB re: 1 meter (RMS). (From Fay, 1984. Reproduced with permission of Fay, R.R., Vol. 1 & 2 221, The Goldfish Ear Codes, Science pp. 952–953, 31 August, 1981.)

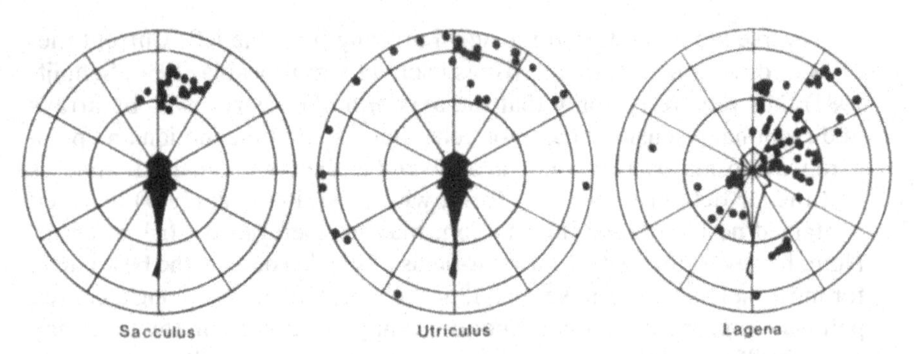

FIGURE 7.6. Distributions of best-direction vectors for 135 units of the saccular (n = 22), lagenar (n = 85), and utricular (n = 29) branches of the eighth nerve. The view is down onto a northern hemisphere with the fish's ears at the globe's center. The symbols locate the points at which best-direction axes penetrate the northern hemisphere. The radial lines are azimuths, with 0° to the right, and 90° straight ahead. The concentric circles are equal elevations at 30° intervals. Straight up is defined as 90° elevation. Sensitivity is not represented. Present data do not allow a determination of whether northward or southward motion from the globe's center is excitatory. (From Fay, 1984.)

motional input. The saccular units respond best to motions along a single axis with an azimuth of about 75° (90° defined as straight ahead), and an elevation of about 50° (90° defined as straight up). The utricular units respond best in the horizontal plane, with a wide range of azimuths. The lagenar units are generally more dispersed, but tend to cluster in a vertical plane with a wide range of elevations. These results demonstrate that peripheral nerve fibers respond as if their inputs (most likely from a group of hair cells) were equivalent to a single hair cell with a particular three-dimensional axis of best sensitivity. It would seem that if these afferents are "labeled" in the brain according to their orientations, then the goldfish has the ingredients for a "place principle" for analyzing the axis of any motional input to the ears.

Very little is known about central processing of sound in the fish auditory system and less about directional coding in the CNS. However, Horner et al (1980) reported an interesting experiment on the cod in which units of the medulla and mesencephalon (torus semicircularis) were monitored during underwater sound stimulation and anodal electrical block of a portion of the peripheral saccular input. Binaural interaction was demonstrated in some of the cells contacted at both sites, but evidence for ascending inhibition in these cells was seen only in the midbrain. Fay et al (1982) looked at midbrain-evoked potentials in the goldfish to both pressure and particle motion stimuli. An intriguing finding was that swimbladder deflation caused a large loss in pressure sensitivity across the entire frequency range of hearing, but resulted in an increase in vibration-evoked potentials.

This suggests an inhibitory interaction between directional (particle motion) and nondirectional (pressure) information at the midbrain level. This interaction effect was larger at medial midbrain sites than at more lateral sites. Future studies should concentrate on central responses recorded under conditions of combined (and well specified) pressure and displacement stimulation.

Other Phenomena of Directional Hearing

Cocktail Party Effect

Another aspect of directional hearing, which has been well documented in man, may be termed the "cocktail party effect," or the ability to use the directional information of a "signal" to differentiate it from noise emanating from different spatial locations. Humans seem to accomplish this through binaural interaction in which the interaural time and intensity differences characteristic of a signal from a given source location are used as a basis for signal detection in a noisy environment. This could be viewed as an example of directional or space-domain filtering. This type of filtering could operate in any detection system with directional characteristics, and there is good behavioral evidence that it operates among fishes (Chapman, 1973; Chapman and Johnstone, 1974).

In these experiments, a sound source producing a tone signal was displaced in the horizontal plane relative to that producing a masking noise, and the signal-to-noise ratio (S/N) at threshold was measured as a function of the angle between them. A separation of 45° and greater increased signal detectability by 6 to 8 dB relative to the 0° case (see Fig 7-3B). The current interpretation of these results is that the partial independence of inputs emanating from different directions is due not to binaural processing, but to the fact that signals with different axes of particle motion stimulate somewhat different populations of directional receptor cells, and thus different peripheral auditory neurons. Buwalda (1981) has also found evidence for this segregation of information in psychophysical experiments using synthesized standing wave sound fields in the lab. It appears, then, that enhanced signal detectability in fishes is not only accomplished by filtering mechanisms in the frequency domain (Fay & Coombs, 1983; Fay and Ream, 1986), but also by filtering in the space domain.

Reflex Orientation

The study of Moulton and Dixon (1967) presented earlier showed that directional tail-flip reflexes of the goldfish were dependent on an intact labyrinth. This response is common in fishes and some amphibians and is believed to be mediated by Mauthner's cells (M-cells), comparatively large, paired neurons originating in the medulla and projecting down the spinal cord to synapse with contralateral body musculature. There is abundant input to the M-cell from the otolith organs, lateral line system, and from other sensory modalities. Saccular input via large diameter fibers produces both excitation and inhibition, which are mediated electrically as well as chemically (Furukawa, 1966). The M-cells mediate directional startle and escape responses to mechanical input such as vibration (Eaton and Kimmel, 1980). Since the saccular fibers providing input most likely respond to sound pressure, we may be confident that sound creates a large input to the system. Blaxter et al (1981) and Blaxter and Hoss (1981) studied sound-evoked startle responses in herring shoals, which may be mediated by the M-cells. They conluded that while both sound pressure and particle motion are probably necessary to provide the directional information, the pressure waveform (mediated by the utricle in this species) triggers the response. It is not yet clear how directional information is coded in the M-cell's afferent input and how this information is processed. An important question concerns the relation between directional hearing as discussed above, and reflex orientation mediated by the M-cells.

Cues for the Structure of Acoustic Environments

Underwater acoustic environments can be reverberant since the bottom and the surface are excellent sound reflectors. Waveforms detected in such environments contain information about reflecting surfaces and about the spatial relations among sound sources, reflectors, and listeners (Supra et al, 1944). A simple stimulus for studying the detection and processing of this type of information is "repetition noise," or the addition of a noise signal to its delayed repetition (Bilsen and Ritsma, 1970). When the "echo" is comparable in amplitude to the original signal (within 15 dB), and when delay times (t) are in the 1 to 50 ms range, humans judge repetition noise to have a pitch that is equal in frequency to 1/t. These coloration attributes of sound convey some information on the proximity of reflectors, and on the spatial relations among sources, reflectors, and receivers.

Fay et al (1983) have extensively studied the processing of repetition noise by the goldfish. The major finding is that goldfish can discriminate a change in echo delay for noise of about 5% even when the echo is considerably attenuated relative to the direct sound. Neurophysiological studies showed that the information used in these discriminations is most likely the details of the sound's waveform as represented in the intervals between phase-locked spikes evoked in the saccular nerve. Although this acute sensitivity to echo delay [rivaling that of man (Yost et al, 1978)] certainly exists for the goldfish, we do not yet know whether it provides useful information for acoustic orientation in natural environments. Tavolga (1976) has speculated that this type of processing may underly the ability of the sea catfish to detect and avoid obstacles by using a simple form of echolocation. Recently, Rogers (1986) speculated that fish detect the presence of other fishes by processing ambient noise scattered by their swimbladders. This suggests a passive form of echolocation, possibly based on the mechanisms for repetition noise processing, which has been studied in detail by Fay (1983).

Acoustic Intensity Determination

Schuijf and Hawkins (1983) have shown that cods are able to discriminate differences in sound source distance at a range of several meters in a free-field. They speculate that the animal uses as information the relative phase and amplitudes of pressure and particle motion, since in the near-field these relations are a function of distance (Schuijf and Buwalda, 1980). We note that the ability to code the amplitudes and phases of the sound pressure and particle motion waveforms using directional receptors provides the fishes with the information required for determining underwater acoustic intensity (Fay, 1986). Fishes appear to be unique among vertebrates in having the information required for such acoustic signal processing coded in fibers of the eighth nerve.

Directional Hearing in Anurans

Acoustic communication in anurans involves sounds having relatively shorter wavelengths than those in fish. Despite this, the wavelengths of the communicative sounds tend to be longer than the interaural distance for most of these animals. Sound localization, therefore, is a difficult problem that may necessitate specialized adaptations in the auditory periphery or the central nervous system, or both.

The sounds that anurans produce can generally be classified according to their communicative function, i.e., mating (or advertisement) calls, territorial calls, release calls, warning calls, and etc. These calls have distinct spectral and temporal characteristics. The mating calls, which are species specific and produced by males, play the most important role in the reproductive behavior, and they have been carefully analyzed for many species. The spectral energy in the mating calls of a given species of North American hylid or ranid frog typically has a bimodal distribution: a low-frequency (100–1,000 Hz) peak and a broad high-frequency (1.2–5 kHz) peak. The temporal envelope of the calls, or of individual pulses in the croaks, has a fairly rapid rise time. These acoustical features of the calls are used to provide the cues for locating the sound source.

Gravid female frogs have been shown to be able to locate the source of mating calls in both horizontal and vertical planes (Feng et al, 1976; Rheinlaender et al, 1979, Gerhardt and Rheinlaender, 1982; Passmore et al, 1984). In the green tree frog, females take individual hops into the direction of the sound source (Fig. 7-7). Lateral head scanning often occurs prior to the jumps. The accuracy of horizontal head orientation and jump has been carefully measured (Rheinlaender et al, 1979). The mean angle of the head relative to the sound source at the end of head scanning is

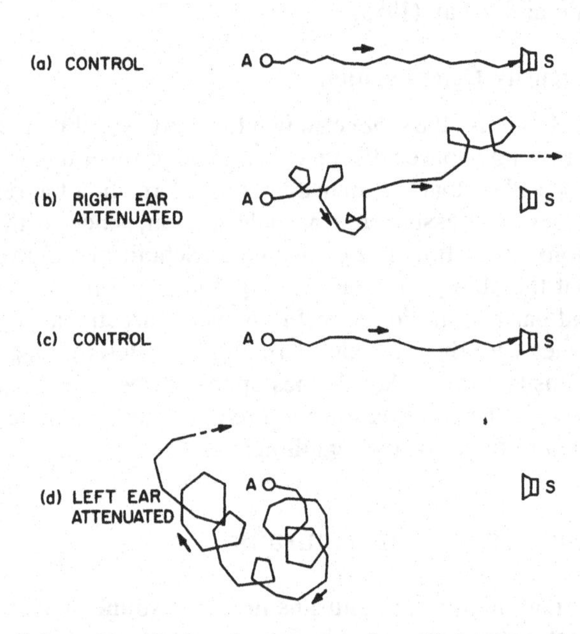

FIGURE 7.7. Orientation of a female green tree frog (*Hyla cinerea*) in response to playback of mating calls. The arrows indicate the direction of movement in successive leaps. (**a**) Control trial in which both eardrums were untouched. (**b**) Pattern of movement when the animal's right eardrum was coated with a thin layer of silicone grease. (**c**) Control trial following removal of the coating from the right eardrum. (**d**) Pattern of movement when the animal's left eardrum was coated with a thin layer of silicone grease. (From Feng et al, 1976.)

8.4°, whereas the mean jump angle is 11.8°. When jumping occurs without head scanning, the accuracy deteriorates with a mean jump angle of 17.6°. To locate an elevated sound source, the head is additionally tilted upward before and after the lateral scan (Gerhardt and Rheinlander, 1982). In the vertical plane, sound localization is not as precise (Passmore et al, 1984). Male frogs presumably can also locate a sound source, since in a breeding pond or in typical territorial encounters they maintain certain spatial relationships.

Given the relatively long wavelengths of frog calls and the short interaural distance, what may be the acoustic cues for sound localization? Blocking the sound input to one ear of tree frogs renders the animal incapable of locating a sound source (Fig. 7-7), suggesting that binaural processing is involved for sound localization (Feng et al, 1976). Interaural time (phase or arrival time) and intensity differences are probably the primary cues. For small animals such as tree frogs, however, the time difference is small, and unless the nervous system is extremely sensitive to small interaural time differences, this may not be very useful. Physically, the interaural intensity difference is also small, unless there is special adaptation in the auditory periphery. Frogs have no pinna to generate directional information, and for sound frequencies below 5 KHz, the tympanic membrane itself cannot provide sufficient directionality. The head-shadowing effect is poor owing to the shallowness of the head and the small head size at low sound frequencies. How is it then frogs can accurately locate a sound source?

Mechanical measurements of the frog's eardrum (Chung et al, 1978, 1981; Pinder and Palmer, 1983) have shown that the displacement amplitude of the frog's eardrum depends on the sound direction. The sound originating from the ipsilateral sound field is more effective than that from the contralateral field to stimulate the eardrum to vibrate at high frequencies (1,100–2,300 Hz) examined. An eighth nerve study (Feng, 1980) in the northern leopard frogs *(Rana p. pipiens)* reveals quantitatively how the responses of single primary auditory fibers depend on sound direction at various frequencies. In response to low-frequency (65–450 Hz) tones, individual eighth nerve fibers show a "figure 8" directivity pattern (directional characteristic in decibels): sound from the two sides of the animal has equal effectiveness in exciting the fibers but sound in the frontal anterior (or posterior) field is relatively ineffective (Fig. 7-8a). At higher frequencies (500–2,100 Hz), the directivity pattern is ovoidal: ipsilateral sound excites the fibers rigorously but excitation is progressively reduced as the sound is rotated toward the frontal field and the contralateral sound field (Fig. 7-8b). Thus the directivity pattern of the frog's ear is frequency dependent.

The above eighth nerve data show that the ear of *Rana p. pipiens* does not behave like a simple pressure receiver since it is not omnidirectional, and the suggestion has been made that the anuran ear may function to some degree as a pressure gradient receiver (Strother, 1959; Chung et al, 1978; Fletcher and Thwaites, 1979; Rheinlaender et al, 1979). A pressure gradient receiver is one that responds to a difference in pressure at two nearby points in space (Fig. 7-9) and not simply to the scalar magnitude of pressure as most "usual" pressure microphones do. The response to gradients is achieved essentially by suspending a diaphragm so that sound may impinge on both of its sides. Clearly, if the pressure signal is equal on both sides, there is no tendency for particles to move across the diaphragm and thus no response. However, if there is a pressure difference across the diaphragm, the particles tend to move across the diaphragm, and thus no response. However, if there is a pressure difference across the diaphragm, the particles tend to move across the diaphragm, and bring it into motion. The pressure gradient receiver therefore responds in proportion to the acoustic particle velocity in a direction perpendicular to the diaphragm plane. This receiver is inherently directional in that its output goes to zero when the diaphragm plane is oriented parallel to the di-

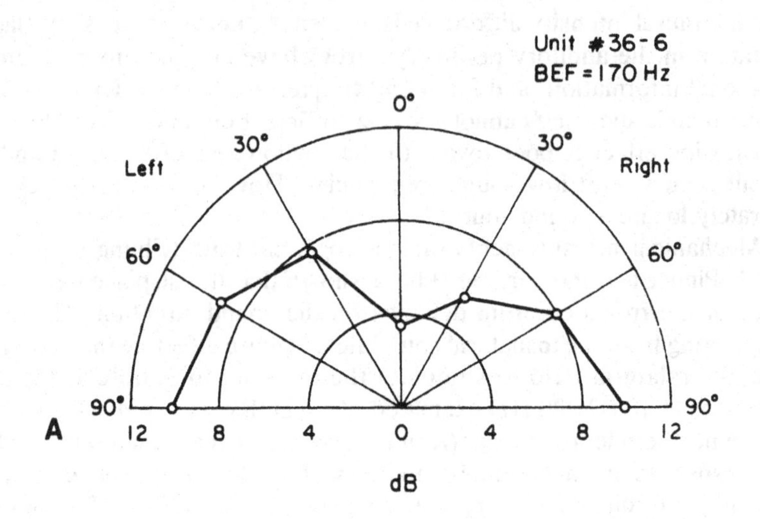

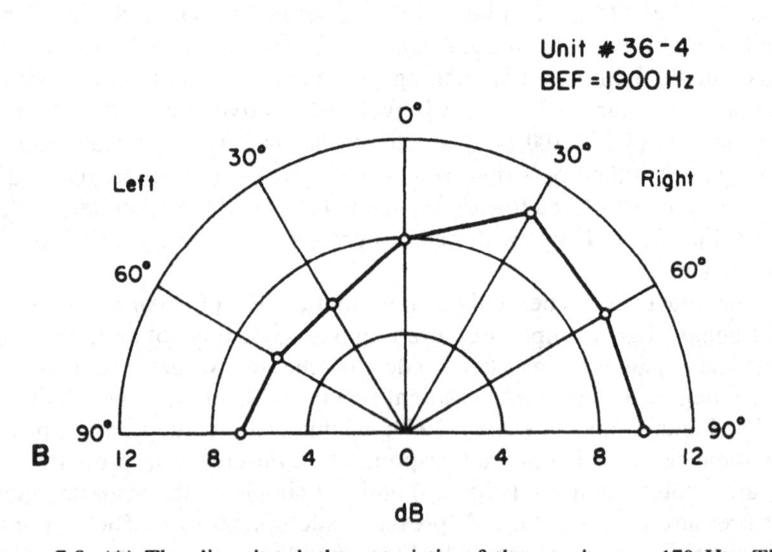

FIGURE 7.8. (**A**) The directional characteristic of the receiver at 170 Hz. This curve was obtained by measuring the firing rates of a primary eighth nerve fiber to a constant tone intensity at the right eardrum with varying incident angles at the horizontal plane. (**B**) The directional characteristic of the receiver at 1,900 Hz. This curve was obtained by measuring the firing rates of a primary eighth nerve fiber to a constant tone intensity at the right eardrum with varying incident at the horizontal plane. (From Feng, 1980. Reproduced with permission from Feng, 1980, J. Acoust. Soc. Anes.)

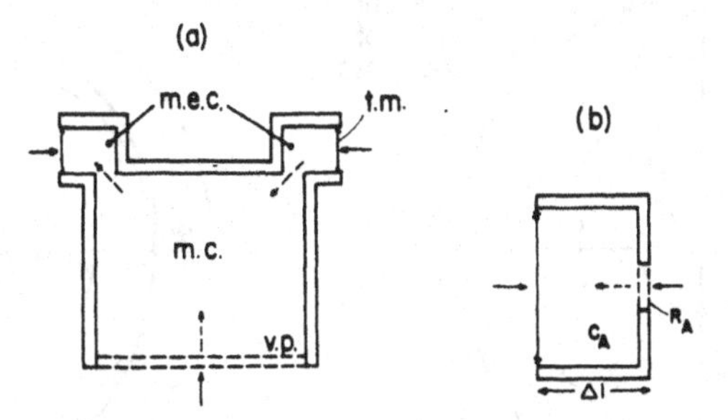

FIGURE 7.9. Schematic diagram illustrating the primary and secondary acoustic inputs to the frog's left ear. (a) The primary acoustic input impinges directly on the external surface of the tympanic membrane (t.m.). The secondary acoustic input impinges on the internal surface of the left tympanic membrane and is derived from two acoustic pathways: one derives from the right tympanic membrane, which is transmitted through the middle ear cavities (m.e.c.) and the mouth cavity (m.c.) to the internal surface of the left tympanic membrane, and the other is mediated by the vocal pouch (v.p.), or other tissues, through the mouth cavity to the internal surface of the left tympanic membrane. (b) A model of combination pressure and pressure-gradient receiver for the frog's left ear (figure modified from Beranek, 1954). C_A represents the compliance of the acoustic cavity where R_A represents the acoustic resistance of the secondary pathways, -1 = linear distance between the primary (left) and secondary (right) sources. Arrows in both figures represent the proposed acoustic pathways. (From Feng and Shofner, 1981.)

rection of particle motion and is maximal when the particle motion and the diaphragm are perpendicular. Theoretically, a pure pressure gradient receiver should produce a "figure 8" (dipole) directivity pattern at all frequencies (Beranek, 1954). In contrast, a pure pressure receiver is essentially omidirectional, since the diaphram is exposed to changes in sound pressure on only one side and thus responds only to the scalar pressure.

A detailed analysis of the frog's ear suggests that it behaves as a combination of a pressure and a pressure-gradient receiver (Feng and Shofner, 1981). A schematic diagram of the layout of a frog's ears and a combination receiver are shown in Figure 7-9 (Beranek, 1954). The combination receiver is depicted as a transducer having a cavity (with a compliance C_A) at the back side of the diaphragm, which has an opening to the outside through an acoustic resistance R_A. The directional characteristics of such a receiver are dependent on the acoustic resistance (R_A) and compliance (C_A) as well as δL, or more precisely by a unitless constant $B = \delta L/cC_AR_A$, where c is the speed of sound in the air. The directivity patterns of a combination receiver at various B values are given in Figure 7-10. For a pure pressure gradient receiver, $R_A = O$; B therefore becomes infinity and the pattern is that of a "figure 8" whereas for a pure pressure receiver $R_A = \infty$, B becomes O and an omnidirectional pattern is obtained. The frog's two ears are coupled via the middle ear and mouth cavities but the degree of

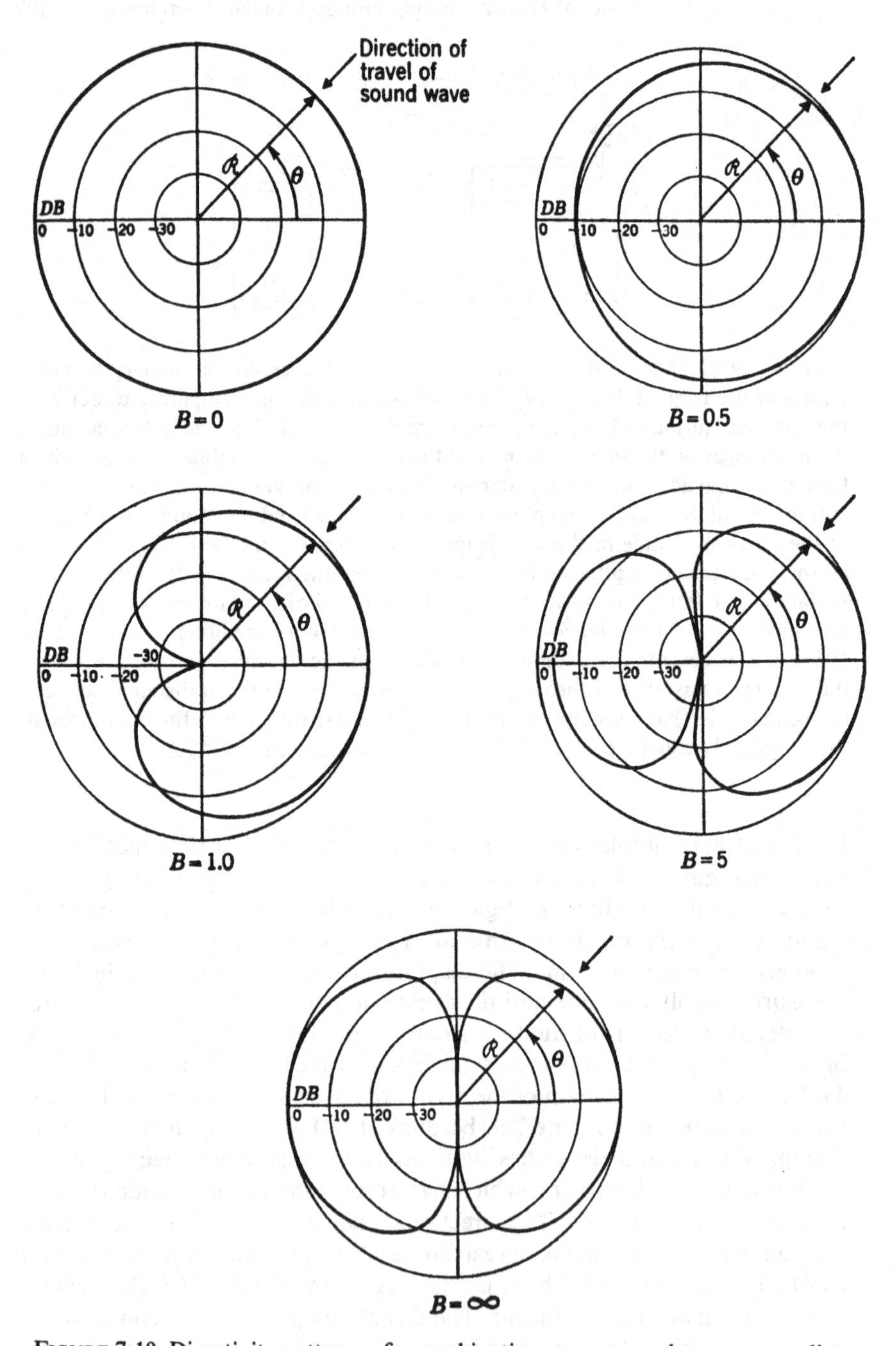

FIGURE 7.10. Directivity patterns of a combination pressure and pressure gradient microphone at various coupling constant (B) values (see text). The force R (which is proportional to particle velocity at any given frequency) acting on the diaphragm is a cosine function of the incident angle 0, i.e., R = 20 log [(1 + B cos 0)/(1 + B)]. (From Beranek, 1954. Reproduced with permission from Beranek, 1954.)

coupling, or R_A, varies with frequency (Feng, 1980). Thus the frog's ear mimics a combination receiver; namely, in addition to the acoustic input impinging on the external side of the tympanic membrane, there is a secondary input arriving from the contralateral ear or other structures onto the internal side of the tympanic membrane. These two inputs are integrated, giving rise to a frequency-specific directivity pattern. Opening of the mouth cavity, which allows the sound to impinge on both the internal and external sides of the tympanic membrane with equal effectiveness, i.e., R_A approaching O, alters the directivity pattern of the ear so that it produces a "figure 8" pattern uniformly, independent of stimulus frequency (Fig. 7-11a; Feng and Shofner, 1981). At high frequencies the secondary input is primarily derived from the contralateral ear, as blocking the contralateral input results in an omnidirectional pattern (Fig 7-11b; Feng and Shofner, 1981). The secondary pathway for low-frequency sounds, however, is not so well understood. Low-frequency sound apparently can penetrate through various tissues and excite the ear without transmitting through the tympanic membrane (Lombard and Straughan, 1974; Wilczynski et al, 1982). Blocking of secondary input from the contralateral ear alone does not alter the directivity pattern of the ear at low frequencies (Feng and Shofner, 1981; Wilczynski and Capranica, 1981), suggesting this pathway is not the major contributing factor to the "figure 8" directivity pattern at these frequencies. This notion is supported by the fact that interaural coupling is poor at low frequencies (Feng, 1980; Vlaming et al, 1984). It is likely that for low-frequency sound, the secondary input is derived from sound transmission through the nostril and vocal pouch (Fletcher and Thwaites, 1979; Feng and Shofner, 1981; Pinder and Palmer, 1983), mouth (Vlaming et, 1984), or by bone conduction or other pathways (Wilczynski et al, 1982; Chung et al, 1981). Palmer and Pinder (1984) recently modeled the frog ear as two coupled eardrums with an additional sound pathway through the mouth cavity. Their model can account for the directionality of the ear under free- and closed-field conditions. Further modification of the model by Aertsen and his colleagues (1985), using three coupled linear oscillators, provides an even better fit of the empirical data at intermediate and high frequencies.

In a preliminary study, Rheinlaender et al (1981) showed that the directional properties of the ear can be altered by changing the volume of the mouth cavity. However, it is not clear if these animals use this strategy actively in localizing a sound source.

Through special mechanical (or acoustical) adaptation, the frog's ear acquires an auditory receiver that is directional. Thus, even though physically the interaural intensity difference at the two ears is small, the auditory periphery at the two sides "sense" a difference of 3 to 8 dB at high frequencies attributed to the asymmetrical directional responses to this frequency range around the frog's longitudinal body axis. Such a difference in perceived intensities can be translated into an average of 0.3- to 4.8-

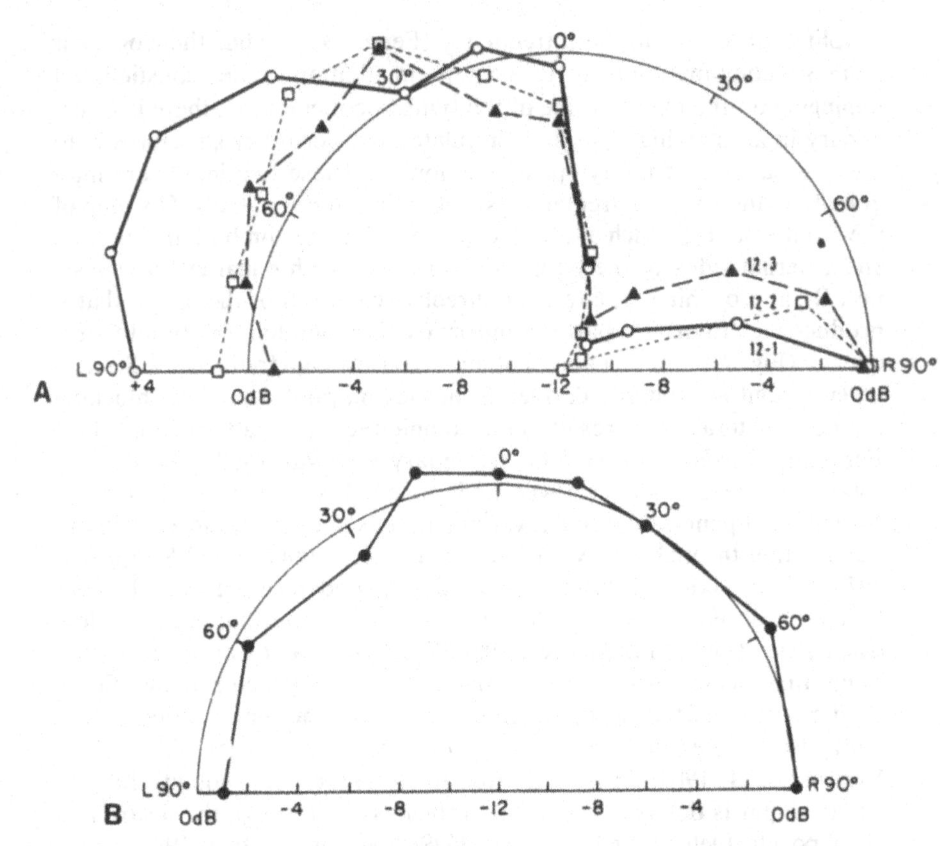

FIGURE 7.11. (A) "Figure 8" directivity patterns of the frog's ear at 415 Hz, 665 Hz, and 1435 Hz obtained from directional responses of units 12-3, 12-2 and 12-1, respectively, from a frog that had its mouth forced open during the recording session. The directional responses were studied for each unit at the unit's best excitatory frequency (BEF). (B) Omnidirectional directivity pattern of the frog's ear at 1,700 Hz obtained from directional responses of unit 3 to 10 from a frog that had its contralateral ear coated with silicone rubber cement prior to the recording session. (From Feng and Shofner, 1981. Reproduced with permission from Feng, 1981 Hearing Res., Elsevier Science Publishers.)

ms difference in excitation time on the two nerves, since a 1-dB change in sound intensity elicits an average shift of 0.1 to 0.6 ms in firing latencies of individual eighth nerve fibers (Feng, 1982). Binaural centers in the frog brain can readily process interaural intensity and time differences of such magnitudes (Bibikov, 1977; Feng and Capranica, 1976, 1978; Kaulen et al, 1972.)

Single-unit recordings from the midbrain torus semicircularis show that almost all neurons in this region exhibit strong responses to changes in sound direction (Feng, 1981; Carlisle and Pettigrew, 1984). The majority of directionally sensitive neurons show ovoidal directivity patterns (or sigmoidal directional response) to all stimulating frequencies with best response to sound from the contralateral field. A few neurons show V- or inverted V-shape directional response to varying sound frequencies with about equal response to sound from the right or left side of the animal and a minimal or maximal response to sound from the frontal field (Feng, 1981). A systematic spatiotopic map, however, has not been conclusively demonstrated in the frog's auditory midbrain. Pettigrew and his colleagues (Pettigrew et al, 1978, 1981; Pettigrew and Carlisle, 1984) indicate that a space map may be present on the basis of gross-evoked potential but electrolytic markings show that their recording sites are not confined to the torus semicircularis.

That directional responses of torus neurons are considerably sharper than those of primary afferents and that response patterns are not frequency dependent as they are for the primary afferent suggests that binaural processes are involved in transforming the directional responses of the auditory periphery in the CNS. This notion is supported by the fact that unilateral eighth nerve ablation results in a decrease in directional responses of torus neurons (Feng, 1981). These results therefore suggest that both the directionality of the peripheral receiver and the central binaural processing are involved in sound localization.

The predominant presence of neurons with sigmoidal directional response implies that a sound from the left side of the animal will maximally excite a large population of torus neurons on the right and that most torus neurons on the left will be minimally excited. Thus the relative excitation levels (van Bergeijk, 1962) on the two sides can be used to indicate the sidedness of the source or the extent the source deviates from the longitudinal body axis of the frog. Alternatively, the absolute firing rates of individual neurons can represent the absolute sound source direction. If, indeed, the animal employs the latter strategy, directional coding will be most precise (and least ambiguous) for sound coming from the frontal field between L30° and R30° owing to the steepness of the response curve in this region of the sound field. Coding of sound direction at both tail ends of the sigmoidal curve will be ambiguous or less precise. Thus, sound localization may be more accurate at these azimuth angles, as was shown for owls (Knudsen et al, 1979) and monkeys (Brown et al, 1982). But azimuth-dependent localization acuity has not been demonstrated behaviorally in frogs. It is interesting to note that when a female frog scans her head, her head and jump orientations are much more accurate than if no head scan was done. The small number of toral neurons with V- or inverted V-shaped directional response can presumably further contribute in directional hearing by signalling the animal that the source is located in the frontal field. The minimal or maximal response of these neurons occurs to frontal stimulation between L15° and R15° in the azimuth, which corresponds to the acuity of horizontal head and jump orientation.

Directional Hearing in Avians

Most birds are soniferous, and acoustic communication plays important roles in behaviors ranging from reproduction, territorial defense, pair-bond maintenance, social and parental-young interactions, to feeding behavior. Sound localization is an integral part of these behaviors. Natural avian sounds are more complex than anuran sounds in terms of their amplitude and frequency modulation characteristics. Species variations, geographical and individual, have been noted. The frequency range of avian sounds surpasses that of anuran sounds but seldom exceeds 10 kHz. The physical and neural bases of sound localization have been reviewed recently (Knudsen, 1980; Konishi, 1983; Lewis, 1983; Lewis and Coles, 1980), and therefore we shall primarily extend these earlier reviews and provide the current views of localization mechanisms in birds.

Despite the small head size, birds, like the anuran counterpart, are able to locate a sound source with an accuracy of just a few degrees (Engelmann, 1928; Gatehouse and Shelton, 1978; Jenkins and Masterton, 1979; Knudsen et al, 1979; Konishi, 1973a,b; Norberg, 1978; Payne, 1971; Shalter, 1978). This ability is a result of a number of factors, i.e., evolution of (1) sounds with optimal locatability (Marler, 1955, 1959), (2) peripheral receivers that are adapted for sound localization (Norberg, 1977; Coles et al, 1980; Hill et al, 1980; Moiseff and Konishi, 1981a), and (3) specialized neural substrates for detecting minute differences in sound intensities and times at the two ears (Moiseff and Konishi, 1981b, 1983; Sullivan and Konishi, 1984; Konishi et al, 1985).

The acuity of sound localization has been carefully evaluated in the barn owl in both horizontal and vertical planes. Owls rely on passive sound localization for catching prey at dusk and can accurately determine the azimuth and elevation of a sound source without moving their heads (Payne, 1971; Konishi, 1973a,b; Knudsen and Konishi, 1979; Knudsen et al, 1979). Localization of tones is most accurate for the frequency range of 4 to 8.5 kHz and comparatively poorer at lower and higher frequencies. The owl's best angular acuity is 1° to 2° in both planes. Other bird species have not been as carefully examined as in the barn owl. The angular acuity of chickens is estimated to be about 4° (Engelmann, 1928).

Most birds have small heads with interaural distances of no more than 2 cm. While the upper cut-off frequency ($f<10$ kHz or >2.5 cm) of their audible range extends beyond that of anurans, the head-shadowing effect is still not pronounced except at the higher end of the audible range. In Japanese quails the maximum head shadow is 6 dB at 8 kHz (Coles et al, 1980). Yet microphonic or gross eighth nerve potential measurements show that it can vary by as much as 25 dB with a 360° horizontal rotation of the sound source in various bird species (Schwartzkopff, 1950, 1962; Payne, 1971; Knudsen and Konishi, 1978; Coles et al, 1980). The reason behind this is that the two ears are coupled via the interaural air passages

(Schwartzkopff, 1962; Payne, 1971; Rosowski and Saunders, 1980; Hill et al, 1980) and thus each ear acts as a push-pull or pressure gradient receiver (Lewis and Coles, 1980; Knudsen, 1980). However, Moiseff and Konishi (1981a) showed that the two ears in owls are not tightly coupled at the frequency range (6–8 Khz) that is important for sound localization in this species. The coupling is reasonably tight at frequencies below 4 kHz to permit the ear to operate as a pressure gradient receiver, but owls poorly localize tones at these frequencies (Payne, 1971; Konishi, 1973a). Thus, the angular acuity in owls seems to rely more heavily on neural adaptations in the brain.

From the above discussion, species variation is clearly evident, and for different species the degree of ear coupling at various frequencies may also vary. Since the directivity pattern conforms to the cardioid pattern at 315 Hz and to a "figure 8" pattern at 5 kHz in quails (Fig. 7-12), and since coupling is frequency dependent, the quail auditory periphery presumably operates also as a combination pressure-pressure gradient receiver (see previous discussion) instead of a pure pressure gradient receiver (Lewis, 1983). The precise acoustical characteristic of the ear, however, has yet to be conclusively determined in various bird species.

In owls the primary physical cues for sound localization have been studied in great detail. These studies have involved interaural differences in time and spectrum, or frequency-dependent intensity (Knudsen, 1980). A unilateral ear plug to the left ear leads the owl to orient above and slightly to the right of the target whereas plugging the right ear causes the owl to orient below and to the left of the target (Knudsen and Konishi, 1979). These experiments demonstrate that (1) owls integrate the signals from both ears to locate a sound source, (2) barn owls have asymmetrical

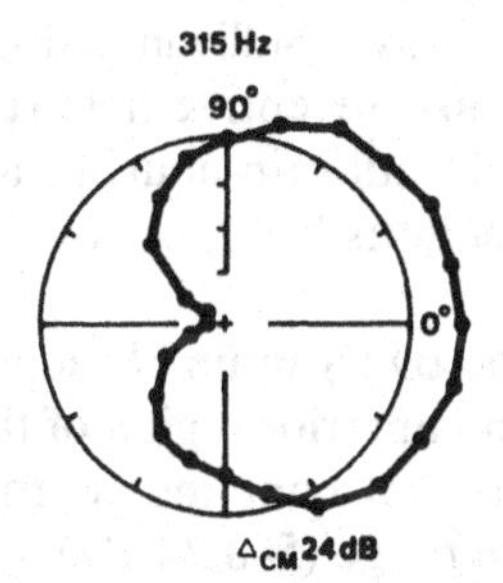

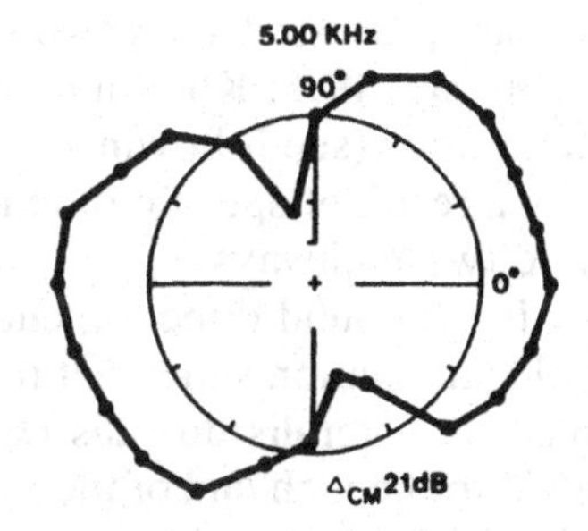

FIGURE 7.12. (Left) The skull of the quail at the level of the middle ear (columella arrowed) to show the extreme trabeculation of the skull. (Right) The cochlear microphonic responses of the right ear plotted on polar coordinates for two frequencies. Maximum interaural difference values (2 cm) are given. When the left ear was blocked, these interaural intensity differences fell to less than 5 dB. (From Coles et al, 1980.)

ears, creating a vertical disparity (additional of the horizontal disparity) in the two ears at a frequency range (4–9 KHz) at which the facial ruff provides an effective reflector. The importance of ear asymmetricity in sound localization has earlier been evaluated (Payne, 1971; Norberg, 1977, 1978), but the theory of sound localization differs between various laboratories. Based on elegant neurophysiological and behavioral data, Konishi and his colleagues (1983) suggested that localization involves a bicoordinate process; i.e., localization in the horizontal plane primarily involves binaural differences in on-going time in the microsecond range, whereas in the vertical plane it involves interaural intensity differences of high-frequency components of sound. By combining single-unit data from the owl auditory midbrain using both free-field and dichotic stimulations, Moiseff and Konishi (1981b) showed that the horizontal angular selectiveness in unit response is attributed to the dichotic microsecond difference of ongoing time and not to the interaural difference of sound envelope or to the interaural intensity difference. Removing the facial ruff feathers, the acoustic-reflective component of the auditory periphery at high frequencies renders the owl incapable of localizing sounds in elevation, but the owl's azimuth localization acuity remains (Knudsen and Konishi, 1979), showing that the intensity cue is most important for elevation angular acuity. Their theory, while differing from all the earlier theories (Payne, 1971; Pumphrey, 1948; Norberg, 1978; Knudsen, 1980), currently represents the most tenable model of sound localization for owls.

Current research has focused on how these interaural cues are processed at lower auditory centers and how these cues are separated from one another. If central binaural neurons were to process interaural time information without intensity interference, these two parameters would be independent of one another. Neurons with such characteristics are shown to be present in the owl's lower auditory centers (Moiseff and Konishi, 1983). It is now firmly established that interaural time and intensity differences are processed along two separate pathways (Sullivan and Konishi, 1984; Sullivan, 1985; Konishi et al, 1985). Thus, receptive field response characteristics (see following discussion) of midbrain neurons are presumably a result of specific functional connections from, or convergence of, these two pathways.

How is the sound direction encoded in the owl's brain? Studies using free-field stimulation show that the lateral and anterior region of the mesencephalicus lateralis dorsalis (MLD) of the owl contains neurons that are tuned to the high end of the owl's audible range (5–8.7 kHz) and have sharply delineated auditory receptive fields (Knudsen and Konishi, 1978a,b). Furthermore, these neurons (Fig. 7-13), as well as those in the owl's optic tectum (Knudsen, 1982), are arranged topographically according to the azimuths and elevations of their receptive fields. Sound azimuths are represented in the horizontal plane with the right side of the structure representing sound field between R15° to L60° and the left side

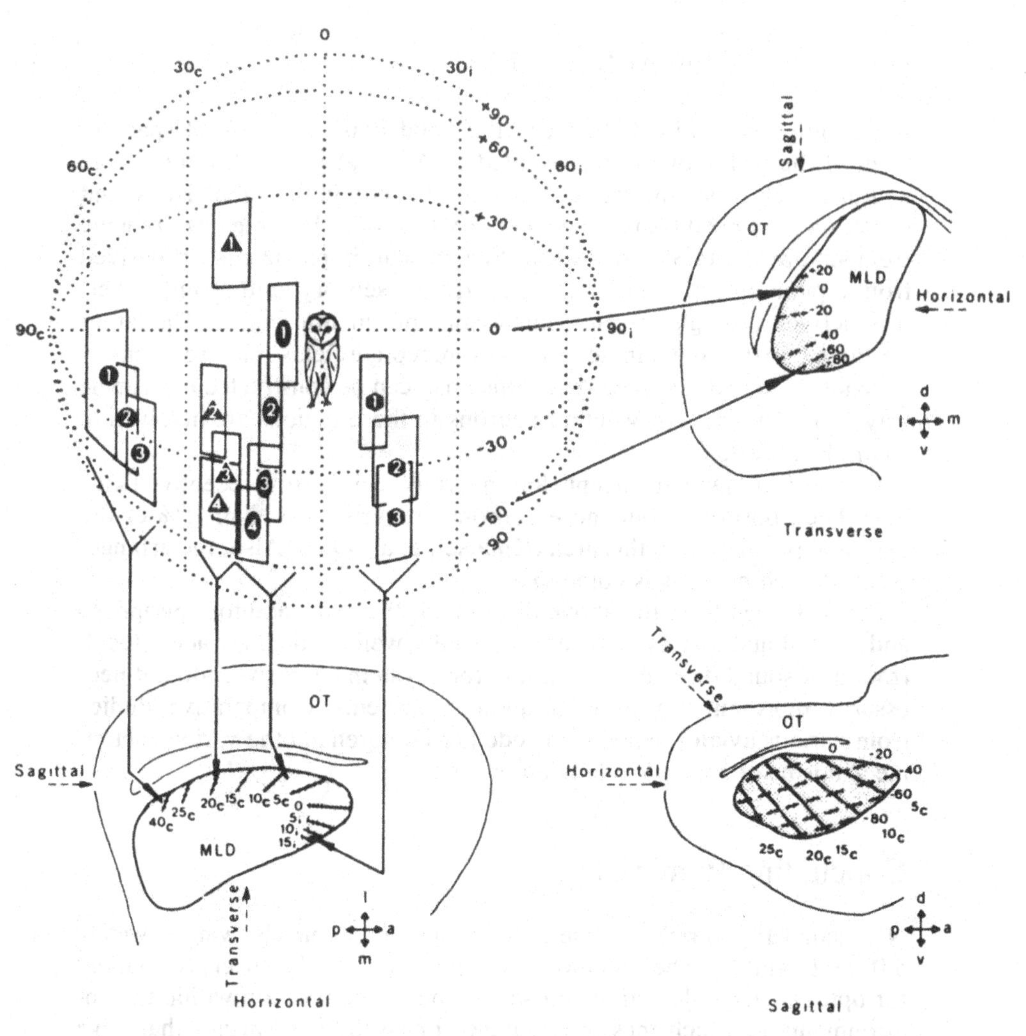

FIGURE 7.13. A neural map of auditory space. Upper left: coordinates of auditory space are depicted as a globe surrounding the owl. Projected onto the globe are the best areas (solid- lined rectangles) of 14 units that were recorded in four separate penetrations. The large numbers backed by the same symbols (dark diamonds, triangles, etc.) represent units from the same penetration; the numbers themselves denote the order in which the units were encountered. Penetrations were made with the electrode oriented parallel to the transverse plane of the mesencephalicus lateralis dorsalis (MLD) at positions indicated in the horizontal section by solid arrows. Below and to the right of the globe are illustrated three histological sections through MLD in the horizontal, transverse, and sagittal planes. The stippled portion of MLD corresponds to the space-mapped region. Isoazimuth contours, based on field centers, are shown as solid lines in the horizontal and sagittal sections. On each section dashed arrows indicate planes of the other two sections. Solid, crossed arrows to the lower right of each section define the orientation of the section: a, anterior; d, dorsal; l, lateral; m, medial; p, posterior; v, ventral; OT, optec tectum. (From Knudsen and Konishi, 1978a. Reproduced with permission from Knudsen and Konishi, *Space*, A Neural Map of Auditory Space in the Owl'', Vol. 200 pp. 795–797. Fig. 2. 19 May 1978.)

representing sound fields between L15° and R60°. Thus 30° of space in front of the owl is overly represented on MLD at both sides, a possible neural correlate having optimal behavioral acuity in this area. The sound elevation is mapped transversely in the MLD with "best regions" ranging from 40° upward to 80° downward. Such a spatiotopic map is not derived from topographic projection of inputs from a sensory epithelium; rather, it is derived through precise comparisons of binaural inputs. The larger portions of MLD contain neurons with receptive fields that are intensity dependent, so that no single spatiotopic map can be constructed in a simple way. It is not yet clear whether neurons in these regions are involved in sound localization.

In field L of owl's telencephalon, neurons with narrow receptive fields have been observed, but these neurons comprise less than 15% of the neuronal population in this area (Knudsen et al, 1977). Also, the arrangement of such neurons is complex.

As evidenced from the above discussion, the owl's auditory periphery and central auditory systems are especially well adapted for acute localization of sound direction. Their auditory systems, however, do not necessarily represent typical avian auditory systems. Comparative studies from various avian species are needed to comprehend the various neural mechanisms of sound localization in birds.

Concluding Remarks

Nonmammalian vertebrates are a diverse group of animals living in widely different habitats. Their auditory systems appear to be specially adapted for optimal signal detection and sound source localization within the environments to which they are adapted. From the few species that have been studied so far, it is obvious that the strategies employed for detection and localization differ profoundly from one class to another, and even for members of the same order. For biologists and sensory physiologists, these animals offer valuable opportunities to study diverse adaptations of the auditory system and to further our understanding of mechanisms for directional hearing and sound localization among all animals.

Acknowlegments. The authors wish to thank G. Neuweiler for his critical comments on the manuscript. Research reported here from the authors' laboratories was supported by the National Science Foundation (NS 82-04160 to A.S.F. and BNS 81-11354 to R.R.F.), the National Institutes of Health, National Institute of Neurological and Communicative Disorders and Stroke (NS-15268 to R.R.F.), and a Research Career Development Award to R.R.F.

References

Aertsen, A.M.H.J., Vlaming, M.S.M.G., Eggermont, J.J., Johannesma, P.I.M. (1985). Directional hearing in the grassfrog (*Rana temporalis* L.) II. Acoustics and modelling of the auditory periphery. Hear. Res. In press.

Beranek, L.L. (1954). Acoustics. New York: McGraw Hill.

Berg, A. van den, Schuijf, A. (1983). Discrimination of sounds based on the phase difference between particle motion and acoustic pressure in the shark. *Chiloscyllium griseum*. Proc. Roy. Soc. (*Lond.* B) 218, 127–134.

Bergeijk, W.A. van (1962). Variation on a theme of Bekesy: a model of binaural interaction. J. Acous. Soc. Am. 34, 1431–1437.

Bergeijk, W.A. van (1964). Directional and nondirectional hearing in fish. In: Tavolga, W.N. (ed.). Marine Bio-Acoustics, Vol. I. pp. 281–300. Oxford: Pergamon Press.

Bergeijk, W.A. van (1967). The evolution of vertebrate hearing. In: Contribution to Sensory Physiology Vol. 2. Neff, W.D. (ed.). pp. 1–49. Berlin: Springer-Verlag.

Bibikov, N.G. (1977). Dependence of the binaural neurons reaction in the frog torus semicircularis on the interaural phase difference. Sechenov Physiol. J. USSR 63, 365–373.

Bilsen, F., Ritsma, R. (1970). Repetition pitch and its implications for hearing theory. Acustica 22, 63–68.

Blaxter, J.H.S., Denton, E.J., Gray, J.A.B. (1981). Acousticolateralis system in clupeid fishes. In: Hearing and Sound Communication in Fishes Tavolga, W.N., Popper, A.N., Fay, R.R. (eds.). pp. 39–59. Berlin: Springer-Verlag.

Blaxter, J.H.S., Hoss, D.E. (1981). Startle response in herring: The effect of sound stimulus frequency, size of fish and selective interference with the acousticolateralis system. J. Mar. Biol. Assoc. UK 61, 871–879.

Bleckmann, H. (1986). Prey identification and prey localization in surface-feeding fish and fishing spiders. In: Atema, J., Fay, R., Popper, A., Tavolga, W. (eds.). Sensory Biology of Aquatic Animals. New York: Springer Verlag. In press.

Bleckmann, H., Schwartz, E. (1982). The functional significance of frequency modulation within a wave train for prey localization in the surface feeding fish *Aplocheilus lineatus* (Cyprinodontidae). J. Comp. Physiol. 145, 331–339

Brown, C.H., Schessler, T., Moody, D., Stebbins, W. (1982). Vertical and horizontal sound localization. J. Acous. Soc. Am. 72, 1804–1811.

Buwalda, R.J.A. (1981). Segregation of directional and nondirectional acoustic information in the Cod. In: Tavolga, W., Popper, A.A., Fay, R. (eds.). Hearing and Sound Communication in Fishes. pp. 139–172. New York: Springer.

Carlisle, S., Pettigrew, A.G. (1984). Auditory responses in the torus semicircularis of the cane toad, *Bufo marimus*. II. Single unit studies. Proc. Roy. Soc. (Lond. B) 222, 243–257.

Chapman, C.J. (1973). Field studies of hearing in teleost fish. Helgolander wiss. Meeresunters *24*, 371–390.

Chapman, C.J., Hawkins, A.D. (1973). A field study of hearing in the cod *(Gadus morhua)*. J. Comp. Physiol. 85, 147–167.

Chapman, C.J., Johnstone, A.D. (1974). Some auditory discrimination experiments on marine fish. J. Exp. Biol. *61*, 521–528.

Chapman, C.J., Sand, O. (1974). A field study of hearing in two species of flatfish-*Pleuronectes platessa* (L) and *Limanda limanda* (L), Comp. Biochem. Physiol. *47*, 371–385.

Chung, S.H., Pettigrew, A., Anson, M. (1978). Dynamics of the amphibian middle ear. Nature 272, 142–147.

Chung, S.H., Pettigrew, A.G., Anson, M. (1981). Hearing in the frog: dynamics of the middle ear. Proc. Roy. Soc. *(Lond.)* 212, 459–485.

Coles, R.B., Lewis, D.B., Hill, K.G., Hutchings, M.E., Gower, D.M. (1980). Directional hearing in the Japanese quail *(Conturnix coturnix japonica)* II. Cochlear physiology. J. Exp. Biol. 86, 153–170.

Coombs, S., Janssen, J., Webb, J. (1986). Diversity of lateral line systems: Evolutionary and functional considerations. In: Atema, J., Fay, R., Popper, A., Tavolga, W. (eds.). Sensory Biology of Aquatic Animals. New York: Springer Verlag. In press.

Corwin, J.T. (1981) Peripheral auditory physiology in the lemon shark: Evidence of parallel otolithic and non-otolithic sound detection. J. Comp. Physiol. 142:379–390.

Denton, E.J., Gray, J.A.B. (1983). Mechanical factors in the excitation of clupeid lateral lines. Proc. Roy. Soc. *(Lond.* B) 218, 1–26.

Denton, E.J., Gray, J.A.B. (1986). Mechanical factors in the excitation of the lateral lines of fishes. In: Atema, J., Fay, R., Popper, A., Tavolga, W. (eds.). Sensory Biology of Aquatic Animals. New York: Springer Verlag. In press.

de Vries, H. (1950). The mechanics of the labyrinth otoliths. Acta Oto-Laryngol. *38*, 262–273.

Dijkgraaf, S. (1960). Hearing in bony fishes. Proc. Roy. Soc. (Lond. B.) 152, 51–54.

Dijkgraaf, S. (1963). The functioning and significance of the lateral-line organ. Biol. Rev. 38, 51–105.

Durlach, N.I. (1972). Binaural signal detection: Equalization and cancellation theory. In:Foundations of Modern Auditory Theory. Vol. 2 Tobias, J. (ed.). pp. 369–458. New York: Academic Press.

Eaton, R.A., Kimmel, C.B. (1980). Directional sensitivity of the Mauthner cell system to vibrational stimulation in zebrafish larvae. J. Comp. Physiol. 140, 337–342.

Engelmann, W. (1928). Untersuchungen uber die Schallokalisation bei Tieren. Z. Psychol. 105, 317–370.

Fay, R.R. (1978). Sound detection and sensory coding by the auditory systems of fishes. In: Mostofsky, D. (ed.). The Behavior of Fish and Other Aquatic Animals. pp. 197–231. New York: Academic Press.

Fay, R.R. (1981). Coding of acoustic information in the 8th nerve. In: Tavolga, W., Popper, A., Fay, R. (eds.). Hearing and Sound Communication in Fishes. New York: Springer Verlag.

Fay, R.R. (1984). The goldfish ear codes the axis of acoustic particle motion in three dimensions. Science 225, 951–953.

Fay, R.R. (1986). Peripheral adaptations for spatial hearing in fishes. In: Atema, J., Fay, R., Popper, A., Tavolga, W. (eds.). Sensory Biology of Aquatic Animals. New York: Springer Verlag. In press.

Fay, R.R., Coombs, S.L. (1983). Neural mechanisms of sound detection and temporal summation. Hear. Res. 10, 69–92.

Fay, R.R., Hillery, C., Bolan, K. (1982). Representation of sound pressure and particle motion information in the midbrain of the goldfish. Comp. Biochem. Physiol. 71A, 181–191.

Fay, R.R., Olsho, L.W. (1979). Discharge patterns of lagenar and saccular neurons of the goldfish eighth nerve: Displacement sensitivity and directional characteristics. Comp. Biochem. Physiol. 62, 377–386.

Fay, R.R., Patricoski, M. (1980). Sensory mechanisms for low frequency vibration detection in fishes. In: Buskirk R., (ed.). Abnormal Animal Behavior Preceding Earthquakes. Conference II. U.S. Geological Survey Open File Report, pp. 80–453.

Fay, R.R., Popper, A.N. (1975). Modes of stimulation of the teleost ear. J. Exp. Biol. 62, 379–387.

Fay, R.R., Ream, T. 1986. Acoustic response and tuning in saccular nerve fibers of the goldfish *(Carassius auratus)*. J. Acoust. Soc. Am. 79, 1883–1895.

Fay, R.R., Yost, W.A., Coombs, S.L. (1983). Psychophysics and neurophysiology of repetition noise processing in a vertebrate auditory system. Hear. Res. 12, 31–55.

Feng, A.S. (1980). Directional characteristics of the acoustic receiver of the leopard frog *(Rana pipiens)*: A study of eighth nerve auditory responses. J. Acous. Soc. Am. 68, 1107–1114.

Feng, A.S. (1981) Directional response characteristics of single neurons in the torus semicircularis of the leopard frog *(Rana pipiens)*. J. Comp. Physiol. 144, 419–428.

Feng, A.S. (1982). Quantitative analysis of intensity-rate and intensity-latency functions in peripheral auditory nerve fibers of northern leopard frogs *(Rana p. pipiens)*. Hear. Res. 6, 242–246.

Feng, A.S., Capranica, R.R. (1976). Sound localization in anurans. I. Evidence of binaural interaction in dorsal medullary nucleus of bullfrogs *(Rana catesbeiana)*. J. Neurophysiol. 39, 871–881.

Feng, A.S., Capranica, R.R. (1978). Sound localization in anurans II. Binaural interaction in superior olivary nucleus of the green treefrog *(Hyla cinerea)*. J. Neurophysiol. 41, 43–54.

Feng, A.S., Gerhardt, H.C., Capranica, R.R. (1976). Sound localization behavior of the green treefrog *(Hyla cinerea)* and the barking treefrog *(H. gratiosa)*. J. Comp. Physiol. 107, 241–252.

Feng, A.S., Shofner, KW.P. (1981). Peripheral basis of sound localization in anurans. Acoustic properties of the frog's ear. Hear. Res. 5, 201–216.

Fletcher, N.H., Thwaites, S. (1979). Physical models for the analysis of acoustical systems in biology. Q. Rev. Biophys. 12, 25–65.

Frisch, K. von, Dijkgraaf, S. (1935). Konnen Fische die Schallrichtung wahrnehmen? Z. Vergl. Physiol. 22, 641–655.

Frisch, K. von (1938). The sense of hearing in fish. Nature 141, 8–11.

Furukawa, T. (1966). Synaptic interaction at the Mauthner cell of goldfish. Prog. Brain Res. 21A, 44–70.

Gatehouse, R.W., Shelton, B.R. (1978). Sound localization in bobwhite quail *(Colinus virginianus)*. Behav. Biol. 22, 533–540.

Gerhardt, H.C., Rheinlaender, J. (1982). Localization of an elevated sound source by the green treefrog. Science 217, 663–664.

Gorner, P. (1973). The importance of the lateral line system for the perception of surface waves in the clawed toad, *Xenopus laevis*. Experientia 29, 295–296.

Harris, G.G., van Bergeijk, W.A. (1964). Evidence that the lateral-line organ responds to near-field displacements of sound sources in water. J.Acoust. Soc. Am. 34, 1831–1841.

Hawkins, A.D., Horner, K. (1981). Directional characteristics of primary auditory neurons from the codfish ear. In: Tavolga, W. Popper, A., Fay, R. (eds.). Hearing and Sound Communication in Fishes. pp. 311–328. New York: Springer Verlag.

Hawkins, A.D., Sand, O. (1977). Directional hearing in the median vertical plane by the cod. J. Comp. Physiol. 122, 1–8.

Hill, K.g., Lewis, D.B., Hutchings, M.E., Coles, R.B. (1980). Directional hearing in the Japanese quail. I. Acoustic properties of the auditory system. J. Exp. Biol. 86, 135–151.

Horner, K., Sand, O., Enger, P. (1980). Binaural interaction in the cod. J. Exp. Biol. 85, 323–332.

Jenkins, W.M., Masterton, R.B. (1979). Sound localization in pigeon *(Columbia livia.)* J. Comp. Physiol. Psychol. 93, 403–413.

Kalmijn, A. (1986). Hydrodynamic and acoustic detection in aquatic vertebrates. In: Atema, J., Fay, R., Popper, A., Tavolga, W. (eds.). Sensory Biology of Aquatic Animals. New York: Springer Verlag. In press.

Kaulen, R., Lifschitz, W., Palazzi, C., Adrian, H. (1972). Binaural interaction in the inferior colliculus of the frog. Exp. Neurol. 37, 469–480.

Kleerekoper, H., Chagnon, E.C. (1954). Hearing in fish with special reference to *Semotilus atromaculatus* (Mitchill). J. Fish Res. Bd. Cand. 11, 130–152.

Knudsen, E.I. (1980). Sound localization in birds. In: Comparative Studies of Hearing in Vertebrates. Popper, A.N., Fay, R.R. (eds.). pp. 289–322. New York: Springer Verlag.

Knudsen, E.I. (1982). Auditory and visual maps of space in the optic tectum of the owl. J. Neurosci. 2, 1177–1194.

Knudsen, E.I., Konishi, M. (1978a). A neural map of auditory space in the owl. Science 200, 795–797.

Knudsen, E.I., Konishi, M. (1978b). Space and frequency are represented separately in auditory midbrain of the owl. J. Neurophysiol. 41, 870–884.

Knudsen, E.I., Konishi, M. (1979). Mechanisms of sound localization in the barn owl *(Tyto alba)*. J. Comp. Physiol. 133, 13–21.

Knudsen, E.I., Konishi, M. (1980). Monaural occlusion shifts receptive field locations of auditory midbrain units in the owl. J. Neurophysiol. 44, 687–695.

Knudsen, E.I., Blasdel, G.G., Konishi, M. (1979). Sound localization by the barn owl *(Tyto-alba)* measured with the search coil technique. J. Comp. Physiol. 133, 1–11.

Knudsen, E.I., Konishi, M., Pettigrew, J.D. (1977). Receptive fields of auditory neurons in the owl. Science 198, 1278–1280.

Konishi, M. (1973a). Locatable and nonlocatable acoustic signals for barn owls. Am. Nat. 107, 775–785.

Konishi, M. (1973b). How the owl tracks its prey. Am. Sci. 61, 414–424.

Konishi, M. (1983). Neuroethology of acoustic prey localization in the barn owl. In: Neuroethology and Behavioral Physiology. Huber, F., Markl, H. (eds.). pp. 303–317. Berlin: Springer-Verlag.

Konishi, M., Sullivan, W.E., Takahashi, T. (1985). The owl's cochlear nuclei process different sound localization cues. J. Acous. Soc. Am. 78, 360–364.

Lewis, D.B. (1983). Directional cues for sound localization. In: Lewis, B. (ed.). Bioacoustics: A Comparative Approach. pp. 233–260. London: Academic Press.

Lewis, D.B., Coles, R. (1980). Sound localization in birds. Trend. Neurosci. 3, 102–105.

Lombard, R.E., Straughan, I.R. (1974). Functional aspects of anuran middle ear structures. J. Exp. Biol. 61, 71–93.

Marler, P. (1955). Characteristics of some animal calls. Nature 176, 6–8.

Marler, P. (1959). Developments in the study of animal communication. In: Bell, P.R. (ed.). Darwin's Biological Work. pp. 150–206. Cambridge: Cambridge University Press.

McCormick, C.A. (1983). Structure and function of the lateral line system. In: Northcutt, R.G., Davis, R.E. (eds.). Fish Neurobiology and Behavior. pp. 179–214. Ann Arbor, Mich.: University of Michigan Press.

McCormick, C.A., Braford, M. (1986). Central connections of the octavolateralis system. In: Atema, J., Fay, R., Popper, A., Tavolga, W. (eds.). Sensory Biology of Aquatic Animals. New York: Springer Verlag. In press.

Moiseff, A., Konishi, M. (1981a). The owl's interaural pathway is not involved in sound localization. J. Comp. Physiol. 144, 299–304.

Moiseff, A., Konishi, M. (1981b). Neural and behavioral sensitivity to binaural time differences in the owl. J. Neurosci. 1, 40–48.

Moiseff, A., Konishi, M. (1983). Local anesthetics demonstrate the separation of time and intensity processing by the owl's auditory system. Soc. Neurosci. Abst. 9:212A.

Moulton, J.M., Dixon, R. (1967). Directional hearing in fishes. In: Tavolga W.N., (ed.). Marine Bio-acoustics, Vol. II. pp. 187–232. Oxford: Pergamon Press.

Myrberg, A., Gordon, C., Klimley A. (1976). Attraction of free ranging sharks by low frequency sound, with comments on its biological significance. In: Schuijf, A., Hawkins, A. (eds.). Sound Reception in Fish. pp. 205–228. Amsterdam: Elsevier.

Nelson, D.R., Gruber, S.H. (1963). Sharks: Attraction by low frequency sounds. Science 142, 975–977.

Norberg, R.A. (1977). Occurrence and independent evolution of bilateral ear asymmetry in owls and implications on owl taxonomy. Phil. Trans. Roy. Soc. (*Lond.* B) 282, 325–410.

Norberg, R.A. (1978). Skull asymmetry, ear structure and function, and auditory localization in Tengmalm's owl, *Aegloius funereus* (Linne). Phil. Trans. Roy. Soc. *(Lond. B)* 282, 325–410.

Northcutt, G. (1980). Central auditory pathways in anamniotic vertebrates. In: Popper, A., Fay, R. (eds.). Comparative Studies of Hearing in Vertebrates. pp. 79–118. New York: Springer Verlag.

Palmer, A.R., Pinder, A.C. (1984). The directionality of the frog ear described by a mechanical model. J. Theor. Biol. 110, 205–215.

Partridge, B. (1981). Lateral line function and the internal dynamics of fish schools. In: Tavolga, W., Popper, A., Fay, R. (eds.). Hearing and Sound Communication in Fishes. pp. 515–522. New York: Springer-Verlag.

Passmore, N.I., Capranica, R.R., Telford, S.R., Bishop, P.J. (1984). Phonotaxis in the painted reed frog *Hyperolius marmoratus).* The localization of elevated sound source. J. Comp. Physiol. 154, 189–197.

Payne, R.S. (1971). Acoustic location of prey by barn owls *(Tyto alba).* J. Exp. Biol. 54, 535–573.

Pettigrew, A., Chung, S.-H., Anson, M. (1978). Neurophysiological basis of di-

rectional hearing in amphibia. Nature 272, 138–142.
Pettigrew A.G., Anson, M., Chung, S.-H. (1981). Hearing in the frog: a neurophysiological study of the auditory response in the midbrain. Proc. Roy. Soc. (Lond. B), 212, 433–457.
Pettigrew, A.G. and Carlisle, S. (1984) Auditory response in the torus semicircularis of the cane toad *(Bufa marimis).* 1. Field potential studies. Proc. Roy. Soc. Lond. (B), 222, 231–242.
Pinder, A.C., Palmer, A.R. (1983). Mechanical properties of the frog ear: vibration measurements under free- and closed-field acoustic conditions. Proc. Roy. Soc. (Lond. B), 219, 371–396.
Popper, A.N. (1977). A scanning electron microscopic study of the sacculus and lagana in the ears of fifteen species of teleost fishes. J. Morph. 153, 397–418.
Popper, A.N. and Coombs, S.L. (1982). The morphology and evolution of the ear in Actinopterygian fishes. Amer. Zool. 22:311–328.
Popper, A.N., Rogers, P., Saidel, W., Cox, H. (1986). The role of the fish ear in sound processing. In: Atema, J., Fay, R., Popper, A., Tavolga, W. (eds.). Sensory Biology of Aquatic Animals, New York: Springer Verlag. In press.
Pumphrey, R.J. (1948). The sense organs of birds. Ibis 90, 171–199.
Pumphrey, R.J. (1950). Hearing. Symp. Soc. Exp. Biol. *4,* 3–18.
Reinhardt, F. (1935). Uber Richtungswahrnehmung bei Fischen, besonders bei der Elritze (*Phoxinus laevis* L.) und beim Zwergwels (*Amiurus nebulosus* Raf.). Z. Vergl. Physiol. 22, 570–603.
Rheinlaender, J., Gerhardt, H.C., Yager, D.D., Capranica, R.R. (1979). Accuracy of phonotaxis by the green treefrog *(Hyla cinerea).* J. Comp. Physiol. 133, 247–255.
Rheinlaender, J., Walkowiak, W., Gerhardt, H.C. (1981). Directional hearing in the green treefrog: a variable mechanism? Naturwissen-schaften 68, 430–431.
Rogers, P. (1986). What are fish listening to?—A possible answer. J. Acoust. Soc. Am., 79, S22 (abstract).
Rosowski, J.J., Saunders, J.C. (1980). Sound transmission through the avian interaural pathways. J. Comp. Physiol. 136, 183–190.
Sand, O. (1981). The lateral line and sound reception. In: Tavolga, W., Popper, A., Fay, R. (eds.). Hearing and Sound Communication in Fishes. pp. 459–480 New York: Springer Verlag.
Schuijf, A. (1981). Models of acoustic localization. In: Tavolga, W., Popper, A., Fay, R. (eds.). Hearing and Sound Communication in Fishes. p. 267. New York: Springer Verlag.
Schuijf, A., Siemelink, M.E. (1974). The ability of cod *(Gadus morhua)* to orient towards a sound source. Experientia 30, 773–774.
Schuijf, A. (1975). Directional hearing of cod *(Gadus moorhua)* under approximate free field conditions. J. Comp. Physiol. 98, 307–332.
Schuijf, A., Buwalda, R.A. (1975). On the mechanism of directional hearing in cod (*Gadus morhua* L.). J. Comp. Physiol. 98, 333–343.
Schuijf, A., Buwalda, R.A. (1980). Sound localization: A major problem in fish acoustics. In: Popper, A., Fay, R. (eds.). Comparative Studies of Hearing in Vertebrates. pp. 43–78. New York: Springer Verlag.
Schuijf, A., Hawkins, A.D. (1983). Acoustic distance discrimination by the cod. Nature 302, 143–144.
Schwartz, E. (1967). Analysis of surface-wave perception in some teleosts. In:

Cahn, P.H. (ed.). Lateral Line Detectors. pp. 123–154. Bloomington, Ind.: Indiana University Press.

Schwartzkopff, J. (1950). Beitrag zum Problem der Richtungshorens bei Vogeln. Z. Vergl. Physiol. 32, 319–327.

Schwartzkopff, J. (1962). Zur Frage des Richtungshorens von Eulen *(Striges)*. Z. Vergl. Physiol. 45, 570–580.

Shalter, M.D. (1978). Localization of passerine seeet and mobbing calls by goshawks and pygmy owls. Z. Tierpsychol. 46, 260–267.

Strother, W.F.J. (1959). The electrical responses of the auditory mechanisms in the bullfrog *(Rana catesbeiana)*. J. Comp. Physiol. Psychol. 52, 157–162.

Sullivan, W.E. (1985). Classification of response patterns in cochlear nucleus of barn owl: Correlation with functional response properties. J. Neurophysiol. 53, 201–216.

Sullivan, W.E., Konishi, M. (1984). Segregation of stimulus phase and intensity coding in the cochlear nucleus of the barn owl. J. Neurosci. 4, 1787–1799.

Supra, M., Cotzin, M., Dallenbach, K. (1944). Am. J. Psychol. 57:133.

Tavolga, W.N. (1976). Acoustic obstacle detection in the sea catfish *(Arius felis)*. In: Schuijf, A., Hawkins, A.D. (eds.). Sound Reception in Fish. pp. 185–204. Amsterdam: Elsevier.

Vlaming, M.S.M.G., Aertsen, A.M.H.J., Epping, W.J.M. (1984). Directional hearing in the grass frog *(Rana temporalis L.)* : I. Mechanical vibrations of tympanic membrane. Hear. Res. 14, 191–201.

Walker, T. (1967) History, histological methods, and details of the structure of the lateral line of the Walleye surf perch. In: Cahn, P. (ed.) Lateral Line Detectors. Bloomington: Indiana University Press.

Wilcox, S. (1986). Surface wave reception in invertebrates and vertebrates. In: Atema, J., Fay, R., Popper, A., Tavolga, W. (eds.). Sensory Biology of Aquatic Animals. New York: Springer Verlag. In press.

Wilczynski, W., Capranica, R.R. (1981). A study of the mechanism underlying the directional sensitivity of the anuran ear. Soc. for Neurosci. Abst. 7:147A.

Wilczynski, W., Resler, C., Capranica, R.R. (1982). Relative sensitivity of tympanic and extratympanic sound transmission in the leopard frog, *Rana pipiens*. Soc. Neurosci. Abst. 8:941A.

Yost, W.A., Hill R., Perez-Falcon, T. (1978). Pitch and pitch discrimination of broad band signals with rippled power spectra. J. Acoust. Soc. Am. 63, 1166–1173.

8

Directional Hearing and Sound Localization in Echolocating Animals

JAMES A. SIMMONS

Most animals appear to use their hearing for multiple purposes. These include detecting and identifying significant sounds in the environment (frequently as indicators of the presence of a predator or prey), social communicating (to attract or identify other members of the species, even particular individuals, and to determine the behavioral state of these others for aggressive encounters, mating, or other purposes), and perceiving the locations of various sources of sound, often to guide the orientation of the primary spatial modality, vision. Animals which use their *hearing* as the primary mode of spatial perception for locating and identifying objects provide valuable examples of the ultimate capacity of the auditory system to process spatial information conveyed in sounds. This is especially true for the localization of sound sources in three dimensions by animals who move about freely in the air or water, where a premium is likely to be placed on vertical as well as horizontal localization.

Two important groups of mammals have evolved the unique ability to induce objects to "make" sounds to reveal their positions. Both bats in air and porpoises in water emit sonar signals that reflect back from objects situated nearby (Griffin, 1958). The positions of objects can then be perceived by localizing the sources of echoes in horizontal and vertical directions and by timing the delay of echoes after emissions to determine distance (Ajrapetjantz and Konstantinov, 1974; Evans, 1973; Murchison, 1980; Nachtigall, 1980; Norris, 1968, 1969; Novick, 1977; Schnitzler and Henson, 1980; Wood and Evans, 1980). The active use of sounds explicitly to achieve spatial perception of objects, rather than the localization of sounds which come from a variety of environmental sources and over whose composition the animal has no control, reveals the mechanisms underlying perception of auditory space and the influences of the acoustic features of signals on sound localization. In particular, the examples provided by echolocation illustrate the important distinction between sensitivity and acuity in directional hearing.

When echolocating, a bat or a porpoise emits brief ultrasonic sounds and listens for echoes of these sounds returning from objects in the vicinity.

The maximum operating range of echolocation is determined by the reflective characteristics of targets and by the propagation of sound through the medium in which the animal lives (Au, 1980; Griffin, 1971; Lawrence and Simmons, 1982a; Murchison, 1980; Pye, 1980). The operating range for spherical targets measuring 1 cm or so in diameter, similar in size to the insects that bats prey upon, is 3 to 5 m (Kick, 1982). For porpoises under water, which unfortunately have proved more difficult to observe in the act of using echolocation for natural tasks, the operating range for detecting steel spheres several centimeters in diameter is 70 to 75 m (Murchison, 1980). Once a target is detected, the animal moves nearer, eventually intercepting the target after tracking it during the approach (Griffin, 1958; Novick, 1977). The behavior of echolocating animals leaves no doubt that the target's location in three dimensions is perceived; echolocating animals usually change the pattern of emission of sonar signals as they move closer to the target, indicating that the target's distance and its horizontal and vertical directions are tracked.

Echolocation provides explicit examples of directional hearing and sound localization in an animal's normal behavior. The sonar signals used by various species of bats and porpoises are *designed to achieve spatial perception* as their sole function, and the perceptual distinction between the directional dependence of detection and the accuracy of localization of targets appears directly in the animal's behavior. The capacity to *detect* a target depends on being able to generate and project a sonar signal that will travel to the target and still be audible on its return to the ears. The capacity to *localize* a target depends on the quite different ability to emit a signal that can convey information about the target's position, relative to the animal's own position, and then to extract that information from the echo after reception by the ears. The directionality of hearing is associated both with sensitivity for the detection of targets in different directions and with the acuity of representation of the target's position from the detailed information extracted from echoes.

Directionality of Echolocation Signals

The functional significance of the directional sensitivity of hearing is illustrated by the directional characteristics of echolocation sounds themselves. The sonar signals of porpoises are clicks—brief acoustic transients containing a broad range of frequencies occurring at the same instant. For example, the echolocation sounds of the bottle-nosed porpoise, *Tursiops truncatus,* contain only two or three waves and have durations of 20 to 250 μs (Ajrapetjantz and Konstantinov, 1974; Au, 1980; Evans, 1973). They contain frequencies from 20 to about 150 kHz. The frequency of peak energy is usually between 50 and 100 kHz, but the spectrum is broad and thus not well represented by this one value. The sonar signals of other toothed whales (Cetacea) are similar in design; they differ primarily in their frequency range. Thus, the clicks emitted by the small Amazon River dolphin, *Inia geoffrensis,* have a broad range of spectral maxima from 50 to 100 kHz, whereas those of the large killer whale, *Orcinus orca,* have a spectral peak around 10 to 20 kHz (Evans, 1973). Although cetaceans have been observed to emit relatively long-duration constant-frequency (CF) sounds as well as clicks (Altes and Ridgeway, 1980; Evans, 1973), there is as yet no direct evidence that these signals are used for echolocation by cetaceans as they are by bats.

The echolocation sounds of bats (Microchiroptera) consist either of broadband, short-duration, frequency-modulated (FM) signals or narrowband, long-duration CF signals. Many species of bats transmit compound sonar sounds containing both a CF and an FM component (Griffin, 1958; Novick, 1977; Schnitzler and Henson, 1980). The FM component commonly contains several harmonics as well. For example, the big brown bat *Eptesicus fuscus* emits FM sounds sweeping from about 55 to 25 kHz in the first harmonic, from about 110 to 50 kHz in the second harmonic, and even some third-harmonic energy from 110 to 75 kHz. The greater horseshoe bat, *Rhinolophus ferrumequinum,* emits a long duration CF signal at 83 kHz followed by an FM sweep from 83 down to about 70 kHz. Bats generally change the signals that they emit in different situations (Pye, 1980; Simmons et al, 1979).

The sounds that echolocating animals project outward into the environment to produce echoes from objects are transmitted in a broad beam toward the animal's front. This beam is aimed primarily by moving the head. In the porpoise, *Tursiops,* the overall strength of transmitted sonar clicks is projected in a beam with a 3-dB width of about 11° in the horizontal plane and 12° in the vertical plane (Au, 1980). (This width specifies the angular size of the beam from one edge to another, with the edges defined as the directions in which the signal strength is 3 dB weaker than in the direction of maximum strength.) The 10-dB width of the *Tursiops* emission beam is about 20° in horizontal and vertical dimensions. In the little brown bat, *Myotis,* which emits FM signals having about as broad a bandwidth as the clicks of porpoises, the emitted sonar sounds have a 10-dB beamwidth of about 60° in the horizontal plane and about 40 to 60° in the vertical plane (Shimozawa et al, 1974). For the big brown bat, *Eptesicus fuscus,* the emitted sound has a 10-dB beamwidth of about 90° at 30 kHz (see Fig. 8-1; Simmons, 1969). The horseshoe bat, *Rhinolophus,* emits its 83-kHz CF signals with a 10-dB beamwidth of about 60° in the horizontal plane and 90° in the vertical plane (Schnitzler and Grinnell, 1977). There is thus a striking difference in the sharpness of the directional beams for the sonar emissions of porpoises compared with bats; bats broadcast their signals over a considerably broader front than porpoises do.

The most important characteristic of the directional sensitivity of *hearing* is already apparent from the fact that the sonar *sounds* of bats and por-

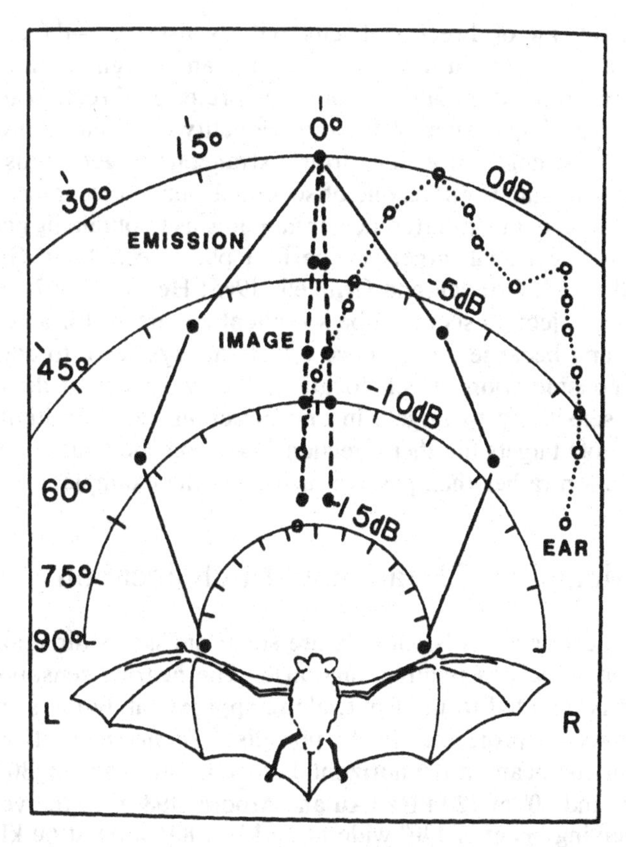

FIGURE 8.1. The directionality of sonar emissions (solid curve), the directionality of the external ear (dotted curve), and the acuity of horizontal-angle discrimination (dashed curve) for the big brown bat, *Eptesicus fuscus*. For discrimination, the bat's performance is expressed in decibels of percent correct responses for comparison with the other curves. The ability of the bat to detect sonar targets from echoes coming from different directions is substantially less directional than the angular resolution provided in the acoustic image that the bat perceives.

poises are beamed in a directional manner. This makes echolocation as a whole a directional process. The sensitivity of an echolocating animal to a *target* is determined by the intensity of the sounds being transmitted, by the reflective acoustic properties of the target and by the propagation of sound to and from the target. A bat or a porpoise is able to detect a target at a greater distance if it is located straight ahead than if it is off to one side because the sound reaching the target to produce an echo is strongest straight ahead. By focusing the energy of the emitted sound in one preferred direction, the animal achieves a longer operating range for echolocation in that direction.

The converse of directional sensitivity is also true—the animal is less sensitive to targets in directions other than straight ahead; therefore, echoes returning from objects not in the preferred direction are relatively weaker. The importance of the directionality of sonar emissions for reducing the strength of echoes from extraneous targets, thus preventing interference, stands out as one of several aspects of echolocation related as much to avoiding interference or jamming as to obtaining good acoustic images of individual targets in the first place (Au, 1980; Griffin, 1958; Grinnell, 1967; Grinnell and Grinnell, 1965; Henson, 1967). Because interference rejection seems to be so salient a concern for an echolocating animal, and because interference effects always seem to come up early in any consideration of echolocation, the two faces of directionality—greater sensitivity to targets in one direction, less sensitivity to (more rejection of) targets in other directions—are somehow more obvious when echolocation rather than passive hearing is the example.

Directionality of Hearing and Echolocation

In echolocating animals not only are sonar emissions directional, but the reception of echoes is directional also. The hearing sensitivity of these animals is directed to the front, along approximately the same axis that the sounds are projected. In the porpoise, *Tursiops*, the 10-dB width of the receiving beam in the horizontal plane is about 100° at 30 kHz, 60° at 60 kHz, and 40° at 120 kHz (Au and Moore, 1984). In the vertical plane the receiving beam is 140° wide at 30 kHz, 80° wide at 60 kHz, and 30° wide at 120 kHz. Thus, porpoises hear sounds having a broader directionality than that of the sounds they broadcast, although at the higher ultrasonic frequencies used for echolocation, the directionality of emissions and of reception become more alike. One significant potential cause of this discrepancy is that porpoises may use their hearing to detect sounds other than echoes over a broader front than for which they detect echoes themselves.

In the horseshoe bat, *Rhinolophus*, the 10-dB directional receiving beam is 60° wide in the horizontal plane and 80° wide in the vertical plane, at the 83 kHz frequency of the bat's CF signals (Grinnell and Schnitzler, 1977). The receiving beamwidths given above were all measured behaviorally. In bats that emit FM sonar sounds, the only data presently available on the directionality of hearing have been obtained using acoustical measurements or physiological rather than behavioral responses. As indicated by evoked potentials in the auditory midbrain, the 10-dB directional receiving beamwidth for *Myotis* is approximately 90° in the horizontal and vertical planes at frequencies predominantly used for echolocation (Grinnell and Grinnell, 1965; Shimozawa et al, 1974). In *Eptesicus* the receiving 10-dB beamwidth is about 75° in the horizontal plane (see Fig 8-1; unpublished data). In other species using FM sounds the directionality both of emissions and of hearing has not been determined.

Since both the sonar emissions and the auditory sensitivity of echolocating animals are directional, the sensitivity of the animal to a sonar *target* will be directional, too. The directional sensitivity to targets is compounded from the directionality of emissions and of hearing. In porpoises the directional beam for the emissions is noticeably sharper than for reception, so the overall directionality of echolocation is largely due to this characteristic of the sounds being sent out. The 10-dB width of the directional beam for echolocation as a whole in *Tursiops* would be about 15°, depending on which frequencies in the emissions were being considered. For the horseshoe bat, *Rhinolophus*, the 10-dB width of the beam of sensitivity to sonar targets is about 40° (Grinnell and Schnitzler, 1977). This refers to the directionality of echolocation at the 83 kHz frequency of the bat's CF signals. In the FM bat, *Myotis*, the combined directionality of emissions and reception results in a directional beam about 30° to 40° wide for sensitivity to targets. In *Eptesicus* the combined directionality (Fig. 8-1) would yield a beam with a 10-dB width of 50°.

Echolocating bats appear to use a substantially broader beam to detect sonar targets than do porpoises. Generally, in bats the directionality of emissions and of reception are similar, whereas in porpoises the directionality of emissions appears conspicuously sharper than that of hearing. Except for the special case of ear movements in bats that emit long CF signals (see below), most bats aim both the emitted sound beam and the receiving beam when pointing the head to track a target. The ears are kept relatively stationary on the head, so the bat steers its whole echolocation system when it moves the beam of emissions and reception in tandem. Porpoises emit their sounds through the front surface of the head and may be able to alter the direction in which emissions are pointed without moving the head (Au, 1980; Evans, 1973). If so, this might account for the relative sharpness of the directional beam for the emissions. Porpoises may also be able to change the directionality of emissions.

We do not know very much about how porpoises use their sonar to perform tasks in their natural environment. Is echolocation used by porpoises to find prey (fish) as it so predominantly is by bats? If echolocation is used for different purposes by porpoises and bats, we might expect the directionality of their echolocation systems to differ. For example, obstacle avoidance requires only that sonar targets immediately along the line of movement be detected; therefore, a relatively sharply directional sonar system would seem appropriate for this task. Searching for prey ought to require a broader directional sensitivity to sample a larger volume of space for potential targets. Insectivorous bats seem to search over a conical zone about 120° wide when they are hunting for prey (Griffin et al, 1960; Schnitzler and Henson, 1980). Bats may depend upon echolocation for more aspects of spatial orientation than porpoises, and the hearing of bats, or at least its directional properties, could well be more intimately bound to the requirements of echolocation than in the case of porpoises. More observations are needed about how different species of bats and porpoises use echolocation to interpret the functions of the directionality of echolocation.

Acuity of Localization of Targets

The accuracy with which a porpoise or bat can determine the horizontal and vertical positions of a sonar target is substantially sharper than the width of the directional beam for detecting sonar targets. The porpoise, *Tursiops truncatus*, can perceive a shift of 0.7° in the horizontal position of a sound source generating clicks similar to echolocation sounds. In the vertical plane the porpoise can detect a shift of 0.9° (Renaud and Popper, 1975). The big brown bat, *Eptesicus fuscus*, can perceive a shift of 1.5° in the horizontal direction of a sonar target (Simmons et al, 1983) and a shift of 3° in the vertical direction of a target (Lawrence and Simmons, 1982b). *Eptesicus* can track a moving target with comparable accuracy (Masters, *et al*, 1985). The width of the bat's acoustic image of a target, determined psychophysically, is shown in Figure 8-1.

To achieve an accuracy for target localization in the region of a degree or two by using a sonar system with a directional sensitivity much broader than that requires echoes to be processed to extract from them the directional information they contain, a process beyond that required simply to detect targets. In effect, an acoustic image showing the target's position in relation to other targets and to the orientation of the animal's body must be derived from the signals received at the two ears. Although the directionality of emissions and reception is likely to figure into the process of determining a target's location, it seems less likely that the signal-processing operations that produce acoustic images representing the target's location depend on the directional sensitivity of echolocation for their success. Directional sensitivity as such probably represents a compromise between the need to search for objects over a relatively broad front and the need to isolate individual objects from extraneous objects that produce interfering echoes. These two constraints are concerned with the detection of targets, whereas the acuity of localization refers to the quality of acoustic images obtained after detection has occurred.

The direction of a target could be perceived using three categories of information in echoes. The first, and simplest, category consists of the intensity of the echo as this is determined by the position of the target in the directional beam of the sonar system. To perceive a target's location solely from echo strength, the animal could move the beam up and down and from side to side around the target, observing the direction of aim that yields the strongest echoes (Reynaud and Popper, 1975). As far as is presently known, this is similar to the method that *Rhinolophus* uses to determine the vertical position of a target (Grinnell and Schnitzler, 1977; Neuweiler, 1970; Schnitzler and Henson, 1980). *Rhinolophus* moves its ears in an alternating motion synchronized to the emission of sonar sounds. These movements sweep the directional receiving pattern for each ear vertically over the location of a target; the alternating movements of the ears have the interesting effect of moving one ear's receiving beam upward past the target while the other ear's beam is moving downward. The emitted sound is beamed broadly to the front, and the bat appears to probe along the vertical axis within the emitted beam to locate the target. Since the horseshoe bat's sonar signals contain long-duration CF components, there is a long enough interval of time during which each echo continues to return for ear-movements to pick up changes in echo intensity as the receiving beam pattern sweeps over the target. The ears of *Rhinolophus* do not move horizontally, only vertically, so it is unlikely that the same means is used to perceive a target's horizontal position. Porpoises move the whole head to scan targets and could determine a target's location from the direction yielding the strongest echoes (Reynaud and Popper, 1975).

The second category of information in echoes that could be used to localize targets consists of differences in the echoes received at the two ears. It is conventional to identify interaural intensity and arrival-time cues as the primary basis for horizontal sound localization (see Mills, 1972, for example), and these cues would be available to an echolocating animal (Schnitzler and Henson, 1980; Simmons et al, 1983). Isolation of intensity and time cues for localization depends on experiments using earphones for delivery of controlled stimuli to the ears, and such behavioral experiments have not yet been conducted on porpoises or on bats. One piece of evidence that bears directly upon binaural localization is the observation that the acuity of horizontal localization of targets by *Eptesicus* can be predicted from other data on the acuity of perception of the time of occurrence of echoes (Simmons et al, 1983). The sonar signals of *Eptesicus* are broadband FM sounds, and they would be ideal for target localization using interaural time cues because their broad bandwidth permits the time of arrival of echoes at the two ears to be determined very accurately. Another observation concerns the effects of plugging one or both of the ears of bats while they fly through arrays of obstacles to be avoided (Griffin, 1958; Schnitzler and Henson, 1980). When only one of the two ears of the horseshoe bat, *Rhinolophus*, is plugged, the bat experiences difficulty avoiding obstacles, but when both ears are plugged by about the same amount, the bat is capable of avoiding obstacles virtually as effectively as if neither ear was plugged (Flieger and Schnitzler, 1973). Evidently the level of sound delivered to the ears must not undergo asymmetrical disruptions if the bat is to localize targets. Taken together, these two results seem to indicate that bats rely on the interaural arrival time and intensity of echoes to perceive target direction. It is to be hoped that experiments in which bats are stimulated with echoes through earphones will be conducted in the near future to sort out the detailed roles of binaural time and intensity cues for localization of targets by bats using FM and CF signals.

In porpoises the task of finding out how binaural cues contribute to determining the location of targets is complicated by the difficulty of knowing the path taken by sounds traveling from the water to the inner ear (Reynaud and Popper, 1975). At ultrasonic frequencies it is likely that the two halves of the lower jawbone act as the "external ears" in porpoises (McCormick et al, 1980; Norris, 1968). In any event, no evidence directly bears upon the relative roles of binaural time and intensity cues delivered to the inner ears after reception through the lower jaws or other sound-collecting sites. It seems reasonable to suppose that both cues are used by porpoises, since binaural intensity discrimination presumably is an easy task (Reynaud and Popper, 1975) and since the sonar signals of porpoises are of great bandwidth and could easily be perceived as having a sharply defined time of occurrence.

The third category of information that might be used to localize targets consists of directional dependence not merely of the intensity but of detailed features of the echo waveform reaching the two ears. This directional dependence has variously been expressed in terms of differences in the spectrum of signals reaching the ears from different directions and differences in the time waveform of signals reaching the ears. Cues of this kind are likely to be especially important for localization of targets in the vertical plane (Diercks, 1980; Fuzessery and Pollak, 1984; Grinnell and Grinnell, 1965; Lawrence and Simmons, 1982b). In bats, the directional dependence of echoes stimulating the cochlea originates primarily in the complex directional response characteristics of the external ear (Fuzessery and Pollak, 1984; Grinnell and Grinnell, 1965). The signals transmitted by bats have different directional beams at different frequencies (Shimozawa et al, 1974), so the waveform of the sound actually impinging upon the target will differ according to the location of the target in the emitted beam pattern. There is thus ample opportunity for the location of a target to be represented in the bat's auditory system either by the spectrum of echoes or by the time structure of the waveform of echoes. The only direct evidence for the role of the external ear in localization of targets is the observation that vertical localization is disrupted in *Eptesicus* if the tragus, an important structure of the external ear, is displaced from its normal position (Lawrence and Simmons, 1982b). In porpoises the signals transmitted in different directions have different spectra (Au, 1980), and the sensitivity of hearing for different frequencies also varies according to the direction of the sound (Au and Moore, 1984). There are thus opportunities for the direction of a sonar target to be represented by the waveform of echoes.

The issue of what cues bats and porpoises use to determine the location of sonar targets is primarily a behavioral one and needs to be addressed in further experiments. It seems essential to develop procedures for separate stimulation of the two ears to manipulate binaural cues and cues conveyed by the waveform of echoes. Such work is beginning in several different laboratories at this time. Theoretical work (Altes, 1978) suggests that the traditional cue-by-cue approach to sound localization might best be replaced by a more comprehensive model to guide research on echolocation.

Acknowledgment. The writing of this chapter was supported by National Science Foundation grant BNS 83-02144.

References

Ajrapetjantz, A.I., Konstantinov, A.I. (1974). Echolocation in Nature. 2nd Edition, Report No. JPRS-63328-1, National Technical Information Service, Arlington, VA.

Altes, R.A. (1978). Angle estimation and binaural processing in animal echolocation. J. Acoust. Soc. Am. 67, 1232–1246.

Altes, R.A., Ridgeway, S.H. (1980). Dolphin whistles as velocity-sensitive sonar/navigation signals. In: Animal Sonar Systems, Busnel, R.-G., Fish, J.F. (eds.). pp. 853–854. New York: Plenum Press.

Au, W.W.L. (1980). Echolocation signals of the Atlantic bottlenose dolphin *(Tursiops truncatus)* in open waters. In: Animal Sonar Systems, Busnel, R.-G., Fish, J.F. (eds.). pp. 251–282. New York: Plenum Press.

Au, W.W.L., Moore, P.W.B. (1984). Receiving beam patterns and directivity indices of the Atlantic bottlenose dolphin *Tursiops truncatus*. J. Acoust. Soc. Am. 75, 255–262.

Diercks, K.J. (1980). Signal characteristics for target localization and discrimination. In: Animal Sonar Systems. Busnel, R.-G., Fish, J.F. (eds.). pp. 299–308. New York: Plenum Press.

Evans, W.E. (1973). Echolocation by marine delphinids and one species of freshwater dolphin. J. Acoust. Soc. Am. 54, 191–199.

Fleiger, E., Schnitzler, H.-U. (1973). Obstacle avoidance ability of the bat, *Rhinolophus ferrumequinum*, with one or both ears plugged. J. Comp. Physiol. 82, 93–102.

Fuzessery, Z.M., Pollak, G. (1984). Neural mechanisms of sound localization in an echolocating bat. Science 225, 725–728.

Griffin, D.R. (1958). Listening in the Dark. New Haven, CT: Yale University Press.

Griffin, D.R. (1971). The importance of atmospheric attenuation for the echolocation of bats (Chiroptera). Anim. Behav. 19, 55–61.

Griffin, D.R., Webster, F.A., Michael, C.R. (1960). The echolocation of flying insects by bats. Anim. Behav. 8, 141–154.

Grinnell, A.D. (1967). Mechanisms of overcoming interference in echolocating animals. In: Animal Sonar Systems: Biology and Bionics. Busnel, R.-G. (ed.). Laboratoire de physiologie acoustique, Jouy-en-Josas, France, pp. 451–481.

Grinnell, A.D., Grinnell, V.S. (1965). Neural correlates of vertical localization by echo-locating bats. J. Physiol. (Lond.) 181, 830–851.

Grinnell, A.D., Schnitzler, H.-U. (1977). Directional sensitivity of echolocation in the horseshoe bat, *Rhinolophus ferrumequinum*. II. Behavioral directionality of hearing. J. Comp. Physiol. 116, 63–76.

Henson, O.W., Jr. (1967). The perception and analysis of biosonar signals by bats. In: Animal Sonar Systems: Biology and Bionics. Busnel, R.-G. (ed.). Laboratoïre de physiologie acoustique, Jony-en-Josas, France, pp. 949–1003.

Kick, S.A. (1982). Target-detection by the echolocating bat, *Eptesicus fuscus*. J. Comp. Physiol. 145, 431–435.

Lawrence, B.D., Simmons, J.A. (1982a). Measurements of atmospheric attenuation of ultrasonic frequencies and the significance for echolocation by bats. J. Acoust. Soc. Am. 71, 585–590.

Lawrence, B.D., Simmons, J.A. (1982b). Echolocation in bats: the external ear and perception of the vertical positions of targets. Science 218: 481–483.

Masters, W.M., Moffat, A.J.M., Simmons, J.A. (1985). Sonar tracking of horizontally moving targets by the big brown bat *Eptesicus fuscus*. Science 228, 1331–1333.

McCormick, J.G., Wever, E.G., Ridgway, S.H., Palin, J. (1980). Sound reception in the porpoise as it relates to echolocation. In: Animal Sonar Systems. Busnel, R.-G., Fish, J.F. (eds.). pp. 449–467. New York: Plenum Press.

Mills, A.W. (1972). Auditory localization. In: Foundations of Modern Auditory Theory. Vol. II. Tobias, J.V. (ed.). pp. 303–348. New York: Academic Press.

Murchison, A.E. (1980). Detection range and range resolution of echolocating bottlenose porpoise (*Tursiops truncatus*), In: Animal Sonar Systems. Busnel, R.-G., Fish, J.F. (eds.). pp. 43–70. New York: Plenum Press.

Nachtigall, P.E. (1980). Odontocete echolocation performance on object size, shape and material. In: Animal Sonar Systems. Busnel, R.-G., Fish, J.F. (eds.). pp. 71–95. New York: Plenum Press.

Neuweiler, G. (1970). Neurophysiologische Untersuchungen zum Echoortungssystem der Grossen Hufeisennase *Rhinolophus ferrumequinum*. Z. Vergl. Physiol. 67, 273–306.

Norris, K. (1968). The evolution of acoustic mechanisms in odontocete cetaceans. In: Evolution and Environment. Drake, E.T. (ed.). pp. 297–324. New Haven, CT: Yale University Press.

Norris, K. (1969). The echolocation of marine mammals. In: The Biology of Marine Mammals. Andersen, S. (ed.). pp. 391–423. New York: Academic Press.

Novick, A. (1977). Acoustic orientation. In: Biology of Bats, Vol. II. Wimsatt, W.A. (ed.). pp. 73–287. New York: Academic Press.

Pye, J.D. (1980). Echolocation signals and echoes in air. In: Animal Sonar Systems. Busnel, R.-G., Fish, J.F. (eds.). pp. 309–353. New York: Plenum Press.

Reynaud, D.L., Popper, A.N. (1975). Sound localization by the bottlenose porpoise *Tursiops truncatus*. J. Exp. Biol. 63, 569–585.

Schnitzler, H.-U., Grinnell, A.D. (1977). Directional sensitivity of echolocation in the horseshoe bat, *Rhinolophus ferrumequinum*. I. Directionality of sound emission. J. Comp. Physiol. 116, 51–61.

Schnitzler, H.-U., Henson, O.W., Jr. (1980). Performance of airborne animal sonar systems: I. Microchiroptera. In: Animal Sonar Systems. Busnel, R.G., Fish, J.F. (eds.). pp. 109–181. New York: Plenum Press.

Shimozawa, T., Suga, N., Hendler, P., Schuetze, S. (1974). Directional sensitivity of echolocation system in bats producing frequency-modulated signals. J. Exp. Biol. 60, 53–69.
Simmons, J.A. (1969). Acoustic radiation patterns for the echolocating bats *Chilonycteris rubiginosa* and *Eptesicus fuscus*. J. Acoust. Soc. Am. 46, 1054–1056.
Simmons, J.A., Fenton, M.B., O'Farrell, M.J. (1979). Echolocation and pursuit of prey by bats. Science 203, 16–21.
Simmons, J.A., Kick, S.A., Lawrence, B.D., Hale, C., Bard, C., Escudié, B. (1983). Acuity of horizontal angle discrimination by the echolocating bat, *Eptesicus fuscus*. J. Comp. Physiol. 153, 321–330.
Wood, F.G., Evans, W.E. (1980). Adaptiveness and ecology of echolocation in toothed whales. In: Animal Sonar Systems. Busnel, R.-G., Fish, J.F. (eds.). pp. 381–425. New York: Plenum Press.

9

Binaural Hearing in Land Mammals

George Gourevitch

Introduction

The benefits of directional hearing for an animal are obvious. In survival situations where the capture of prey or the evasion of a predator is paramount, sound localization can be the key to success. No less is its importance for communication with other members of the species within diverse parental, courtship, and other social contexts.

Until recently only a dearth of comparative behavioral measurements of directional hearing existed (see Gourevitch, 1980). It is gratifying to find that research interest in this area has been expanding in the past few years. Previously unexamined species, ranging from the elephant to the ferret, have been tested, adding to the comparative panorama of sound localization. Furthermore, by means of behavioral auditory lateralization experiments, the characteristics of the mammalian binaural system underlying localization have also continued to be explored. Finally, of pertinence to a comprehensive understanding of sound localization in animals are the recent, novel measurements of acoustic interaural time and intensity differences made on nonhuman mammals.

Interaural Physical Disparities

The extraction of directional information from acoustic cues is primarily a binaural activity; although it can also be achieved monaurally, it is not as efficient (Butler, 1975). Two broad categories of interaural cues convey this information to the organism—interaural time differences and interaural amplitude differences. The first occurs because sound that originates on one side of the listener must follow a longer path and consequently take a greater time traveling to the distal than to the proximal ear. The second results from the interference of the head and ears of the listener with the propagation of the sound so that a "sound shadow" may develop, i.e., the signal may become attenuated on the far side relative to the near one.

In natural settings an acoustic signal typically gives rise to both time and amplitude differences. Under certain conditions only one of the cues retains sufficient magnitude to serve the listener. Moreover, interaural differences may be altered by factors other than the direction of the impinging sound, e.g., head shape. By making direct physical measurements of interaural differences at the ears of the listener for sounds originating at various azimuths, it has been possible to determine what directional information is available to the binaural system.

Interaural Intensity Differences

Physical measurements of interaural cues in animals have been sparse. Until recently, for example, only interaural intensity differences (IID) have been determined directly on animals, and these on only a few small-headed species, e.g., cats (Bismarck, 1967; Wiener et al, 1966; Harrison and Downey, 1970). Since the head becomes obstructive to the propagation of a tone only when its wavelength is of comparable size, these measurements included high frequencies. The common findings were that small or inconsequential IIDs occurred at low frequencies irrespective of azimuth; that significant IID amplitudes appeared at high frequencies, but varied in an irregular pattern with azimuth; and, finally, that large variability was common among individual subjects, especially at high frequencies. (Most likely, this variability reflected the difficulties in controlling precisely experimental conditions, in particular, the exact placement in the sound field of the head and pinnae of individual animals.)

Recently, a developmental approach to these measurements was undertaken on cats (Moore and Irvine, 1979). As would be expected from physical considerations, kittens exhibited IIDs comparable in magnitude to those of adults, but concentrated at high frequencies (10–20 kHz). In adults, interaural time differences (ITDs) of similar or greater magnitude to those observed in kittens were found over a much wider spectrum ranging from 4 to 20 kHz. Thus, in situations where directional hearing was based on IID, the young could be expected to perform as well as adults only for restricted signals.

It should be noted that in all these findings, not unlike those reported for man (Feddersen et al, 1957), the IIDs do not vary in a simple, systematic fashion with azimuth. Even where relatively large IIDs exist, as with some high frequencies, they may remain constant over a wide range of azimuths (45° or more). At other frequencies IIDs can exhibit great variations with the result that within a 15° azimuthal change, multiple peaks and troughs occur. Thus, IIDs by themselves often do not give unequivocal information on the precise direction of tones. On the other hand, they may still indicate the side of the sound source and thus partially contribute to the localization task. Finally, interaural spectral differences that develop when complex acoustic signals occur, particularly those signals consisting of high-frequency components, provide considerable directional information that can enhance the localization task.

Interaural Time Differences

No physical measurements of the relationship between ITDs and azimuth existed for any nonhuman mammal until quite recently. Even for humans the number of such measurements was limited (Hartley and Fry, 1921; Firestone, 1930; Abbagnaro et al, 1975). Following the lead of Kuhn (1977), who made extensive, direct measurements of ITDs on a human model, Roth et al (1980) undertook comparable measurements on cats.[1] Their results showed, as was the case for humans, that over some segments of the audible spectrum ITDs are dependent on frequency. In man, estimates based on a simple geometric model (typically it is applied to the horizontal plane with the head represented by a circle and the velocity of sound assumed to be a constant; Woodworth, 1938) approximate direct ITD measurements at high but not at low frequencies (Fig. 9-1). In cats the discrepancies between the Woodworth model estimates and directly observed ITDs are considerably greater and exist over a wider spectral range (Fig. 9-2). For example, at a low-incidence angle (15°) the observed ITD for a 400 Hz signal is three times greater than the ITD based on the Woodworth model. Even at high frequencies, e.g., near 6 kHz, where the observed ITDs in man are about the same as predicted by the model, those in cats are more than twice as large (95 μs *v* 43 μs). These disparities occur at all angles of incidence, at all frequencies, and are especially large and stable at low frequencies. Thus, calculation of ITDs based on the ubiquitous geometric model in which velocity of sound is independent of frequency may be, in some instances, inappropriate.

Since for a given angle of incidence any ITD derived from the geometric model depends on the magnitude of the interaural distance, the finding

[1]The ITDs measured in both of these studies were based on the steady-state or phase velocities of the acoustic signals, that is, the propagation velocity of a fixed point of the wave, e.g., the crest. Although in many instances the velocity of sound is constant, under certain conditions, as when a sound encounters an obstruction in its path, dispersion results; that is, the velocity of the sound no longer remains independent of frequency. Thus, under dispersive conditions a compound wave resulting from a burst of a tone or of a complex sound such as noise will lose its spatial distinctiveness and in time spread out. Two different velocities will emerge, the group velocity at which the compound wave propagates and the phase velocity at which individual components propagate. One way to distinguish these two velocities is to imagine that a moving caterpillar has a group velocity, while the ripples traveling along its back have a phase velocity (Stephens and Bate, 1966). Under certain conditions the two velocities may differ appreciably so that an additional localization cue could be available to the listener (Roth et al, 1980). (For an extensive analysis of ITDs see Chapter 1.)

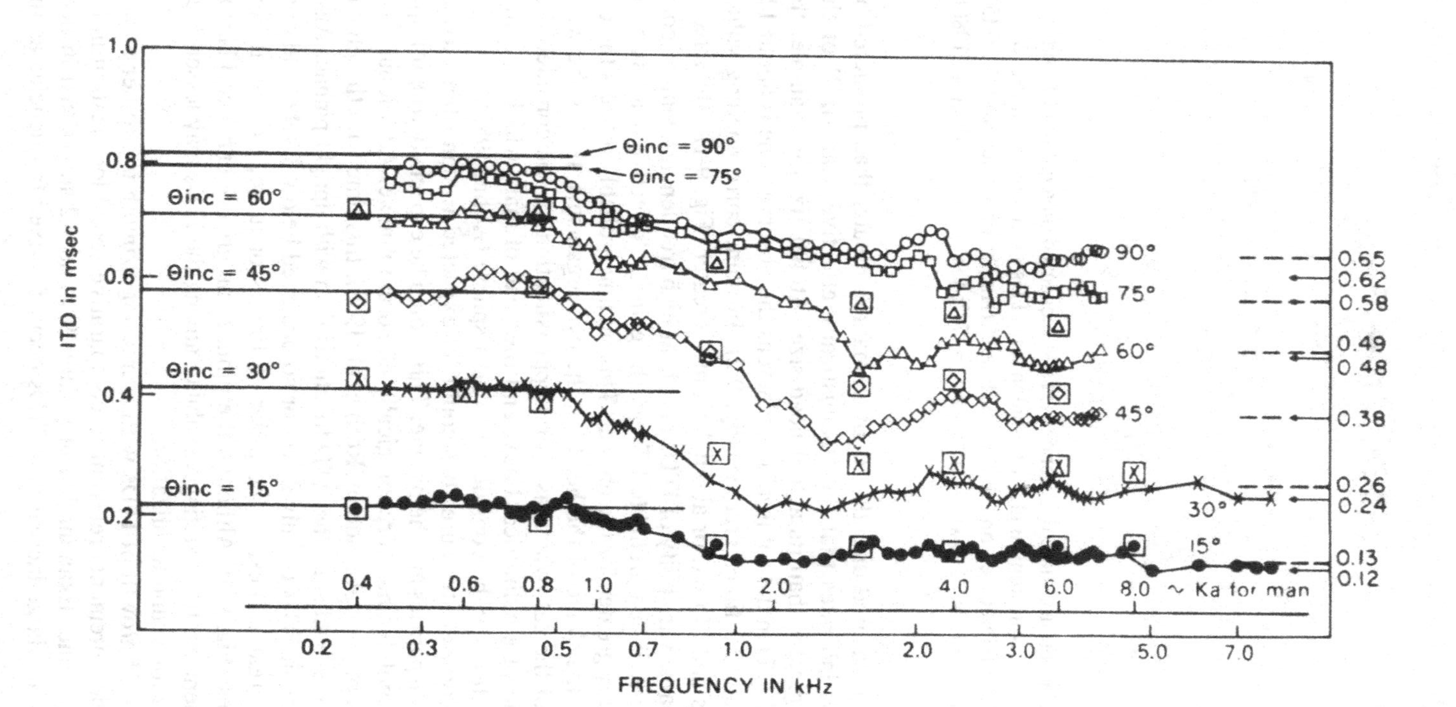

FIGURE 9.1. Interaural time differences (ITDs) measured on a manikin at different frequencies and for different angles of incidence. Dashed lines on the right ordinate indicate ITDs calculated from the Woodworth model (Kuhn, 1977).

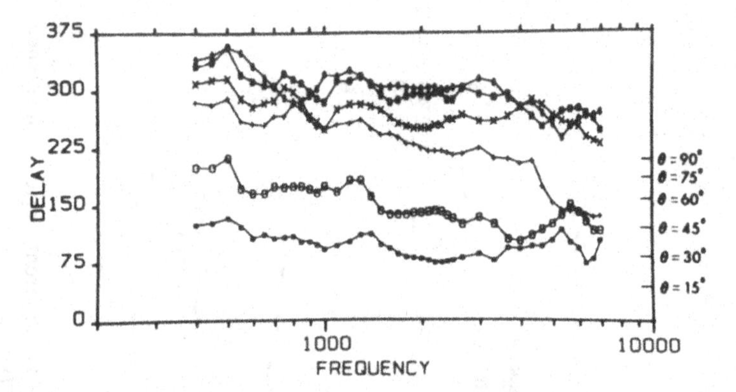

FIGURE 9.2. Interaural time differences (ITDs) measured on a cat at different frequencies and for different angles of incidences. ITDs calculated from the Woodworth model are indicated on the right ordinate. (Angles of incidence: *** = 15°; ooo = 30°; + + + = 45; xxx = 60°; ### = 75°; $$$ = 90° (Roth et al, 1980).

that ITDs measured directly on an animal are greater than predicted by the model, indicates that the "acoustically effective head size" of the animal is greater than its actual head size. At 400 Hz, for example, the calculated ITD in the cat is 48 μs, whereas the observed one is about 129 μs, or about 2.7 times greater (Fig. 9-2). In man, corresponding values are 120 μs and 215 μs, or about 1.8 times greater (Fig. 9-1). (Because of the noticeable variability of ITDs in cats at high frequencies, similar comparisons would be unreliable.) Thus, the gain in "effective head size" is considerably greater for the animal. This finding probably reflects the difference in the shape of the heads of the two organisms, that is, between the head of the cat, which as a first approximation can be considered to be one end of a cylindrical body, and the head of man, which as a first approximation can be considered to be a sphere (Kuhn, 1983).

An anatomical structure that could be expected to contribute to the observed disparities are the pinnae. This would be especially true for species with mobile and morphologically elaborate pinnae. Additional measurements on the cat made by Roth et al (1980), this time with the pinnae retracted, revealed smoother ITD functions than with upright pinnae. Also, decrements in ITDs of only 25 μs or so occurred below 3 kHz, and considerably greater ones, in the order of 100 μs or more, existed at high frequencies (Fig. 9-3). Although the pinnae can greatly enhance ITDs at high frequencies, the resulting variability among the ITDs may lessen their usefulness for sound localization.

The rate of growth of ITDs with increasing azimuth is another aspect of these measurements relevant to comparative sound localization in the horizontal plane. Examination of Figures 9-1 and 9-2 shows that in both man and cats, large increases in ITDs occur for fixed frequencies as the

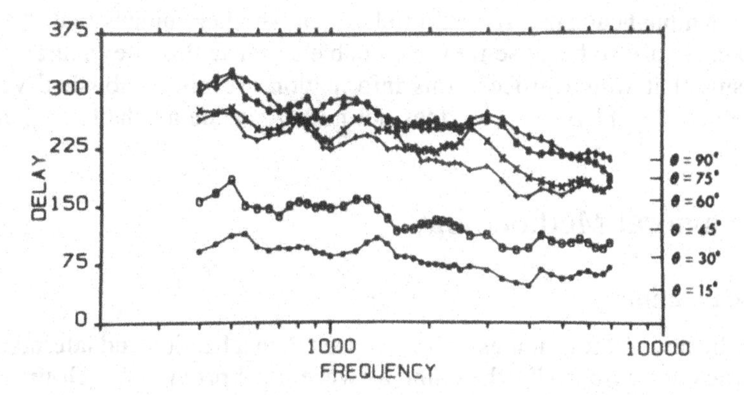

FIGURE 9.3. Interaural time differences (ITDs) measured on a cat with its pinnae retracted. ITDs calculated from the Woodworth model are indicated on the right ordinate: xxx = 15°; ooo = 30°; + + + = 45°; xxx = 60°; ### = 75°; $$$ = 90° (Roth et al, 1980).

angle of incidence changes from the midline to 45°, whereas at higher angles of incidence the growth is very much reduced. The rate of change of ITDs, however, is smaller for the cat than for man. Between 15° and 30°, for example, the ITD for a 400-Hz signal increases by a factor of 1.6 in the cat and by a factor of 1.9 in man. (Similar comparisons at high frequencies are prevented by the irregular changes of ITD with azimuth in the cat.) Thus, it appears that whenever the differential ITD cue is used in localizing sound, e.g., with small-head rotation, its effectiveness is greatest for sources situated near the midline and for signals of relatively low frequency.

From the preceding observations, although only based on two head shapes, it is clear that ITDs are significantly modulated by the physiognomy of the organism. It remains to be seen how ITDs behave with other head shapes, e.g., conical, and with more prominent pinnae.

Although it is clear that ITDs of considerable magnitudes are available not only to man, but also to smaller mammals, there are questions as to how this cue can serve sound localization unambiguously. The uncertainty is most extreme for sound originating at different sites well to the side of the listener, 60° and above, since identical ITDs are commonly found among the different azimuths. Furthermore, to the extent that the magnitude of this cue is frequency dependent, different tones originating from the same site would seem to come from different sites. Thus, ITDs cannot serve as a dependable localization cue for all sounds. Yet, extensive spectral segments exist, especially at middle to low frequencies and at small angles of incidence, where ITDs are independent of frequency and can convey unequivocal directional information. To the extent that the animal uses this cue, he should localize with much greater acuity than estimated from simple head size. It is still to be seen whether animals that are poor localizers are so because they are unable to sense this cue (neural mechanisms that would process this information are not established yet) or whether their physiognomy does not make this cue available to them.

Behavioral Methodology

Localization

The behavioral techniques currently used in localization and lateralization studies are essentially the same as were used previously (Gourevitch, 1980). Although some earlier studies of sound localization used an approach-to-target procedure requiring that the animal proceed from a reference position to the sound source (Casseday and Neff, 1973), most recent investigations used nonlocomotive behavioral procedures.

In the typical arrangement the animal was confined to one location, where it indicated the occurrence of a change in the locus of a sound source (yes-no task; Brown et al, 1980) or the side of the sound source (two-choice task; Heffner and Heffner, 1982, 1984). Positive reinforcement (food or water) was used in these studies. Presentation of stimuli and estimation of thresholds followed classical psychophysical methods such as constant stimuli or a modified method of limits.

A conditioned avoidance technique has also continued to be used in assessing sound localization acuity in mammals (Heffner and Heffner, 1984, 1986). In this procedure a water-deprived animal was trained to lick a spout continuously during which time trials would occur at random. A "safe" trial consisted of brief acoustic signals presented at the rate of 5/sec from a right speaker. A "warning" trial consisted of acoustic signals that were always presented from the left speaker; unless the animal stopped licking the spout during the warning trial, an electric shock followed. This paradigm resulted in the cessation of licking if the animal heard the warning signal. Presentation of "safe" and "warning" trials was randomized. A suppression ratio was devised as a threshold index (Sidman et al, 1966; Smith, 1970). A form of this ratio is given by (S-W)/S, where S is the average time spent licking the spout during safe trials, and W is the average time spent licking the spout during the warning trials. A suppression ratio of 0.5 was usually taken as the threshold (Heffner and Heffner, 1984). Presentation of the stimuli followed a modified method of limits, e.g., the speaker separation was always decreased.

Comparison of directional hearing measurements obtained on animals tested with the avoidance technique with those obtained on many other mammals who were tested with other procedures presents some difficulty since the threshold estimates were derived so differently (percentage correct *v* suppression ratio). It should be noted that the avoidance technique (suppression ratio) and the two-choice technique (percentage correct) applied to the same species yielded comparable thresholds that did not, however, overlap (Heffner and Heffner, 1984).

Since the last review of directional hearing in mammals (Gourevitch, 1980), the principal lateralization studies were conducted on monkeys. The procedure in these experiments was the same as employed in some earlier binaural studies on monkeys (Houben and Gourevitch, 1979). The monkey wore earphones while sitting in a restraining chair. By pressing a center button in front of him the monkey would initiate a trial. The trial consisted of a diotic signal, followed by a silent interval, and then by a dichotic signal. The side of lateralization of the dichotic signal was identified by pulling one of two laterally situated levers. Stimuli were presented according to the method of constant stimuli. Psychometric functions obtained on each animal and for all experimental conditions typically extended from below 60% to above 90% correct. The threshold was taken as the stimulus value corresponding to 75% correct lateralization.

Sound Localization

Large Mammals

A new aspect of directional hearing in animals has been investigated recently, namely, the sound localization acuity of mammals whose heads are of similar size or larger than that of man. The ability of small-headed animals to localize sound, until now the only animals that have been studied, is typically poorer than man's (Gourevitch, 1980). Since these animals have small heads, both ITDs and IIDs are much smaller for comparable signals and azimuths than for man. Consequently, poorer performance would be expected for these animals on the basis of head size. The contrary would be expected for larger headed animals.

Elephants

Heffner and Heffner (1982), using a two-choice procedure, were able to determine the characteristics of sound localization in an Indian elephant (*Elephas maximas*). (The great difficulty in acquiring such animals for research understandably resulted in only one animal being tested. The findings, therefore, are tentative.)

The elephant was trained to initiate a signal from one of two speakers by pressing the center button with its trunk. Pressing the button on the side of the activated speaker was rewarded. To obtain thresholds the speakers were moved toward the 0° azimuth until the animal was unable to distinguish whether the sound emanated from the left or right speaker.

An unexpected finding emerged from a localization screening test in which tones were presented at random from one of two speakers 60° apart,

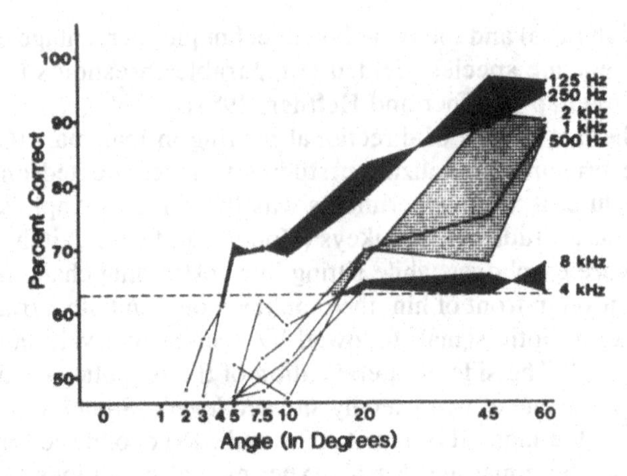

FIGURE 9.4. Pure-tone localization performance by an elephant. (From Heffner and Heffner, 1982; Copyright 1982 by the American Psychological Association. Reprinted by permission.)

30° on each side of the midline: At 4,000 Hz and 8,000 Hz, the elephant was unable to make the discrimination. A minimum audible angle (MAA) function obtained on the elephant confirmed its inability to localize high-frequency tones (Fig. 9-4). In general, the elephant had considerable difficulty localizing sinusoids. Its best performance was at 250 Hz; the MAA was about 13°. At other frequencies, ranging up to 2,000 Hz, MAAs were vastly greater, extending to about 50° at 500 Hz.

The ability of the elephant to localize sound improves considerably when the acoustic signals are not sinusoids, but rather complex sounds such as clicks, broadband and narrow bands of noise. The MAAs, in these instances, range from 1° to approximately 5° (Heffner and Heffner, 1982). An interesting aspect of these findings is the accurate localization of noise bands centered at high frequencies, e.g., 4 kHz and 8 kHz (Fig. 9-5). The apparent inability of the elephant to utilize intensity differences at high frequencies suggests that a time difference cue may be mediating directional information. ITDs are available in the envelopes of the noise signals and are detectable by man (Henning, 1974; McFadden and Pasanen, 1976) and by the monkey (Houben and Gourevitch, 1980). The Heffners concluded that this cue mediates directional information for such signals. Whether the elephant is sensitive to the ongoing time differences at high frequencies is still uncertain, since no auditory lateralization studies have been conducted on the elephant in which this cue was tested in isolation. It is important to note, however, that filtered noise centered at high frequencies possesses low-frequency components, albeit considerably attenuated, that carry interaural temporal disparity information. It was shown in man that despite the low intensity of these components, they

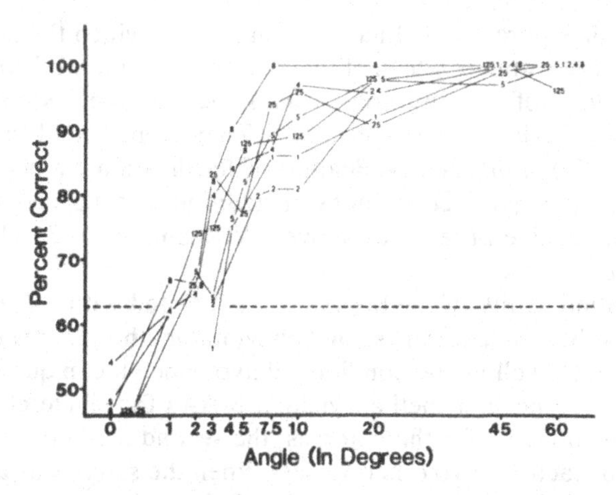

FIGURE 9.5. Noise band localization performance by an elephant. The center frequency of the noise band is indicated for each function; roll off from center frequency at 24 dB/octave. (From Heffner and Heffner, 1982; Copyright 1982 by the American Psychological Association. Reprinted by permission.)

could be effective in mediating the time differences in the waveform of the high-frequency noise bands (Bernstein and Trahiotis, 1982). Thus, even if the elephant was incapable of discriminating time differences of high-frequency signals, its ability to discriminate time differences at some low frequencies, as indicated by its accurate localization of such tones, could serve to localize the high-frequency band-passed signals.

The source for the poor performance in localizing middle- and high-frequency tones is uncertain. The size of the animal implies that the magnitude of interaural differences is great. However, no measurements are available to establish the relationship of these cues with azimuth and the extent of any ambiguities that may exist. Furthermore, the binaural sensitivity of this animal to these cues is also unknown. A great deal of additional information is necessary before localization in the elephant can be well understood.

Ungulates

In contrast to elephants, the interaural distance of the other large mammals whose localization acuity has been examined recently, i.e., horses and cows, is approximately the same as man's (Heffner 1981; Heffner and Heffner, 1984, 1986). Although the shapes of the heads differ among these mammals and, furthermore, the animals have large and mobile pinnae, comparable localization performance could be expected between man and the ungulates.

Using a similar behavioral technique to the one they applied to the elephant, the Heffners found that both cows and horses were generally unable to localize pure tones. In a screening test in which the animals were required to discriminate which of two speakers in front of them, 60° apart, was the source of the pure-tone signals, neither species could achieve 75% correct localization above 1 kHz. They were tested up to 20 kHz (Heffner, 1981). Further examination of localization by horses with an identical screening procedure and with additional animals confirmed their inability to localize pure tones above 1 kHz and up to 25 kHz (Heffner and Heffner, 1986).

Localization acuity of the horse was also tested with a 1-ms click and with a broadband noise. The same behavioral method that was used with the elephant as well as the conditioned avoidance technique was applied in this study. The first method yielded MAAs for single clicks ranging from 31° to about 50° for three horses; the second method yielded MAAs of about 30° each for two other horses. When the signal was a broadband noise instead of a click, the MAA threshold with the first method was about 19° and with the second method near 25° (Heffner and Heffner, 1984). Thus, ungulates, in contrast to elephants, do not exhibit accurate sound localization, even when noise signals are used.

The results obtained from large animals raise more questions than they answer. For example, an expectation would be that even if the binaural sensitivity of these animals, in particular the elephant, to interaural difference cues was considerably less than observed in man, localization accuracy would be comparable (Gourevitch, 1980). Yet, none of these animals localized most pure tones accurately, except at some low frequencies, even when the tones belonged to the most sensitive interval of their audiograms (Heffner and Heffner, 1982, 1983). At this time it appears that sound localization in large animals as a group is especially deficient at high frequencies. Much additional study, however, is required before their directional hearing can be well understood.

Small-Headed Animals

Sound localization has been examined in a heretofore untested carnivore, the domestic ferret (*Mustela putorious*), a member of the weasel family. A left *v* right procedure was used in this study; it was similiar to the one applied to elephants (see above) except that after initiating a trial from a reference position, the animal had to proceed for reinforcement to the speaker that was activated (Kelly and Kavanagh, 1984). Descending series of angular speaker separation were used for threshold testing. The mean MAA obtained on two animals with single clicks was 17°.

The ferret's localization acuity is considerably poorer than the MAA threshold for noise signals of about 5° that were determined for the cat (Casseday and Neff, 1973) and for clicks of 4° to 8° that were found in dogs (Heffner, 1978; Heffner and Heffner, 1984). However, in comparison to the red fox (Isley and Gysel, 1975), the ferret's performance is better.

These results appear to reflect the ferret's head size, which is greater than that of the red fox, but smaller than that of the other carnivores.

Another previously unexamined species on which sound localization thresholds were determined recently is the wild Norway rat (Heffner and Heffner, 1985). The conditioned avoidance technique was used to assess MAAs for 100-ms bursts of broadband noise (100 kHz) and for single 25-μs clicks. The thresholds for three animals ranged between 11° and 13.5° for the noise and 12.5° and 15° for the clicks. These results, although showing somewhat lower thresholds than some earlier reports (Kelly and Glazier, 1978), agree in general with previous findings that many small rodents, including albino rats (Kelly, 1980), are not proficient localizers of sound.

Primates

The mammals whose directional hearing has been examined in great detail during the past few years are Old World monkeys. Previous studies had mostly examined auditory localization and auditory lateralization sensitivity mostly for pure tones (see review, Gourevitch, 1980). More recently, binaural hearing in nonhuman primates has been investigated with complex acoustic signals, e.g., noise, vocalizations.

Localization in the Horizontal Plane

A most extensive examination of MAA thresholds for noise signals was conducted by Brown et al (1980) on two species of Old World monkeys (*M. nemestrina* and *M. mulatta*). The behavioral procedure consisted of having the monkey sit in a chair, which was situated in an anechoic room, and identify a change (yes-no) in the source of the signal. Presentation of the signal locus was according to the method of constant stimuli. The signals consisted of various bandwidths of noise centered at different frequencies, octaves apart, ranging from 250 to 16,000 Hz.

They found that monkeys localized wide-band signals accurately throughout most of the tested spectrum (Fig. 9-6). Localization improved with increasing bandwidth at all frequencies except about 1,000 Hz, where bandwidth change did not alter the accuracy of localization. The lowest MAA thresholds that were obtained occurred at the widest noise bands and were all lower than those obtained with pure tones (Fig. 9-7), except at 1,000 Hz (Brown et al, 1978). The gain in localization accuracy with noise was particularly significant above 2,000 Hz, as was the reduction in variability among animals. At the high frequencies, MAAs were about 5° with wide-band noise, whereas with tones they ranged from 10° to 17°.

Thus, the pattern of directional hearing in monkeys changed significantly when tested with complex signals rather than with sinusoids. The most striking form of this change was in the overall shape of the MAA functions.

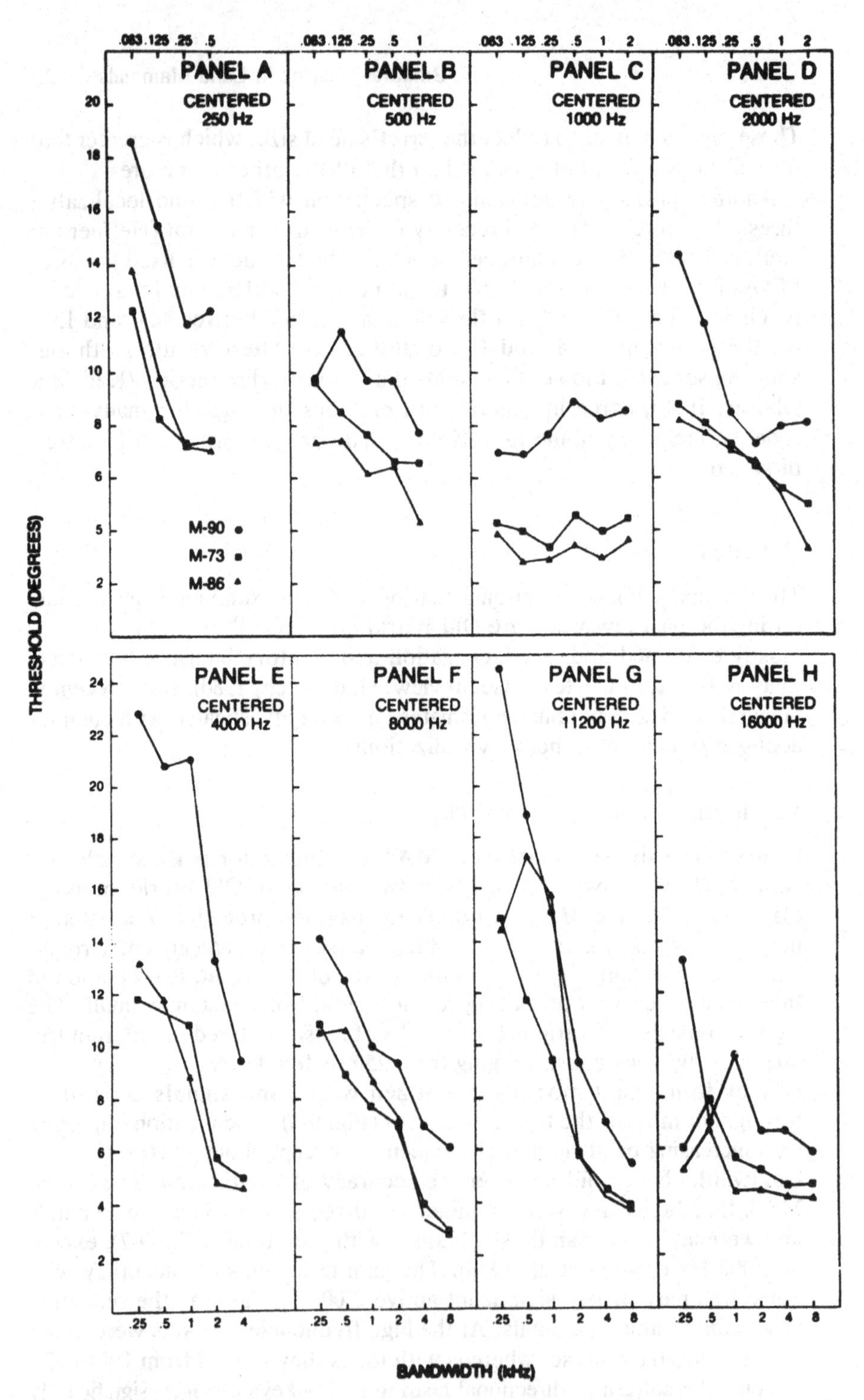

FIGURE 9.6. Minimum audible angle thresholds for bands of noise in Old World monkeys (Brown et al, 1980).

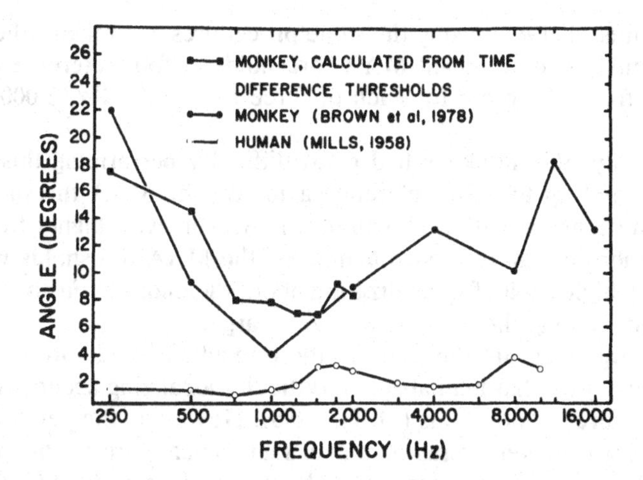

FIGURE 9.7. Minimum audible angles in Old World monkeys and in man (Houben and Gourevitch, 1979).

As can be seen in Figure 9-7, the MAA function for pure tones is V shaped, with the best performance at 1,000 Hz. On the other hand, the MAA function for the widest noise bands decreases from 250 to 1,000 Hz and remains relatively constant thereafter; the overall shape of the function is that of an "L" with a slanted leg. At higher frequencies the function is indeterminate. (Audibility for these animals ceases about an octave above the highest frequency shown in Fig. 9-6; hence, localization accuracy for signals near this limit might decrease rapidly.) It should be noted that with noise bandwidths other than the widest ones, the MAA function retains the general V shape it has for pure tones. Unless the complex signal is broadband, the greatest accuracy for localization seems to be confined to a moderately narrow spectral interval in the neighborhood of 1,000 Hz.

In man there also appears to be an improvement in sound localization when complex signals are used instead of sinusoids. However, this improvement is relatively small. MAAs for pure tones are about 1° to 2° over a wide spectral range (Mills, 1958); for wide-band clicks they are about ¾°, and for high-pass clicks about 1.5° (Banks & Green, 1973). Thus, an important difference exists between the way sound is localized by man and by nonhuman primates. Over a wide part of his audible spectrum man attains a barely changing limit for his localization acuity irrespective of the type of signal or its spectral locus. Nonhuman primates, on the other hand, have distinctly different acuity limits for sinusoids and narrow bands of noise than they do for wide bands of noise.

Lateral Sound Localization

Another aspect of horizontal sound localization examined in animals is "lateral sound localization," i.e., the ability to discern a change in the location of a sound source from one lateral position to another.

Brown et al (1982), using the same procedures as in their other studies with monkeys, determined MAA thresholds at four reference azimuths ranging from 15° to 60° for each of three tones (500 Hz, 2,000 Hz, and 8,000 Hz).

Generally, the monkeys had great difficulty performing this task. At 500 Hz, and up to a 30° reference azimuth, the MAA thresholds were about the same as with a 0° azimuth. However, with higher frequencies and at more lateral reference azimuths, the MAA thresholds were considerably higher than for localization about the midline. Furthermore, the variability among the animals was very large.

To a large measure the trend of these results agreed with those found in humans, i.e., that man also had difficulty localizing accurately around lateral reference positions (Mills, 1958). However, the performance of monkeys was poorer overall than man's and much more erratic, preventing detailed comparisons, except at the lowest frequency. (See Chapter 1 for comparison of human and monkey thresholds at 500 Hz).

It appears that primates cannot discriminate shifts in the directionality of lateral sounds as effectively as they can frontal ones. Whether this feature of horizontal sound localization applies to other species, especially those whose vision is not frontal, remains to be seen. Such information will help to better understand the role of sound localization with that of vision in identifying the position in space of targets of interest.

Localization in the Vertical Plane

Except for man, investigations of sound localization in terrestrial mammals have been confined to the horizontal plane. Localization in the vertical plane by semiarboreal nonhuman primates (macaques) was examined recently (Brown et al, 1982). The method used in this study was the same as for the investigation of horizontal localization. The acoustical signals presented in this study were macaque vocalizations and bands of noise. The results showed that two vocalizations, a "harsh call" with a spectrum ranging from 200 Hz to 2 kHz, and a "clear call" with the fundamental band centered at 820 Hz, and harmonic bands centered at 1,640 Hz and 2,460 Hz, were localized less accurately in the vertical than in the horizontal plane by a factor of about 2.5. Determination of vertical localization acuity with noise bands indicated that performance was as accurate as in the horizontal plane, that is, about 4°, for the widest noise band (125–16,000 Hz). Decrease of the upper cutoff frequency to 8 kHz barely increased the MAA thresholds; but by the time the upper cutoff frequency was reduced to 2 kHz, one animal could not localize the sound at all, and the thresholds for the other two had risen by more than threefold.

Thus, both monkeys and man require high frequencies for localization in the vertical plane (Butler, 1969; Hebrank and Wright, 1974). It appears, however, that monkeys are somewhat more effective than man in using high-frequency information. The vertical and horizontal thresholds of monkeys are about the same for broadband noise (10 kHz), whereas in humans, vertical thresholds are a little more than two times greater than the horizontal thresholds (Hausler et al, 1983).

Lateralization

Sensitivity of the binaural system of Old World monkeys to interaural differences for pure tones was examined previously by Wegener (1974) at one frequency, and by Houben and Gourevitch (1979) over a wide spectral range. These studies showed that ITD (phase) thresholds were considerably higher than those of man. At their lowest thresholds, found to be between 750 and 2,000 Hz, monkeys were about 35 μs less sensitive than man. Moreover, at low frequencies the disparity between man and monkeys increased significantly. It was 120 μs at 250 Hz, the lowest tested frequency. This difference reflects the fact that the drop that occurs in sensitivity to time differences as frequency is decreased, begins for the monkey at a higher frequency than for man. A sharp rise in time difference thresholds for monkeys takes place between 750 and 250 Hz, with no corresponding increase in human thresholds. Only at lower frequencies, i.e., 125 Hz or less, do humans exhibit similar threshold elevation (Klumpp and Eady, 1956).

Houben and Gourevitch (1979) also found close agreement between the MAAs calculated from ITD thresholds and those measured experimentally, indicating that the falling leg of the V-shaped MAA function resulted, in large part, from the way ongoing time differences were processed by the monkey's binaural system (see Fig. 9-7).

More recently, noise signals have been used to investigate processing of ITDs by the binaural system of nonhuman primates. Of the variables affecting binaural time differences sensitivity, signal duration and bandwidth were examined. The method used in these studies is described above (Behavioral Methodology).

As with man (Tobias and Zerlin, 1959; Houtgast and Plomp, 1968), longer signal durations resulted in lowering time difference thresholds for monkeys (Houben and Gourevitch, 1982). However, two significant differences between these two primates emerged. First, the lowest time difference threshold that the monkey attained was about 23 μs in contrast to less than 10 μs in man. Second, the duration over which the monkey could extract time difference information was about 125 ms. At longer signal durations, extending to 500 ms, the thresholds remained relatively constant. In man this process lasts a second or so, although, it becomes less and less efficient for the longer durations (Hafter et al, 1979). Thus, the duration effects in monkeys, although considerable, are not as extreme as in man, both in the time that these effects last and in the sensitivity to ITDs.

The other aspect of binaural processing examined recently in monkeys was that of ongoing time differences in noise bands centered at several points along the low and midrange portions of the spectrum (Gourevitch and Houben, 1983). This study used bands of noise, 50 to 800 Hz wide, and centered at 500 Hz, 1,000 Hz, and 2,000 Hz. The behavioral procedure was the same as in the preceding study. ITD thresholds were found to decrease with the bandwidth for noise centered at 500 Hz, but not decrease appreciably for bands centered at about 1,000 or 2,000 Hz. Moreover, these thresholds were considerably lower than those obtained with pure tones at corresponding center frequencies. For the widest noise band (800 Hz) the thresholds were about 24 μs at about 500 Hz in contrast to 100 μs for a 500-Hz tone (Houben and Gourevitch, 1979).

These results are reflected in the pattern of the MAA functions obtained by Brown et al (1980) for noise bands along similar spectral segments. In particular, no bandwidth effect on MAAs occurred at about 1,000 Hz. Moreover, the MAAs were smaller for all wide-band noise signals than for pure tones; and in many instances, the same was true even for the narrowest bandwidths (Figs. 9-6 and 9-7).

The close correspondence between the results obtained from lateralization and localization studies suggests that monkeys have a very sensitive region in the neighborhood of 1 to 2 kHz in which they localize sound most accurately by means of interaural time processing. At higher frequencies the localization performance may also be mediated, in part, by a time difference cue, namely the ongoing time difference in the envelope of the noise (Henning, 1974; McFadden and Pasanen, 1976). However, measurements of ITD thresholds for noise signals at high frequencies, i.e., above those where the monkey can detect ongoing differences in sinusoids (2,000 Hz), indicate that the monkeys are considerably less sensitive to time differences in this spectral region than in lower ones; thresholds no lower than 60 μs were reported (Houben and Gourevitch, 1980; Gourevitch and Houben, 1983). Thus, it does not appear that this cue serves the localization of wide-band noise signals at high frequencies, since such signals are well localized and would require low time difference thresholds. This cue may still contribute, in part at least, to the localization of narrow high-frequency noise bands, since these are localized much less accurately and would not depend on very fine time discriminations.

It is obvious that considerable differences in directional hearing exist among primates. The most consistent one is the lower sensitivity of nonhuman primates in comparison to man in all binaural hearing discriminations. The weaker performance of the monkey in tasks such as sound localization is understandable, since it is derived, in part, from the lower sensitivity of the monkey's binaural system to interaural cues. On the other hand, the sources of the lower sensitivity to interaural differences are not as evident at this time and requires further investigation.

Conclusion

A striking characteristic of the recent research on directional hearing in nonhuman mammals is that all the measurements support the observation made from earlier studies: directional hearing in animals is not as precise as in man (Gourevitch, 1980). That is not to say that in some species the difference between their best localization performance, or their finest discrimination of interaural disparities, and that of man is not quite small; however, even in these instances, man's performance remains the most acute.

It also appears that the range of stimulus values over which the "best" performance occurs, whether in localizing pure tones or in discriminating time differences, etc., is considerably wider in man than in animals. Thus, the directional hearing of man continues to be foremost among land mammals, a characteristic for which reasons are not obvious.

Lately, physical measurement of interaural differences, a neglected, and yet fundamental aspect of mammalian directional hearing, is being pursued. As a consequence of some of the measurements, the possibility has been raised that the shape of a small-headed animal may be a significant factor in incrementing the accuracy of its localization performance. Although the size of the animal is the principal determinant of the magnitude of interaural cues, clearly other aspects of the anatomy, e.g., particular head shape, pinna mobility, can modify the magnitude of these cues. The extent to which these factors appear in some species and not in others may underlie some of the differences in localization acuity between animals. Thus, physical measurements of interaural cues for the major groups of head size and shape are needed for a better understanding of comparative directional hearing.

Finally, current work, as well as previous behavioral work, showing important differences between man and animal in the sensitivity and response patterns of their binaural systems to interaural cues, suggests that physiological processing may be another differential contributor (not as significant as head size, but probably as much as physiognomy) to the diversity of sound localization performance among terrestrial mammals. For a more accurate estimate of the extent of the variability among mammals in processing binaural cues, which at present rests primarily on lateralization experiments in human and nonhuman primates, further study in animals is needed.

Acknowledgments. Preparation of this chapter was supported in part by National Institutes of Health (NIH) grant RR 08176. The studies of lateralization in monkeys were supported by grants from the National Science Foundation (BNS-7915834) and NIH (RR 08176).

References

Abbagnaro, L.A., Bauer, B.B., Torick, E.L. (1975). Measurements of diffraction and interaural delay of a progressive sound wave caused by the human head-II. J. Acoust. Soc. Am. 58, 693–700.

Banks, M.S., Green, D.M. (1973). Localization of high- and low-frequency transients. J. Acoust. Soc. Am. 53, 1432–1433.

Bernstein, L.R., Trahiotis, C. (1982). Detection of interaural delay in high-frequency noise. J. Acoust. Soc. Am. 71, 147–152.

Bismarck, G.V. (1967). The sound pressure transformation function from free-field to eardrum of chinchilla. MS thesis. MIT. Cambridge, Mass.

Brown, C.H., Beecher, M.D., Moody, D.B., Stebbins, W.C. (1978). Localization of pure tones by Old World monkeys. J. Acoust. Soc. Am. 63, 1484–1492.

Brown, C.H., Beecher, M.D., Moody, D.B., Stebbins, W.C. (1980). Localization of noise bands by Old World monkeys. J. Acoust. Soc. Am. 68, 127–132.

Brown, C.H., Schessler, T., Moody, D.B., Stebbins, W. (1982). Vertical and horizontal sound localization in primates. J. Acoust. Soc. Am. 72, 1804–1811.

Butler, R.A. (1969). Monaural and binaural localization of noise bursts vertically in the median saggital plane. J. Aud. Res. 3, 230–235.

Butler, R.A. (1975). The influence of the external and middle ear on auditory discriminations. In: Handbook of Sensory Physiology, Vol. V/2. Keidel, W.D., Neff, W.D. (eds.). pp. 247–260. New York: Springer Verlag.

Casseday, J.H., Neff, W.D. (1973). Localization of pure tones. J. Acoust. Soc. Am. 54, 365–372.

Feddersen, W.E., Sandal, E.T., Teas, D.C., Jeffress, L.A. (1957). Localization of high frequency tones. J. Acoust. Soc. Am. 29, 988–991.

Firestone, F.A. (1930). The phase difference and amplitude ratio at the ears due to a source of pure tone. J. Acoust. Soc. Am. 2, 260–268.

Gourevitch, G. (1980). Directional hearing in terrestrial mammals, In: Comparative Studies of Hearing in Vertebrates, Popper, A.N., Fay, R.R. (eds.). pp. 357–373. New York: Springer-Verlag.

Gourevitch, G., Houben, D. (1983). Sensitivity of monkeys (*M. nemestrina*) to interaural time disparities in noise bands. 11th Inter. Cong. Acoust, Revue d'Acoustique, 3, 229–302.

Hafter, E.R., Dye, Jr., R.H., Gilkey, R.H. (1979). Lateralization of tonal signals which have neither onsets nor offsets. J. Acoust. Soc. Am. 65, 471–477.

Harrison, J.M., Downey, P. (1970). Intensity changes at the ear as a function of the azimuth of a tone source: A comparative study. J. Acoust. Soc. Am. 47, 1509–1518.

Hartley, R.V.L., Fry, T.C. (1921). The binaural localization of pure tones. Phys. Rev. 18, 431–442.

Hausler, R., Colburn, S., Marr, E. (1983). Sound localization in subjects with impaired hearing. Acta Oto-laryngol (Stockh) Suppl. 400.

Hebrank, J., Wright, D. (1974). Spectral cues used on the localization of sound sources in the median plane. J. Acoust. Soc. Am. 56, 1829–1834.

Heffner, H.E. (1978). Personal communication.

Heffner, H.E., Heffner, R.S. (1984). Sound localization in large mammals: Localization of complex sounds by horses. Behav. Neurosci. 98, 541–555.

Heffner, H.H., Heffner, R.S. (1985). Sound localization in Norway rats (*Rattus norvegicus*). Hear. Res. 19, 151–155.

Heffner, R.S. (1981). Sound localization and the superior olivary complex in horses and cattle. J. Acoust. Soc. Am. 69, S 10.

Heffner, R., Heffner, H. (1982). Hearing in the Elephant (*Elephas maximus*): Absolute sensitivity, frequency discrimination, and sound localization. J. Comp. Physiol. Psychol. 96, 926–944.

Heffner, R.S., Heffner, H.E. (1983). Hearing in large mammals; Horses (*Equus caballus*) and cattle (*Bos taurus*). Behav. Neurosci. 97, 299–309.

Heffner, R.S., Heffner, H.H. (1986). Localization of tones by horses: Use of binaural cues and the role of the Superior Olivary Complex. Behav. Neurosci. 100, 93–103.

Henning, G.B. (1974). Detectability of interaural delay in high frequency complex waveforms. J. Acoust. Soc. Am. 55, 84–90.

Houben, D., Gourevitch, G. (1979). Auditory lateralization in monkeys: An examination of two cues serving directional hearing. J. Acoust. Soc. Am. 66, 1057–1063.

Houben, D., Gourevitch, G. (1980). Lateralization of high frequency noise with interaural time delays by monkeys (*M. nemestrina*). J. Acoust. Soc. Am. 58, S 96.

Houben, D., Gourevitch, G. (1982). Binaural time difference thresholds in monkey and man for signals of different durations. Assoc. Res. Otolaryngol. 52.

Houtgast, T., Plomp, R. (1968). Lateralization threshold of a signal in noise. J. Acoust. Soc. Am. 44, 807–812.

Isley, T.E., Gysel, L.W. (1975). Sound localization in the Red Fox. J. Mammal. 56, 397–404.

Kelly, J.B. (1980). Effects of auditory cortical lesions on sound localization by the rat. J. Neurophysiol. 44, 1161–1174.

Kelly, J.B., Glazier, S.J. (1978). Auditory cortex lesions and discrimination of spatial location by the rat. Brain Res. 145, 315–321.

Kelly, J.B., Kavanagh, G. (1984). Personal communication.

Klumpp, R.G., Eady, H.R. (1956). Some measurements of interaural time differences thresholds. J. Acoust. Soc. Am. 28, 859–860.

Kuhn, G.F. (1977). Model for the interaural time differences in the azimuthal plane. J. Acoust. Soc. Am. 62, 157–167.

Kuhn, G. (1983). Personal communication.

McFadden, D., Pasanen, E.G. (1976). Lateralization at high frequencies based on interaural time differences. J. Acoust. Soc. Am. 59, 634–639.

Mills, A.W. (1958). On the minimum audible angle. J. Acoust. Soc. Am. 30, 237–246.

Moore, D.R., Irvine, D.R.F. (1979). A developmental study of the sound pressure transformation by the head of the cat. Acta Otolaryngol. 87, 434–440.

Roth, G.L., Kochhar, R.K., Hind, J.E. (1980). Interaural time differences: Implications regarding the neurophysiology of sound localization. J. Acoust. Soc. Am. 68, 1643–1651.

Sidman, M., Ray, A.B., Sidman, R.L., Klinger, J.M. (1966). Hearing and vision in neurological mutant mice: A method for their evaluation. Exp. Neurol. 16, 377–402.

Smith, J. (1970). Conditioned suppression as an animal psychophysical technique. In: Animal Psychophysics, Stebbins, W.C. (ed.). pp. 125–159. New York: Appleton-Century-Crofts.

Stephens, R.W.B., Bate, A.E. (1966). Acoustics and Vibrational Physics. pp. 70–75. New York: St. Martin's Press.

Tobias, J.V., Zerlin, S. (1959). Lateralization threshold as a function of stimulus duration. J. Acoust. Soc. Am. 31, 1591–1594.

Wegener, J.G. (1974). Interaural intensity and phase angle discrimination by Rhesus monkey. J. Sp. Hear. Res. 17, 638–655.

Wiener, F.M., Pfeiffer, R.R., Backus, A.S.N. (1966). On the sound pressure transformation by the head and auditory meatus of the cat. Acta oto-laryngol. (Stockh.) 61, 255–269.

Woodworth, R.S. (1938). Experimental Psychology. New York: Holt.

Part IV Problems and Solutions of Directional Hearing in the Real World

As a result of the products of civilization, especially in modern times and in particular cultures, man's real world imposes unique conditions on his directional hearing. This section examines how human spatial hearing is affected by singular surroundings, cultural desires, and medical circumstances.

First among topics discussed in this section is the concern of architectural acoustics in preserving directional information in enclosed environments. This requires identification of how the sounds are physically altered by the environment, e.g., reverberation and, consequently, what changes occur in the perception of the direction of sound. The importance of this research is for the design of enclosed structures ranging from rooms to concert halls. Another aspect of real world listening is that individuals with hearing impairment of one kind or another are afflicted not only by their particular hearing problems, but also by what appears to be much poorer directional hearing than that of normals. Study of various hearing impairments and of associated degraded directional hearing is being followed by attempts at combatting the loss in directional hearing with specially designed hearing aids. Finally, for many people the real world includes listening to music in the home. Thus, the stereo enthusiasts, and even less compulsive listeners, want their audio system to reproduce sound in as natural a way as possible. Many "tricks" have been developed to enhance the "space quality" of normally produced stereophonic music and sounds.

10

Hearing in Rooms

DAVID A. BERKLEY

Up to this chapter discussion has centered on the basic phenomena of directional hearing. A number of subtle and complex experiments designed to elicit and understand these phenomena have been examined. This chapter begins consideration of directional hearing in room environments where normal-hearing human beings (and many other animals as well) naturally apply directional discrimination.

Hearing is one of the most important social senses. Directional hearing is a vital part of the process of distinguishing between desired and undesired sound sources. In conversation the desired talker must be distinguished from several common causes of interference: (1) Other simultaneous talkers, (2) noise sources, such as traffic, music, air conditioning, and (3) the effects of wall reflections in the room, i.e., echoes of all the above sources, including the desired talker. In listening to music, directional hearing plays a different, but also important, role—that of providing a sense of space and surrounding.

The remainder of this chapter discusses the fundamental mechanisms underlying the human ability to achieve directional perception in a room environment and what effect that ability has on perception of sources in such an enclosed space. This chapter concentrates on specific effects produced by rooms, as noted under item (3) above, summarizes the physical basis for room reverberation, discusses perception in such a physical environment in qualitative terms, and, finally, gives some examples of attempts at quantifying these perceptual effects.

Physics of Rooms

Figure 10-1 is a schematic drawing of an idealized rectangular room. Such an idealization yields most of the important effects associated with listening in a real room (Allen and Berkley, 1979; see also Borish, 1984). Two-point pickups can be used as an approximation to two-eared listening;

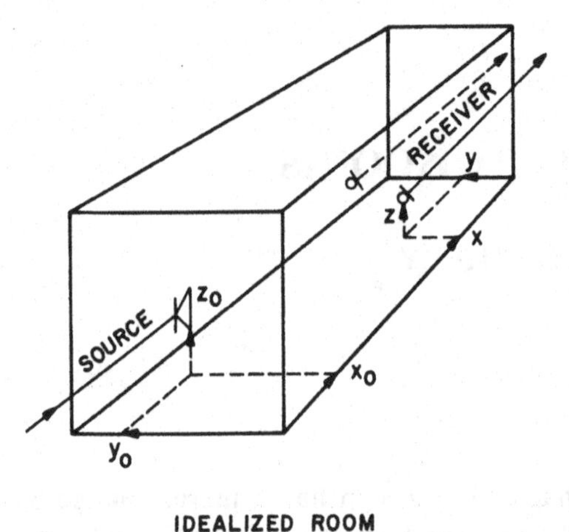

FIGURE 10.1. Idealized rectangular room—two receivers are shown for binaural modeling.

however, as discussed later, this ignores the important effects of the head and outer ear.

A hard acoustic surface acts as an acoustic mirror and the surface may be replaced by an "image" of the actual acoustic source, as shown in Figure 10-2. If there are two opposite surfaces (two opposing walls of a room), each produces images of the other. Figure 10-3 shows a schematic picture of the images produced by a two-dimensional room and it is easy to form a similar model for the three-dimensional room of Figure 10-1. To study sound transmission in such a model, all sources are excited simultaneously, and the free-fields resulting from each source (including the transmission delays) are added at the desired pickup point. For perfectly reflecting walls this model is mathematically exact. If the walls of the room are partially absorbing, this may be approximately taken into

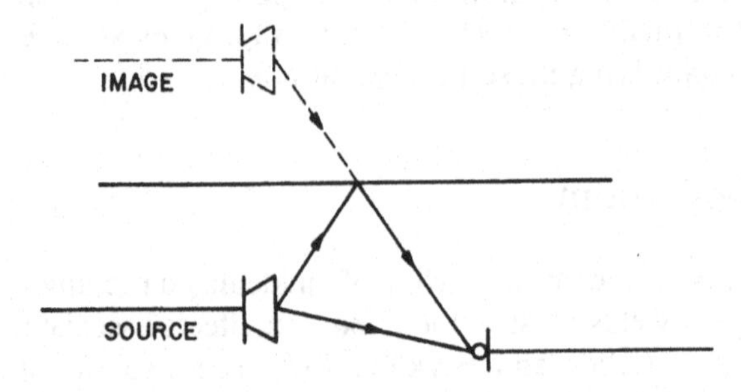

FIGURE 10.2. Schematic illustration of replacement of a wall with an equivalent acoustic image.

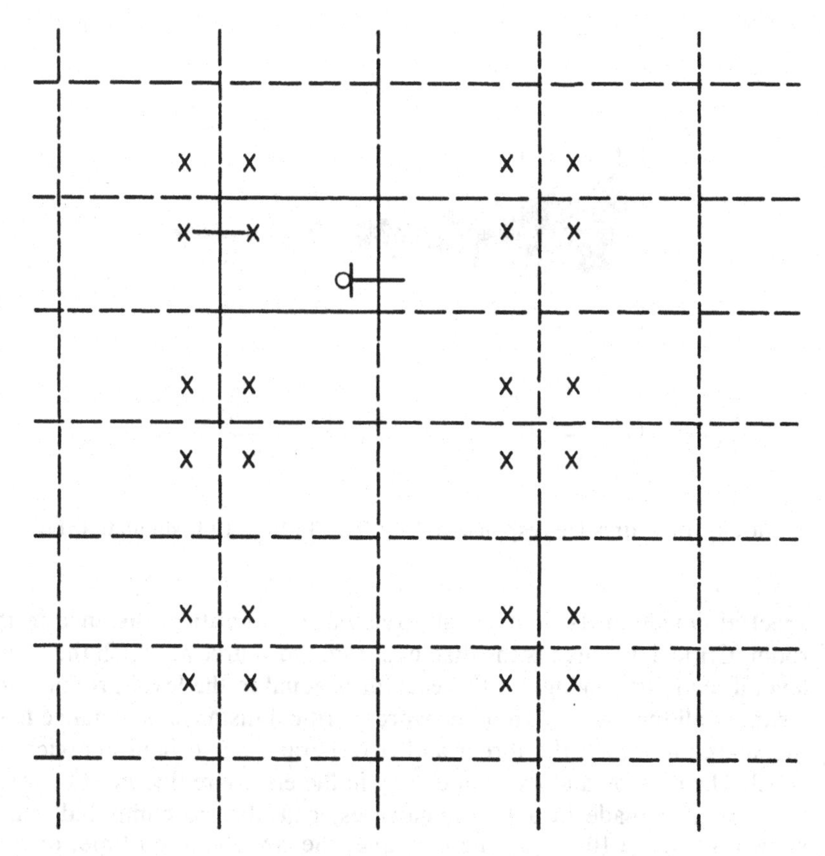

FIGURE 10.3. Image model of a two-dimensional (rectangular) space.

account by attenuating the transmission by the reflection coefficient each time an image wave crosses an image-wall boundary.

If each image is excited by a pulse, the result is the impulse response of the simulated room. Figure 10-4 shows the first 0.25 seconds of the calculated impulse response of a 12 ft × 15 ft × 19 ft (3,420 cu ft) simulated room with a source-receiver spacing of 8.5 ft. A second response can be calculated for a nearby receiver (second "ear") and would look essentially the same. All of the acoustic information about the room, relevant to those two points, is entirely included in the pair of impulse responses. For example, the result of speech excitation of the sources (instead of pulse excitation) can be obtained by "convolving" (see Gold and Radar, 1969) the speech with the impulse responses.

There are several important physical measures that can be derived from the impulse response. The frequency response, shown in Figure 10-5, can be characterized by the root mean square (RMS) of the deviation of the log spectrum from a flat response (Jetzt, 1979). The result, called the

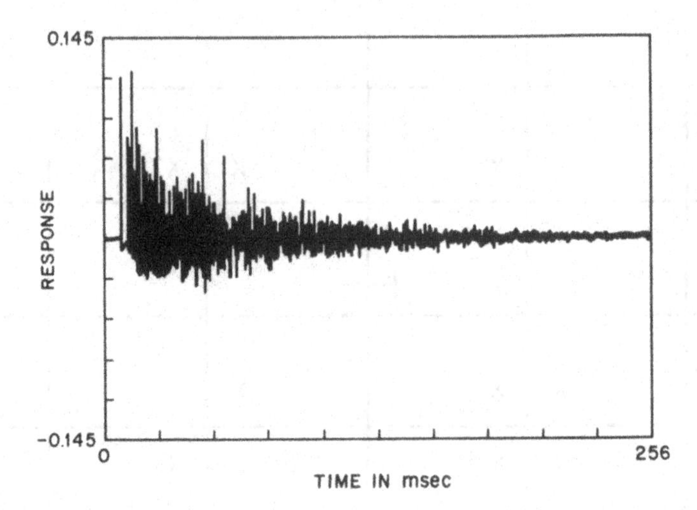

FIGURE 10.4. Impulse response of a 12 ft × 15 ft × 19 ft simulated room.

spectral deviation, is theoretically related to the critical distance in the room. Critical distance is the distance from the source at which the sound level directly impinging on the receiver is equal to the level produced by a sum of all the images. In other words, critical distance is distance from the source at which the direct and reverberant sound field energies are equal. The reverberant decay of energy in the enclosure (the most common measurement made in actual enclosures) can also be computed and is shown in Figure 10-6. For the example, the reverberation time, or time

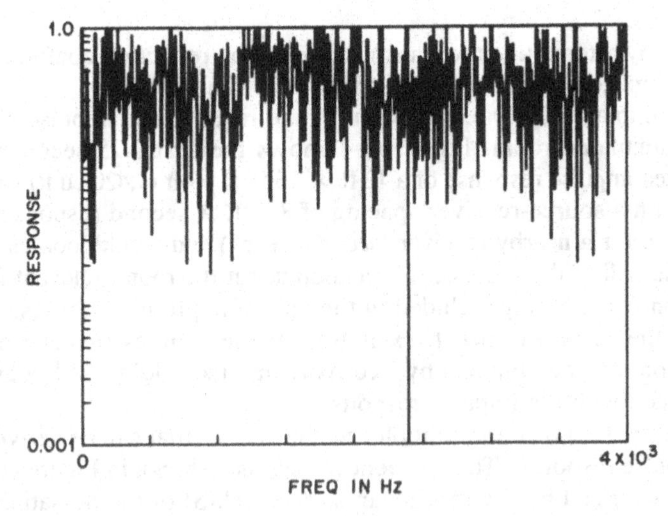

FIGURE 10.5. Frequency response of the simulated room from Figure 10.4.

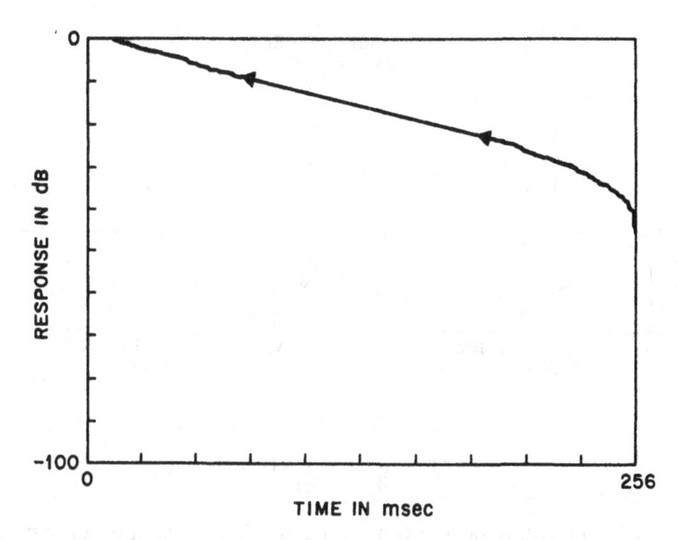

FIGURE 10.6. Reverberant energy decay of the simulated room from Figures 10.4 and 10.5.

for reverberant energy to decay 60 dB, can be extrapolated from the figure to be 0.475 seconds.

Directional Perception in a Reverberant Sound Field

Transients are the primary cue used for perception of direction in a reverberant environment. The precedence effect or "law of the first wavefront" (see Chapter 4) is the means by which listeners avoid chaos and reduce their perception of multiple reflections (reverberation) into a single percept (Mills, 1972, and Blauert, 1983).

The most dramatic demonstration of this effect in a reverberant environment was by Franssen (1963). A sinusoidal signal is reproduced over two separated loudspeakers in a room. Applied first to one loudspeaker, the level in that loudspeaker is gradually reduced (over a number of seconds) and that of the second loudspeaker is raised until all the sound is being produced by the second loudspeaker. A schematic description of a recent demonstration of the effect is shown in Figure 10-7. Subjects listening to this experiment are convinced that the sound continues to come from the first source even when it has totally switched to the second.

Careful listening to the Franssen demonstration and to other source configurations in rooms raises vital distinctions between perception of direction, identification of source, and perception of source quality (e.g., diffuseness and timbre). These, and many other issues related to directional perception, are elegantly treated in the recently translated book *Spatial*

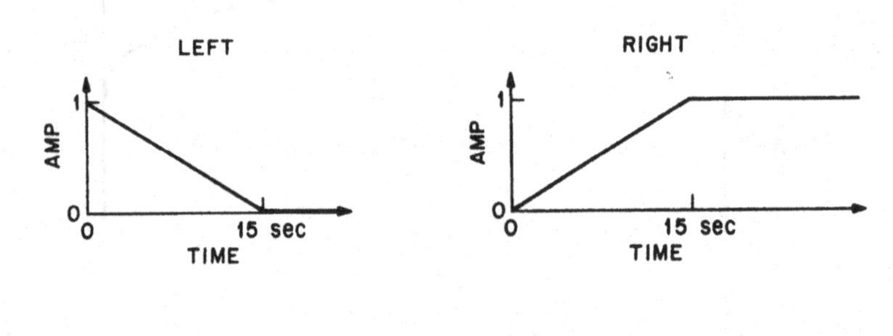

FIGURE 10.7. Schematic description of a simplified version of an experiment by Franssen (1963).

Hearing by Jens Blauert (1983). This book contains not only a summary of the author's original work in the field but also provides a unique review of the literature, including the extensive German literature, in this area.

As seen above, transient signal information mediated by the precedence effect allows identification of sound as coming from a single source even when the actual sound field is reverberant and is impinging on the listener from all directions. The reverberant nature of the field, as mentioned above, does affect the quality of the perceived sound. In the following the perceptual effects of two domains of sound reflections are considered. First, the perception of echo produced by sound arrivals more than about 25 ms after the direct sound, is considered. This is followed by illustrations of the effect of early reflections (less than 25 ms) on perception, namely signal coloration.

If a single reflection of a sound source is heard more than 25 ms after the direct sound reaches the listener (e.g., a reflection from a single wall more than 12 ft away), it is perceived as a discrete echo. This condition is generally avoided in the proper design of acoustic spaces. Significant amounts of energy, arriving in the form of multiple reflections more than 25 ms after the direct sound is heard, are perceived as a time extension of the original sound and contribute to a sense of spaciousness. However, such time extension may degrade the intelligibility of a speech source if the reverberation times approach or exceed 1 s.

Some interesting consequences of precedence can also be demonstrated with such reflections. If a single channel recording of such reverberation is compared with one with the direct sound on one channel and reverberant on a spatially separated second channel, two effects are heard. First, the perceived amount of echo decreases when heard on two channels; second, the direction of the source is slightly offset from the direct sound, and the apparent source is diffused spatially. The decrease in apparent reverberation is important for the ability to communicate in rooms, and the quantitative measurement of this effect is discussed later.

Reflections arriving less than about 25 ms after the direct sound are perceived as changing (or "coloring") the perceived tonal balance of the sound. The physical distinction between early- and late-arriving sound reflections is shown in Figure 10-8. Figure 10-8A displays a single reflection after 50 ms and Figure 10-8B shows a 2 ms reflection. As discussed above, the long-time reflection is perceived as a discrete echo, the rapid spectral variations being ignored by the hearing system, whereas the ear transmits the broader hills and valleys in the frequency response produced by the narrowly separated pulses of Figure 10-8B. Figure 10-9 shows the averaged spectrum of a speech signal before and after passing through a slightly attenuated 1.25-ms reflection. If the direct speech signal and the direct plus reflected speech signal are compared when played over a single loudspeaker, a distinct spectral distortion is heard. However, if the reflection is played over a spatially separated second loudspeaker, the direction of the source is modified by the reflection, but the perceived coloration of the speech sounds is considerably reduced. Such reduction of the distorting effects of short-time echoes plays a significant role in our preference for particular rooms and is also discussed later.

Although simplified, since sound arrivals are far more complex in an actual room, the above examples demonstrate the importance of precedence in understanding directional perception in enclosures. A detailed

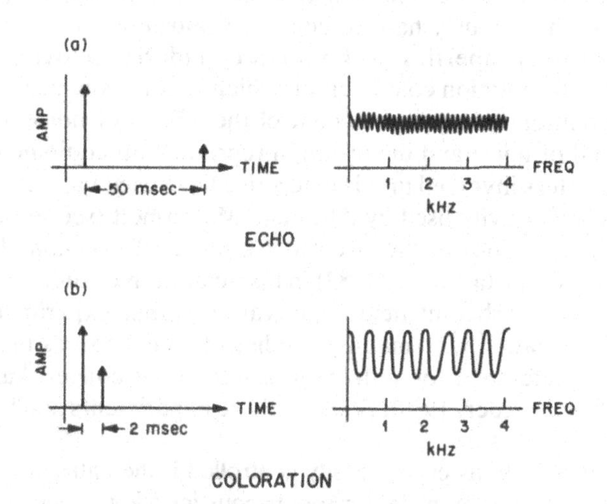

FIGURE 10.8. (a) Time and frequency domain description of a single echo. (b) Time and frequency domain description of coloration.

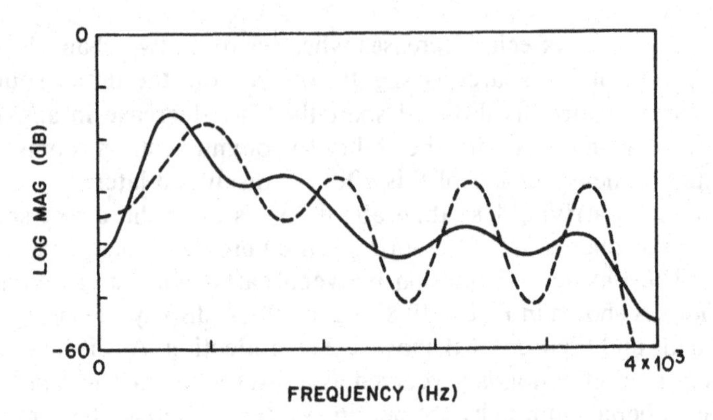

FIGURE 10.9. Average speech spectrum of sentence before (solid curve) and after (dashed curve) passing through a 1.25-ms coloration filter.

understanding of these effects is important in the design of rooms, concert halls, and in reproduction systems (as discussed in a later chapter).

The following section discusses the issues involved in attempts to quantify the effects just discussed.

Quantification of Directional Effects in Rooms

Quantification of directional hearing in rooms is quite difficult. There are two separate areas that cause serious problems. The observed effects are generally multidimensional. That is, more that one observable perceptual effect is present in achievable experiments. Second, physical parameters of rooms are notoriously hard to control. Control of reflections requires variation of the properties of the room boundaries (moving walls and changing their reflection coefficients), which is not easily done. Also, meticulous attention is required to control the effects of noise in the environment and of noise and distortion in reproduction equipment.

Experimenters involved in this enterprise have used various approachs. Simulation was heavily used by Allen and McDermott (see Berkley, 1980). The unique properties of the IRCAM *Espace de Projection (ESPRO)* in Paris was used by Hartmann (1983) in his quantitative studies of direction perception in reverberant fields, and various other experimenters have combined simulation with massive studies of actual spaces in an attempt to quantify preference for reflection properties of concert-hall acoustic spaces (see Schroeder, 1980). Some of these experiments and their results are discussed below.

Boundary reflections can be easily controlled if the entire acoustic space is simulated. The image model for room acoustics allows a simple computer implementation (Allen and Berkley, 1979). The impulse response of the simulated room is computed by summing over the impulsive response of each image at the two receiving locations (the two "ears" of the subject). Test materials are then prepared, as mentioned earlier, by convolving high-quality "dry" (without reverberation) speech samples with the computed impulse responses to form a dichotic pair of samples for presentation over headphones. Impulse responses for a variety of room dimensions and boundary conditions can be easily computed and corresponding test samples produced.

It is known (e.g., see Plenge, 1974) that such reproduction, without consideration of the external ear and head diffraction effect, will lead to perception of source images inside the head. However, qualitatively, this does not affect the variations of echo and coloration perception with room and listening conditions. Experimentation with variations of the simulated room reflection coefficients and listener talker distances confirms the perceptual identification of two dimensions of perception produced by reverberation. To first order, coloration may be accounted for by spectral deviation whereas echo is identified with the simulated room reverberation time.

When listeners are asked for their preferences, responses may also be given in terms of these same two physical variables. Figure 10-10 shows

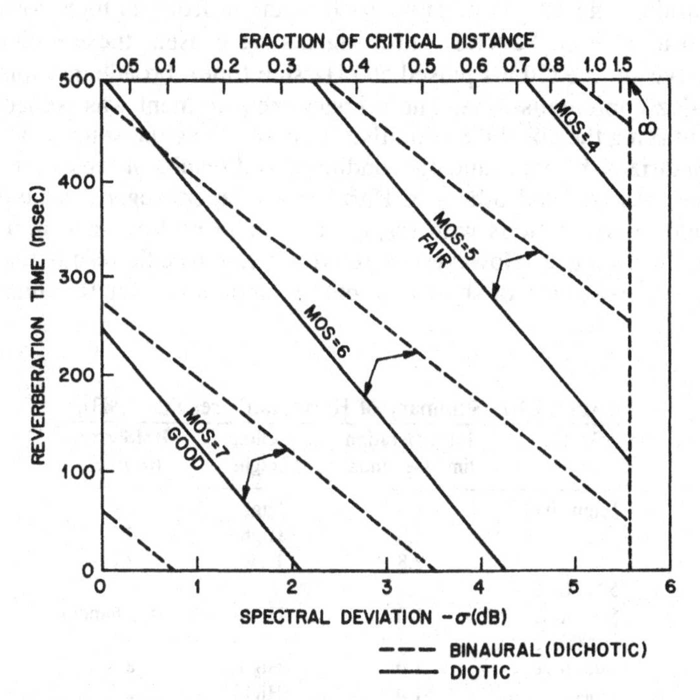

FIGURE 10.10. Isopreference results for simulated rooms with varying reverberation times and spectral irregularity.

a linearized isopreference plot of subject preference (on a nine-point scale labeled excellent through unsatisfactory) for a variety of spectral deviation/reverberation time conditions. As mentioned earlier, the spectral deviation axis can also be interpreted in terms of critical distance, as shown in the alternative labeling of the abscissa at the top of the figure. The effects of directional perception are seen in the differences between the solid curve, representing diotic (signal from one simulated pickup point) presentation of samples, and the dashed curve, representing dichotic (each simulated signal sent to the appropriate ear) presentation.

Figure 10-10 implies that going from diotic to dichotic listening (i.e., introducing directional perception) is equivalent to moving closer to the sound source or treating the room so as to reduce reverberation. This result, of course, says nothing about ability to locate or to accurately point to a source under these conditions. The issue of source localization under reverberant conditions is addressed by the experiments carried out in the IRCAM space where the wall can be moved and sound absorption varied (Hartmann, 1983).

Hartmann studied the ability of subjects to localize the azimuth of impulsive, noise, and tone sources under a variety of reverberant conditions. These conditions included variations in reverberation time and echo conditions. He varied the ESPRO reverberation time from 1 to 5 seconds and also studied the effects of lowering the ceiling from its high position of 11.5 m to 3.65 m. Several source signals were used; these included an impulsive source (50 ms pulsed 500 Hz sine tone), broadband noise, and a 500-Hz continuous tone. The primary measurement was subject error in identifying the absolute azimuthal angle of the actual source. Table 10-1 summarizes the experimental conditions and results of the experiments.

These results, and others in Hartmann's paper, suggest a number of possible interpretations with respect to human performance in a reverberant environment. However, there are several specific results that stand out: (1) Localization of short tone bursts (impulsive signal) is unaffected

TABLE 10-1. Summary of Hartmann's results (1983).

Source type	Reverberation time (seconds)	Ceiling height	RMS error (degrees)
Impulsive	1.0	High	3.4
	4.0	High	3.3
	2.8	Low	2.8
500 Hz Sine tone	1.0	High	12.6 (chance)
5,000 Hz Sine tone	1.0	High	8.6
Noise	1.0	High	2.3
	4.0	High	3.2

by reverberation time, although early echos, under the proper conditions, may improve accuracy; (2) low-frequency simple tones are almost impossible to localize with significant reverberation; higher frequency tones can be poorly localized (although this may be a monaural effect); (3) broadband noise localization, which is the most accurate localization under no-reverberation conditions, is significantly degraded by reverberation but still remains the most accurate among the various sources.

In addition to the results shown in Table 10-1, there are also results on subject bias and precision as well as experiments on spectral complexity, all contributing to an attempt to understand the effects of sound reflections on perception of source position.

Thus far the issues of source preference (and its underlying perceptual basis) and source direction have been addressed. Another question, that of perceived source extent, or diffuseness, has been addressed by a different approach. The primary tool, in roomlike fields, has been to have subjects draw pictures of the shape (and extent) of the sounds perceived under a variety of conditions. The major finding (Wagener, 1971) has been an increase in source extent with increase in the reverberant component relative to the direct sound component at the subjects' position. This is caused by a loss of "coherence," (see Blauert, 1983) between the sound fields at the subjects' left and right ears.

Increase of apparent source diffuseness with decreasing interear coherence is related to an important result in concert hall design (Schroeder, 1980). Schroeder summarizes the results of a series of experiments on concert hall *preference*. In these experiments, long-time reverberation patterns were obtained for a number of different actual concert halls. Then, using computer simulations, a variety of early-echo response patterns were interpolated before the natural reverberation pattern. As the complexity of the early-echo pattern increased, corresponding to more lateral echoes and lower interear coherence, preference for the musical sound of the hall increased. This result has led to a series of new designs for concert-hall ceilings (Schroeder, 1980) that optimize both the density and complexity of the hall's early echo structure.

The above results briefly summarize the state of quantitative understanding of the relationship between physical properties of an enclosure and directional hearing effects therein. To further advance such quantitative research it seems clear that new approaches, such as improved uses of head-related recording techniques (see Blauert, 1983, and Wightman, 1980), will be necessary to increase control over the environment through the use of computer analysis, control, and synthesis.

References

Allen, J.B., Berkley, D.A. (1979). Image method for efficiently simulating small room acoustics. J. Acoust. Soc. Am. 65, 943–950.

Berkley, D.A. (1980). Normal listeners in typical rooms: Reverberation perception, simulation and reduction. In: Acoustical Factors Affecting Hearing Aid Performance. Studebaker, G.A., Hochberg, I. (eds.). Baltimore: University Park Press.

Blauert, J. (1983). Spatial Hearing. Cambridge, Mass: MIT Press.

Borish, J. (1984). Extension of the image model to arbitrary polyhedra. J. Acoust. Soc. Am. 75, 1827–1836.

Franssen, N.V. (1963). Stereophony. Eindhoven: Phillips Tech. Bibl.

Gold, B., Rader, C.M. (1969). Digital Processing of Signals, Chapters 2 and 7. New York: McGraw-Hill.

Hartmann, W.M. (1983). Localization of sound in rooms. J. Acoust. Soc. Am. 74, 1380–1391.

Jetzt, J.J. (1979). Critical distance measurement of rooms from the sound energy spectral response. J. Acoust. Soc. Am. 65, 1204–1211.

Mills, W.A. (1972). Auditory Localization. In: Tobias, J.V. (ed.). Foundations of Modern Auditory Theory, Vol. II. New York: Academic Press.

Plenge, G. (1974). On the difference between localization and lateralization. J. Acoust. Soc. Am. 56, 944–951.

Schroeder, M.R. (1980). Toward better acoustics for concert halls. Phys. Today. 33, 24–30.

Wagener, B. (1971). Spatial distribution of auditory direction in synthetic sound fields (in German). Acustica 25, 203–219.

Wightman, F.L., Kistler, D.J. (1980). A new "look" at auditory space perception. In: Psychophysical, Physiological and Behavioural Studies in Hearing, van den Brink, G., Bilsen, F.A. (eds.). The Netherlands: Delft University Press.

11

Binaural Directional Hearing—Impairments and Aids

H. STEVEN COLBURN, P. M. ZUREK, AND
NATHANIEL I. DURLACH

Introduction

It is apparent, particularly when we close our eyes, that the auditory system creates a spatial representation of the acoustic environment. We are able to monitor sound sources in all directions, determine the nature and positions of the various sound sources, and focus our attention on sounds in a particular direction. In addition, we can determine certain aspects of the environment itself, such as the size and acoustic "liveness" of the space in which the sounds are generated. Although certain aspects of auditory spatial perception and directional hearing can be achieved when listening with a single ear, it is clear that listening is easier and more effective when listening with two ears. This conclusion is supported by subjective reports as well as by objective experiments on both normal and hearing-impaired listeners (e.g., see Durlach and Colburn, 1978; Durlach et al, 1981; Blauert, 1983; Hausler et al, 1983).

In the beginning of this chapter we briefly review some relevant background material on binaural directional hearing. Following that we summarize our picture of the effects of hearing impairments on binaural directional hearing. We then consider various approaches to combatting these effects through the application of hearing aids.

Rather than present a comprehensive overview, we focus on ideas and results that we believe are especially important or interesting. We restrict attention to issues related to response variability and resolution rather than mean responses and subjective attributes and to experiments in which there is an objective criterion for determining correct responses. An excellent and extensive review of directional hearing in normal listeners that includes a much broader class of results is available in Blauert (1983).

Background

Two Conceptualizations of the Binaural System

A natural conceptual model for the binaural hearing system, which is consistent with both subjective impressions and engineering design principles, is based on "spatial filters" that are tuned to various directions. The system is assumed to be able to scan the outputs of the whole filter bank, search for and focus on an output of particular interest, and maintain awareness of all outputs in order to respond to warning signals or important information from a direction that is not currently the one receiving primary attention.

A block diagram of an idealized spatial-filter system is shown in Figure 11-1A. The essence of this conception is that the acoustic field is first processed into separate direction channels, and then each channel is analyzed for its informational content. This system provides sound-source localization (by noting the identity of the channel in which the sound occurs), reduction of interference between sounds in different directions (since such sounds occur in different channels), and overall monitoring of the whole acoustical space (by scanning across all channels).

On the other hand, there is much physiological and psychophysical evi-

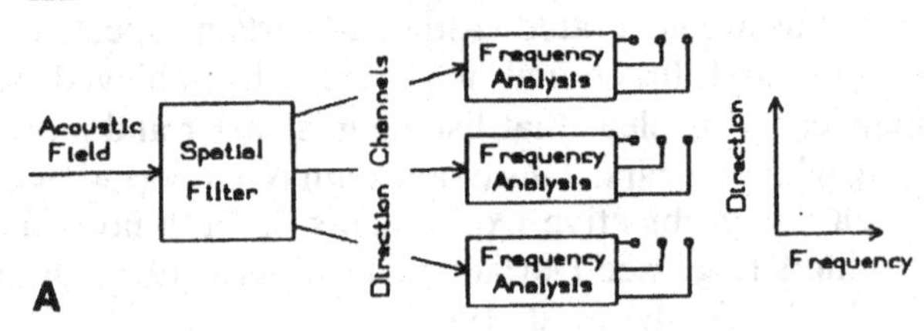

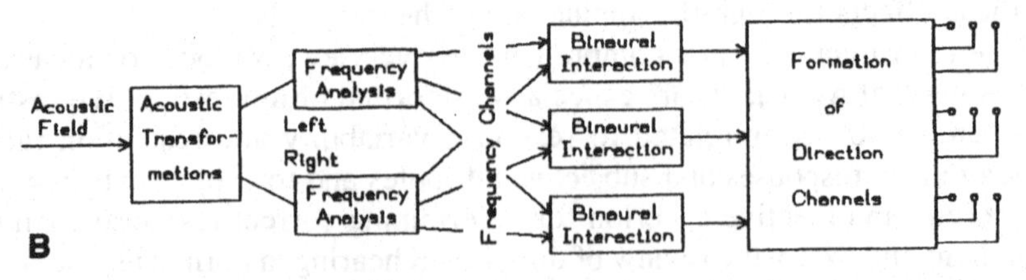

FIGURE 11.1. Two block diagrams illustrating different conceptualizations of the binaural system. (**A**) shows an idealized engineering approach and (**B**) a structure that is more consistent with knowledge of the auditory system.

dence that the peripheral auditory system is organized tonotopically and that a more appropriate model for binaural processing is that shown in Figure 11-1B. In this model the acoustic field is transformed by the head and external ear into two acoustic signals, one for each ear, which are then analyzed into frequency bands prior to binaural interaction or to any directional processing. Since sounds from a particular direction may include the full range of frequencies, the results of the binaural interaction for individual frequency bands must then be recombined in order to determine the properties of the sounds from a particular direction.

It is not clear, at least from a directional-hearing point of view, why the peripheral auditory system is constructed as illustrated in Figure 11-1B. From an engineering standpoint, the natural approach to providing good directional resolution is to design an antenna system that separates the sounds from different directions immediately at the input level. However, the directional resolution provided by an antenna is determined by its size relative to the wavelength. Since, over a substantial portion of the audible spectrum, acoustic wavelengths are relatively large compared with the size of the mammalian body, the peripheral frequency analysis may be related to constraints on the size of a "body-mounted" antenna system. Independent of the causes underlying the development of the system, its characteristics clearly have important implications for the design of aids as well as for the analysis of impairments.

Some Interrelations among Tasks

The study of binaural directional hearing in impaired persons need not consist of replicating in clinical populations all the tests that have already been conducted on normals. It is now believed that many of the results in complex binaural tasks can be predicted in terms of parameters characterizing fundamental components (such as basic sensitivity) for a given listener. Examples of task interrelationships that can be exploited in developing such predictions are presented in the following.

Many of the resolution abilities of the binaural auditory system can be understood in terms of its sensitivity to differences in interaural time (or phase) and interaural intensity. Specifically, one can predict measurements of the horizontal minimum audible angle (MAA), detection thresholds, and sensitivity to interaural correlation by making reasonable assumptions about the use of interaural time and intensity information. The relation of horizontal MAAs to interaural discrimination for tones was delineated by Mills (1958), who showed good correspondence between the MAA and just noticeable differences (JNDs) in interaural time and intensity. For interaural correlation discrimination and binaural detection, the situation is more complicated because the perceptual cue involves a change not in the average values but in the variability of the interaural differences. Furthermore, when one calculates optimum performance for these tasks using interaural JNDs in time and intensity to specify the amount of internal noise, the calculated performance is better than the performance observed empirically. Predictions of correlation JNDs and detection thresholds that are at least roughly comparable to the data can be obtained by postulating that interaural difference samples are temporally averaged (integrated) in the binaural processing. According to recent modeling work by Gabriel (1983) and Zurek et al (1983), available estimates of the degree of binaural temporal integration (Blauert, 1970; Grantham and Wightman, 1978, 1980; and Grantham, 1984) are adequate to make the predictions and data consistent.

Binaural improvements in speech intelligibility can also be explained with reasonable modeling assumptions. Levitt and Rabiner (1967) found that the intelligibility improvements measured with headphone listening could be predicted simply by assuming that the speech-to-noise ratio is effectively increased by the size of the corresponding binaural improvement in detection of a tone in noise. Zurek (1983) extended this type of analysis to the free-field case and computed directional effects for both monaural and binaural listening. Predictions of intelligibility thresholds are derived using physical-acoustics data, binaural-hearing theory, and articulation theory. These predictions, which are reasonably consistent with the data, indicate that most of the intelligibility advantage afforded by the use of two ears can be realized merely by always selecting the better ear. Thus, for example, if one assumes that the target is straight ahead and computes the speech-to-noise ratio required for threshold intelligibility as a function of the azimuthal angle of the noise, one finds that the function achieved by using the left ear when the interference is on the right side and the right ear when it is on the left side is within 2 dB of the function achieved with full binaural processing at all angles. The small magnitude of the improvement resulting from binaural interaction (less than 2 dB at all angles) is due to the relatively small binaural detection advantage at high frequencies where there is a relatively large contribution to intelligibility.

The relation between discrimination of angle and identification of angle appears to be understandable in terms of a model (Searle et al, 1976) originally applied to intensity perception (e.g., Durlach and Braida, 1969; Braida et al, 1984). According to this model, performance in discrimination is limited primarily by sensory factors whereas performance in identification is limited mainly by more central factors, most notably imperfect memory. The only case in which the results of an identification experiment are influenced by sensory factors is when the JND in angle (measured using a discrimination paradigm) is comparable in size to the span of the directions employed in the stimulus set of the identification experiment. To the extent that this model is correct, and to the extent that most localization experiments have used identification paradigms with angular spans that are large relative to the angle JND, most localization experiments reveal memory limitations not sensory ones.

Some Illustrative Background Data

The preceding sections review briefly a number of conceptual and theoretical issues relevant to binaural directional hearing. This section includes some illustrative data on this topic. Comprehensive collections of data are available in Durlach and Colburn (1978) and Blauert (1983).

Angle Discrimination

Binaural directional hearing is generally superior to monaural in locating sound sources. Valid estimates of monaural capabilities are difficult to obtain, however, because they require the use of genuine monaural listeners, i.e., listeners who have only one functional ear and who have been monaural long enough to have adapted to their condition. Several such monaural listeners were tested by Hausler et al (1983) in a study of the MAA. Figure 11-2 shows results from the subject who achieved the best performance and also participated in the largest set of conditions.

For the case of long-duration (1 second) wideband noise, this subject achieved localization performance roughly comparable to the performance

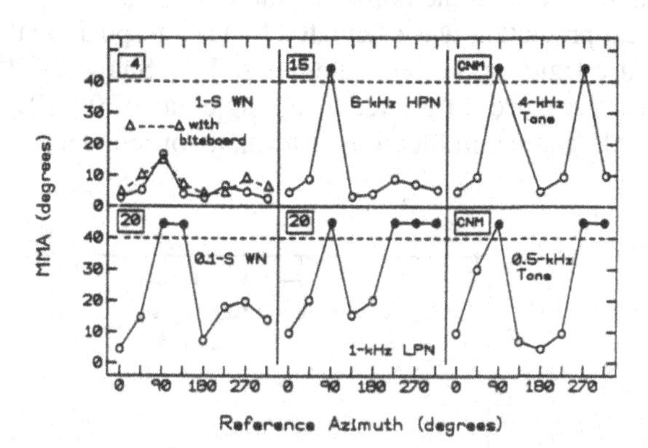

FIGURE 11.2. Localization performance of a monaural listener (adapted from Hausler et al, 1983). All panels show the horizontal minimum audible angle (MAA) as a function of reference azimuth. Different panels correspond to different stimuli: left column shows results for white noise (WN) at durations of 1 and 0.1 seconds, middle column for high-pass noise (HPN) and low-pass noise (LPN), and right column for tones. The listener's normal ear was in the direction of 270°. Unless indicated otherwise, the stimulus duration was 1 second, and the listener's head was not restrained (although the instructions requested that the listener's head remain motionless). The data in the upper-left panel suggest that results would have been similiar if a bite-board had been used (compare dashed and solid lines). The numbers in the upper-left corner of each panel give the vertical MAA (for a reference straight ahead) in degrees. The filled circles indicate that the horizontal MAA could not be measured using angular increments as large as 45°. CNM indicates that the vertical MAA could not be measured using angular increments as large as 45°.

exhibited by the normal binaural subjects in the study; only for sources located opposite the single functional ear was performance significantly below normal. When the bandwidth or duration of the stimulus was decreased, performance was degraded, as one would expect. Except for the surprisingly large vertical MAA obtained, the results for high-pass noise support the notion that spectral information at high frequencies is especially useful for monaural localization. Similarly, the results for tones are consistent with the notion that high-frequency and low-frequency tones can be reasonably well localized in front and back, but not on the sides. Note, however, that since the only monaural localization cue for a tone is its amplitude, randomization of amplitude will reduce monaural performance to chance levels for tones. Note also that the increase in the random fluctuations of the frequency spectrum of the wideband noise stimulus as the duration is decreased from 1 second to 0.1 second may be responsible for the enlarged MAAs obtained with the 0.1-second stimulus.

Monaural and binaural data showing the dependence of the horizontal MAA on frequency for tonal stimuli, along with theoretical predictions for monaural performance, are presented in Figure 11-3. The binaural data are from Mills (1958) and the monaural data are from Hausler et al (1983). The theoretical prediction was calculated by (1) interpolating the directional sound-pressure transformations summarized by Shaw (1974) and (2) assuming that a pressure difference at a single ear of 0.5 dB is necessary for a detectable change in location of a tonal source. These results show

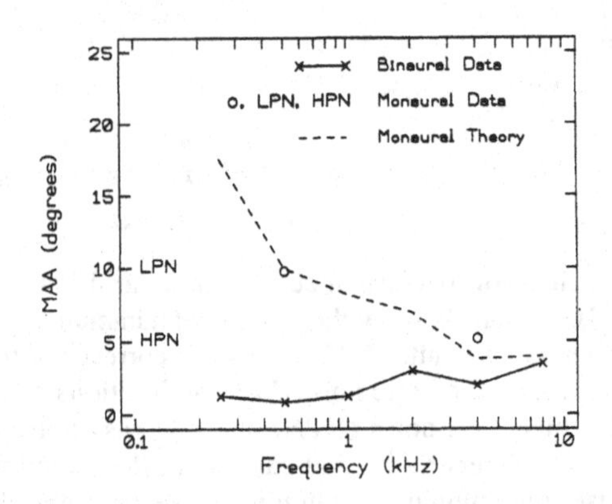

FIGURE 11.3. Minimum audible angle (MAA) in the horizontal plane for a reference location straight ahead as a function of tone frequency. Binaural data are from Mills, 1958; monaural data are from Hausler et al, 1983; and monaural theory is based on head-shadow data of Shaw, 1974, together with the assumption that the monaural MAA corresponds to an incremental intensity change of 0.5 dB. LPN and HPN give values of monaural MAA for low-pass (<1 kHz) and high-pass (>6 kHz) noise.

a substantial binaural improvement at low frequencies where interaural phase differences can be perceived and where the acoustic head shadow is small. At high frequencies, where the interaural phase of pure tones is not perceived and the head shadow becomes increasingly stronger, the difference between binaural and monaural performance decreases toward zero.

Effects of Interaural Level Imbalance

Comparisons of binaural performance between listeners with hearing impairments and listeners with normal hearing are much more complicated when the impairment is asymmetric between the two ears. Confining attention to asymmetries in absolute threshold (as opposed, for example, to asymmetries in pitch perception or temporal resolution), we note that in most binaural experiments with impaired listeners the stimuli have been presented without compensating for threshold asymmetry (e.g., see the review by Durlach et al, 1981). Thus, the observed degradations in performance can be partially due, at least in some cases, to a loudness or sensation-level imbalance, as would occur when a normal-hearing listener with symmetric threshold curves is tested with an intensity imbalance.

Past research on normal listeners (see the review by Durlach and Colburn, 1978) has shown that an intensity imbalance reduces performance below that obtained with equal intensities (at either the higher or lower level of the imbalanced case) in discrimination of interaural time, discrimination of interaural intensity, and binaural masked detection. In addition, such an imbalance produces an asymmetric dependence on interaural time delay (results for positive delays differ from those for negative delays) for interaural time discrimination, but not for interaural intensity discrimination or binaural masked detection. Although these experiments did not provide an opportunity for long-term adaptation to the imbalance, there is no indication that such adaptation occurs in sensitivity tasks (as it does in subjective tasks, such as centering). Data illustrating effects of interaural level imbalance on interaural time discrimination (Hershkowitz and Durlach, 1969; Domnitz, 1973) are shown in Figure 11-4.

Echo Suppression

The third set of data that we consider concerns echo suppression, a phenomenon that is directly related to the reduction in coloration observed with binaural listening. We consider this phenomenon here, even though it has not been studied with impaired listeners, because it represents one of the few cases in which binaural detection is *worse* than monaural detection.

Koenig (1950) noted that room reverberation that was quite apparent monaurally was not noticeable when listening binaurally. Zurek (1979) examined this phenomenon using a simplified laboratory analog to deter-

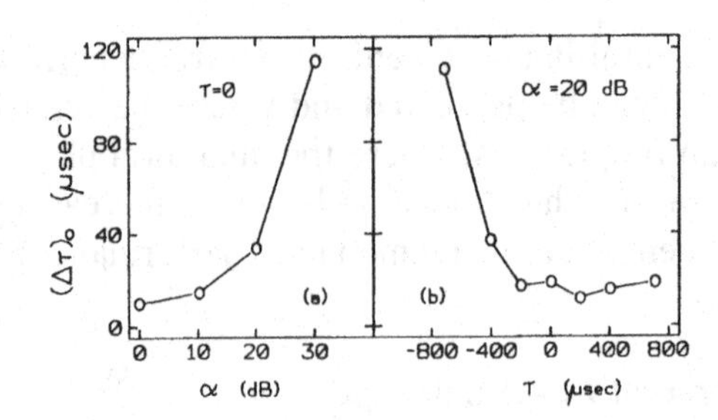

FIGURE 11.4. Effect of level imbalance on interaural time discrimination for a 500-Hz tone. (A) shows the dependence of the interaural time JND $(\Delta\tau)_0$ on the interaural amplitude ratio α for a reference delay $\tau = 0$ (Hershkowitz and Durlach, 1969). (B) shows the dependence of $(\Delta\tau)_0$ on τ for $\alpha = 20$ dB (Domnitz, 1973. Reproduced with permission from Domnitz, 1973, American Institute of Physics [JASA].)

mine the detection threshold of a synthetic echo (the target) in the presence of a synthetic direct signal (the masker). The direct signal was diotic (as if the source were straight ahead), the echo had an interaural delay τ but no interaural amplitude difference (head shadow was ignored), and the detection threshold of the echo was measured as a function of the echo delay δ.

Average data on this function obtained from three subjects are shown in Figure 11-5A for two conditions: a diotic echo ($\tau = 0$) and a dichotic

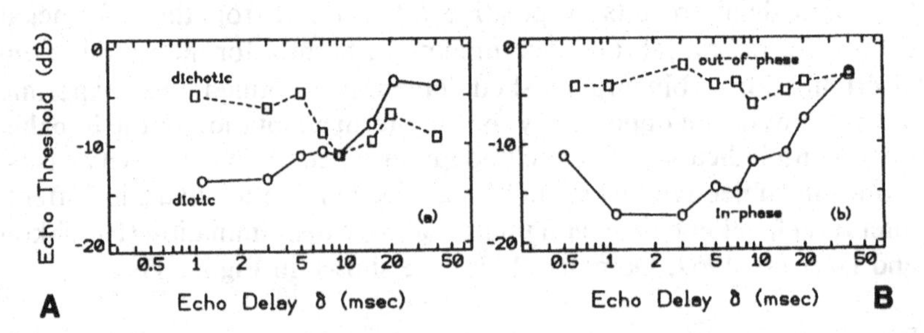

FIGURE 11.5. Measurements of binaural echo suppression (adapted from Zurek, 1979). Stimuli are 5-kHz low-pass noise bursts with 250-ms durations. (A) shows data for the detection of a synthetic echo in the presence of a synthetic direct signal as a function of the delay δ of the echo. The direct signal (masker) was always diotic. The data labeled diotic are obtained from a detection experiment in which the echo (target) was also diotic; the data labeled dichotic are from one in which the echo (target) had an interaural time delay of 500 μs. (B) shows data for the case in which the masker consists of independent noise signals to the two ears, and the target signal to each ear is constructed by delaying the masking signal to that ear by δ. The in-phase case corresponds to adding the target to the masker in both ears; the out-of-phase case corresponds to adding it in one ear and subtracting it in the other.

echo ($\tau = 500\ \mu s$). As δ decreases below 9 ms, the echo threshold for the diotic case becomes as much as 10 dB lower than the echo threshold for the dichotic case. In other words, since the echo threshold for the diotic case is roughly equal to the echo threshold that would be achieved monaurally, these data imply that binaural listening degrades the ability to detect the echo (or, equivalently, reduces the subjective coloration of the signal).

An even larger effect is shown in Figure 11-5B. In this experiment two independent noises serve as the binaural masker. The signal, a delayed version of each masker, was either added to both ears or added to one ear and subtracted from the other. The average results show the diotic thresholds to be as much as 15 dB lower than dichotic thresholds, with this difference decreasing, as before, at longer signal delays.

Hearing Impairments

The effect of hearing impairments on binaural hearing and auditory spatial perception is a large and not yet well-understood topic. A detailed review of this topic, organized according to auditory task, is available in Durlach et al (1981). A further review, in which the discussion is organized according to diagnostic category and includes more recent results, is available in Colburn (1982). In general, although the degradation of directional hearing is not well specified by conventional clinical examinations, the prevalence of patient complaints about the disturbing effects of background noise, as well as the results of objective experiments, suggest that the degradation of directional hearing is a common and important consequence of hearing impairments. In the following comments, we consider gross diagnostic categories and summarize briefly our integrated impressions of the main thrusts of the reports in this area. For details, the reader should consult the reviews.

Conductive Impairments

Results for subjects with substantial conductive losses (greater than about 35 dB HL) are summarized in Table 11-1. It is immediately apparent that these subjects have considerable difficulty in most binaural tasks. Furthermore, this difficulty is not a simple consequence of the lower effective sound level; performance is not significantly affected by changes in the stimulus level when the level is above threshold. Note also that unilateral losses disrupt performance more than bilateral losses, especially in tests of interaural intensity discrimination and binaural detection (masking level differences, MLDs). In bilateral cases, the MLDs are good in spite of substantial losses. Interaural parameter discrimination is fair to poor in all cases except that of intensity with wide-band stimuli and bilaterally symmetric losses. Horizontal angle discrimination is poor in all cases.

TABLE 11-1. Lateralization and localization in listeners with *conductive* losses greater than 35 dB HL.

	Unilateral	Bilateral
Interaural time JND		
Wideband	Poor	Poor
Narrowband	Fair to poor	Fair to poor
Interaural intensity JND		
Wideband	Fair to poor	Good
Horizontal angle JND		
Wideband	Poor	Poor
Source angle identification		
Wideband	Poor	Fair to poor
Masking-level differences	Poor (with asymmetries)	Good

Since subjects with bilaterally symmetric losses have good MLDs, good interaural intensity JNDs, and poor interaural time JNDs, it appears that these subjects use interaural intensity differences to detect the antiphasic target (e.g., Durlach, 1964). Note also that asymmetries in the dependence of MLDs on interaural time delay have been observed for subjects with unilateral conductive losses, although such asymmetries have not been seen in normal subjects with interaural level imbalance.

The results summarized in Table 11-1 are consistent with the idea (Roser, 1966; Hausler et al, 1983) that conductive losses reduce the air-conducted component of the cochlear stimulus, that bone-conducted sounds stimulate both cochleas roughly equally, and therefore that conductive losses lead to a lack of cochlear isolation (in the sense that the stimulus to each ear affects both cochleas). Zurek (1986) described such a mechanism quantitatively and showed that it can produce a variety of substantial and interesting effects (including transformations between interaural time delays and interaural intensity differences).

Cochlear Impairments

Results for subjects with hearing losses of cochlear origin (again greater than 35 dB HL) are shown in Table 11-2. It is remarkable that these subjects generally perform better in these tests than subjects with conductive losses. In fact, interaural time and intensity JNDs in the normal range have been measured (Hausler et al, 1983) for subjects with cochlear losses as large as 80 dB HL throughout the audible range. Note also that wideband stimuli give generally good performance whereas narrowband stimuli give relatively poor performance on these tests. The MLDs were measured with narrowband targets and thus we would expect the effective stimulus to be narrowband, at least in normal listeners. Although the MLD was normal for subjects with noise-induced hearing losses, the individual

TABLE 11-2. Lateralization and localization in listeners with *cochlear* losses greater than 35 dB HL.

	Unilateral	Bilateral		
	Meniere's	General	Noise-Induced	Presbycusis
Interaural time JND				
Wideband	Good	Good	Good	Good
Narrowband	Poor (with asymmetries)	Good to CNM	Fair to poor	Poor
Interaural intensity JND				
Wideband	Fair	Good	Good	Good
Narrowband	Poor	—	Poor	Poor
Interaural correlation JND				
Narrowband	—	—	Fair to CNM	Poor
Horizontal angle JND				
Wideband	Good	Good in front Variable on sides	Good in front Variable on sides	Good in front Variable on sides
Masking-level difference	Poor (with asymmetries)	Fair to Poor	Good	Fair

CNM, performance was so poor it could not be measured.

thresholds were abnormal for these patients. Meniere's patients, the only tabulated category of unilateral cochlear-loss patients, show results that depend on the bandwidth of the stimulus and show left-right asymmetries reminiscent of those measured in normal listeners with an interaural intensity imbalance. Performance by patients with bilateral cochlear losses depends on the etiology of the loss. For example, the limited data available on MLDs suggest that performance distinctions might be made between noise-induced and presbycusic losses.

Retrocochlear Impairments

The two general categories of subjects for which most information is available are acoustic neuromas (i.e., vestibular schwannoma with auditory nerve involvement) and brainstem lesions of various kinds, including multiple sclerosis. The predominant characteristic of these categories is the large variability in binaural performance from subject to subject without any obvious correlation with other available information. For example, compare the following two patients, both of whom had unilateral acoustic neuromas at the time of testing (later surgically confirmed). One (Florentine et al, 1979) had no binaural abilities at all, even though the hearing loss was less than 30 dB HL. The other (Hausler et al, 1983) exhibited normal performance in all binaural abilities tested (interaural time and intensity discrimination and angle discrimination).

Subjects with brainstem lesions typically exhibit no hearing loss; the nature of their impairments suggests that they are subjects with true binaural impairments as opposed to impairments that are basically monaural with consequent degradation in binaural performance. Available evidence from these subjects indicates that interaural time and intensity JNDs can be degraded independently, a result that has implications for hearing theory and for the design of hearing aids.

Aids to Restore Directional Hearing

Directional hearing in normal listeners is achieved primarily through the use of two spatially separated sensors (the ears) and appropriate processing of the two signals derived from these sensors (binaural interaction). Hearing impairments can degrade directional hearing by eliminating the use of one ear, by reducing cochlear isolation, by destroying the "fidelity" of the neural information in one or both of the monaural channels, or by interfering more centrally with the binaural-interaction process. Aids designed to restore directional hearing will differ depending on the nature of the impairment. In this section we first describe some of the current approaches to improving directional hearing and then speculate on paths that future research might follow.

Current Approaches

Attempts to combat degraded directional hearing have focused on binaural aids, the "CROS family" of aids, and directional-microphone aids.

There is an extensive history of laboratory research and clinical evaluations of binaural aids (see, for reviews, Markides, 1977, and Libby, 1980). Work on binaural aids has generally been restricted to the case in which the aid is composed of two single-microphone monaural aids with independent processing of the two received signals. Furthermore, the design and fitting of these aids have generally been treated as though the binaural auditory system consisted of two independent channels. Little attention has been given to the relations between the impairments in the two ears or to the state of the listener's binaural processing abilities. Aside from a few studies concerned with the effects of splitting the spectrum between the two ears (e.g., see Franklin, 1969, 1975; Rosenthal et al, 1975; Haas, 1982), it has been implicitly assumed that the optimum binaural aid consists of two optimum monaural aids—one that is optimum for the left ear and one that is optimum for the right ear. Although it is conceivable that this assumption is roughly correct for a substantial segment of the impaired population [and some recent data obtained by Levitt and his associates (personal communication) are consistent with this assumption], it is an assumption that must be treated with caution. In general, the sensations experienced by a listener who has significant residual hearing in both ears involve the processing performed by both ears and the binaural interaction between the peripherally processed signals in the two channels, and the design and fitting of aids (monaural as well as binaural) should take serious account of these factors. The only case in which such consideration is irrelevant and the complexities introduced by the availability of two ears and by binaural interaction can be totally eliminated is that of the unilaterally deaf subject.

For some types of impairments it makes eminent sense, from a psychoacoustic point of view, to use binaural amplification with independent processing. Such is likely to be the case, for example, with symmetric sensorineural losses of cochlear origin. As discussed previously, binaural capabilities appear to survive such losses relatively well. If the binaural abilities are normal, and the monaural degradations are symmetric, then one can be fairly confident that some benefit can be gained by symmetric independent processing.

In general, the benefits obtained from binaural amplification will depend on a multitude of factors in addition to the intrinsic auditory abilities of the user. Other factors to be considered include the ease of using a binaural aid, the goodness of fit of each aid individually, and the nature of the acoustic environment. Relatively little benefit is expected in environments with no interference or reverberation or with extremely complex interference and reverberation.

Many types of CROS aids have been considered (see, for example,

Harford and Dodds, 1974). The MULTI-CROS aid constitutes a useful illustration. If we disregard the aspects concerned with "earmold plumbing" and frequency-gain characteristic, this aid can be described as a two-microphone monaural aid with user-controlled switches. The two microphones are placed on opposite sides of the head, and, depending on the setting of two switches, either the two channels are summed and fed to one ear, or one or the other of the two channels is used alone (hence the acronym for Contralateral Routing Of Signals). The aid thus permits a unilaterally deaf user to eliminate or to capitalize upon the head-shadow effect, depending on the environmental situation. The advantages for speech intelligibility to be expected from such an aid under a variety of idealized conditions can be deduced from the study by Zurek (1983) mentioned previously. Although there have been attempts to evaluate certain aspects of user satisfaction for some of the CROS-family aids by means of questionnaires (Gelfand, 1979; Gelfand and Silman, 1981), we know of no direct performance evaluation of MULTI-CROS aids.

A third current approach is to use a directional microphone with its output signal presented to one ear. Directional microphones achieve directionality by delaying and subtracting the sound pressures sensed at two closely spaced points. In free space, directional microphones typically have a frequency-independent cardioid pattern with a null at 180 degrees (behind the listener). When an aid with such a microphone is placed on the head, however, the directivity pattern is complicated by acoustic scattering by the head. Evaluations of such microphones in hearing aids (e.g., Hawkins and Yacullo, 1984; Sung et al, 1975) show, as expected, that the benefits are greatest when reverberation is small relative to the direct signal (at the listener's location) and when the target and interference are favorably located with respect to the directivity pattern. A directional microphone, under the best of conditions, can provide an effective increase in signal-to-noise ratio of about 10 dB. However, the range of source locations and the realistic percentage of environmental conditions for which a benefit of that size can be achieved are rather limited. A rough calculation, assuming that a single interference source may appear at any horizontal azimuth with equal probability, predicts an average improvement in effective speech-to-noise ratio of about 3 dB in an *anechoic* environment.

Research Directions

We conclude this chapter by considering possible aids for two extreme but illustrative cases of hearing impairment. In the first, we assume that the listener has only one functional ear, and we focus on the problem of providing good directional hearing for a monaural listener. In the second, we assume that the listener has relatively good monaural hearing in both ears but severely degraded binaural processing, and we focus on the problem of combatting this central degradation.

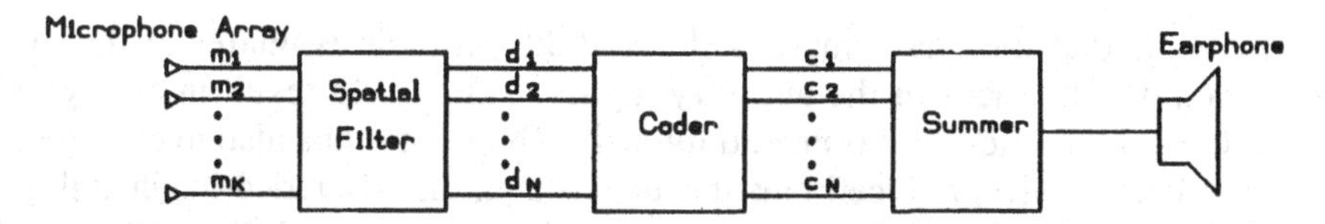

FIGURE 11.6. Idealized monaural aid. The microphone array and spatial filter use parallel processing to achieve simultaneous directional signals d_1,--, d_N. The coder then transforms these to signals C_1,--,C_N which can be resolved after summation and presentation to a single ear.

A conceptual structure for an idealized monaural aid (which exploits the spatial-channel concept discussed earlier) is shown in Figure 11-6. In this aid the sound field is processed to form directional signals d1, d2, etc. The coder then translates the various directional signals into signals that, when summed for monaural presentation, permit the listener to monitor all directions or to focus on the direction of choice.

There are two major problems that must be overcome to realize such an aid. The first (the *resolution* problem) concerns the achievement of good directionality within the geometric constraints imposed by the need for the aid to be wearable. The second (the *coding* problem) concerns the establishment of a transformation that allows the ear both to keep the channels perceptually separated and to evaluate the signals in each channel.

Approaches to the resolution problem include application of optimum linear array processing, and directional microphones, as well as exploitation of the head shadow. Included with optimum linear array processing are techniques in which the processing adapts to the environment. Further approaches to the resolution problem involve processing that is nonlinear (e.g., Kaiser and David, 1960; Mitchell et al, 1971). In the Kaiser-David scheme, for example, an attempt is made to enhance signals from straight ahead relative to signals from other directions by temporally gating the sum of the signals picked up by two symmetrically placed microphones according to the value of the cross-correlation. When the correlation is near unity, the sum signal is presented to the single ear; when the correlation differs from unity, the sum signal is attenuated.

Approaches to the coding problem include manipulation of frequency spectra and recruitment of senses other than hearing. In effect, the coding transformation must provide a subjective variable for monaural listening that plays the same role as lateralization in binaural listening. Unfortunately, the frequency coding of direction evident in monaural localization, although sufficient to provide some information on the direction of an isolated source, is not adequate to prevent cross-masking among channels.

Consider finally the problem of designing a two-channel output aid for listeners who have defective binaural processing. One model of these listeners is that they have increased variability in their estimates of the interaural differences in the auditory inputs. This would result in enlarged values of the interaural time and intensity JNDs and thus also in enlarged correlation JNDs, reduced binaural unmasking, and degraded intelligibility performance in binaural situations. One obvious approach to combatting this increased internal interaural noise is to magnify the interaural difference (Durlach and Pang, 1986). In principle, such magnification could be achieved either by use of an artificially enlarged head (the "fat-head solution") or by electronic signal processing. In the latter case, the dependence of the magnification on frequency could be designed according to the frequency profile of the impairment rather than imposed by physical acoustics.

This approach is limited by three major difficulties. First, little is currently known about the ability to adapt to magnified interaural differences. Adaptation to transformations that only require a relabeling of stimuli, such as rotation of the interaural axis, appears to be possible (e.g., see Held, 1955). However, adaptations to transformations that require new processing (such as an increased internal interaural delay to compensate for increased headwidth) is less clear (e.g., see Wein, 1964). Second, the ability to tolerate the increased dynamic range associated with magnification of interaural intensity differences is highly restricted in many patients, particularly those who exhibit recruitment (and for whom amplitude compression might be more suitable). Third, and finally, magnification of interaural differences implies an increase in the phase-ambiguity problem for narrowband signals.

The extent to which either type of aid discussed above can be realized in a practical device that is useful to a significant number of impaired listeners remains to be seen. It is clear that the restoration of normal binaural benefits to a monaural listener is still a long way off.

Acknowledgments. The preparation of this chapter was supported by the US Public Health Service (National Institute of Neurologic and Communicative Disorders and Stroke grants NS10916 and NS21322).

References

Blauert, J. (1970). Zur Traegheit des Richtunghoeren bei Laufzeit- und Intensitaets-Stereophonie [On the persistence of directional hearing in connection with time- and intensity-difference stereophonic sound], Acustica 23, 287–293; also Int. Audiol. (1972) 11, 265–270.

Blauert, J. (1983). Spatial Hearing. Cambridge, Mass: MIT Press.

Braida, L.D., Lim, J.S., Berliner, J.E., Durlach, N.I., Rabinowitz, W.M., Purks, S.R. (1984). Intensity perception. XIII. Perceptual anchor model of context-coding. J. Acoust. Soc. Am. 76, 722–731.

Colburn, H.S. (1982). Binaural interaction and localization with various hearing impairments. In: Binaural Effects in Normal and Impaired Hearing. Pedersen, O.J., Poulsen, T. (eds.). Scand. Audiol. Suppl. 15.

Domnitz, R. (1973). The interaural time jnd as a function of interaural time and interaural amplitude. J. Acoust. Soc. Am. 53, 1549–1552.

Durlach, N.I. (1964). Note on binaural masking-level differences at high frequencies. J. Acoust. Soc. Am. 36, 576–581.

Durlach, N.I., Braida, L.D. (1969). Intensity Perception. I. Preliminary Theory of Intensity Resolution. J. Acoust. Soc. Am. 46, 372–383.

Durlach, N.I., Colburn, H.S. (1978). Binaural phenomena. In: Handbook of Perception, Vol. 4. Carterette, E., Friedman, M. (eds.). New York: Academic Press.

Durlach, N.I., Pang, X.D. (1986). Interaural magnification. J. Acoust. Soc. Am. 80, 1849–1850.

Durlach, N.I., Thompson, C.L., Colburn, H.S. (1981). Binaural interaction in impaired listeners—A review of past research. Audiology 20, 181–211.

Florentine, M., Thompson, C.L., Colburn, H.S., Durlach, N.I. (1979). Psychoacoustical studies of a patient with a unilateral vestibular schwannoma. J. Acoust. Soc. Am.—Speech Communication Papers, 579–582.

Franklin, B. (1969). The effect on consonant discrimination of combining a low-frequency passband in one ear with a high-frequency passband in the other ear. J. Aud. Res. 9, 365–378.

Franklin, B. (1975). The effect of combining low- and high-frequency passbands on consonant recognition in the hearing impaired. J. Speech Hear. Res. 18, 719–727.

Gabriel, K.J. (1983). Binaural interaction in hearing-impaired listeners. Ph.D. Thesis, MIT, Cambridge, Mass.

Gelfand, S.A. (1979). Usage of CROS hearing aids by unilaterally deaf patients. Arch. Otolaryngol. 105, 328–332.

Gelfand, S.A., Silman, S. (1981). Use of CROS and IROS hearing aids by patients with high-frequency hearing loss. Ear Hear. 3, 24–29.

Grantham, D.W. (1984). Discrimination of dynamic interaural intensity differences. J. Acoust. Soc. Am. 76, 71–76.

Grantham, D.W., Wightman, F.L. (1978). Detectability of varying interaural temporal differences. J. Acoust. Soc. Am. 63, 511–523.

Grantham, D.W., Wightman, F.L. (1980). Detectability of a pulsed tone in the presence of a masker with time-varying interaural correlation. J. Acoust. Soc. Am. 65, 1509–1517.

Haas, G.F. (1982). Impaired listeners recognition of speech presented dichotically through high- and low-pass filters. Audiology 21, 433–453.

Harford, E., Dodds, E. (1974). Versions of the CROS hearing aid. Arch. Otolaryngol. 100, 50–57.

Hausler, R., Colburn, H.S., Marr, E. (1983). Sound localization in subjects with impaired hearing. Acta Oto-Laryngol. Suppl. 400.

Hawkins, D.B., Yacullo, W.S. (1984). Signal-to-noise ratio advantage of binaural hearing aids and directional microphones under different levels of reverberation. J. Speech Hear. Disord. 49, 278–286.

Held, R. (1955). Shifts in binaural localization after prolonged exposures to atypical combinations of stimuli. Am. J. Psychol. 68, 526–548.

Hershkowitz, R.M., Durlach, N.I. (1969). Interaural time and amplitude jnds for a 500-Hz tone. J. Acoust. Soc. Am. 46, 1464–1467.
Kaiser, J.F., David, E.E. (1960). Reproducing the cocktail party effect. J. Acoust. Soc. Am. 32, 918.
Koenig, W. (1950). Subjective effects in binaural hearing. J. Acoust. Soc. Am. 22, 61–62.
Levitt, H., Rabiner, L.R. (1967). Predicting binaural gain in intelligibility and release from masking for speech. J. Acoust. Soc. Am. 42, 820–829.
Libby, E. (1980). Binaural Hearing and Amplification. Chicago: Zenetron.
Markides, A. (1977). Binaural Hearing Aids. New York: Academic Press.
Mills, A.W. (1958). On the minimum audible angle. J. Acoust. Soc. Am. 30, 237–246.
Mitchell, O.M.M., Ross, C.A., Yates, G.H. (1971). Signal processing for a cocktail party effect. J. Acoust. Soc. Am. 50, 656–660.
Rosenthal, R.D., Lang, J.K., Levitt, H. (1975). Speech reception with low-frequency speech energy. J. Acoust. Soc. Am. 57, 949–955.
Roser, D. (1966). Directional hearing in persons with hearing disorders. J. Laryngol. Rhinol. 45, 423–440.
Searle, C.L., Braida, L.D., Davis, M.F., Colburn, H.S. (1976). A model for auditory localization. J. Acoust. Soc. Am. 60, 1164–1175.
Shaw, E.A.G. (1974). Transformation of sound pressure level from the free field to the eardrum in the horizontal plane. J. Acoust. Soc. Am. 56, 1848–1861.
Sung, G.S., Sung, R.J., Angelelli, R.M. (1975). Directional microphone in hearing aids: Effects on speech discrimination in noise. Arch. Otolaryng. 101, 316–319.
Wien, G. (1964). A preliminary investigation of the effect of headwidth on binaural hearing. S.M. Thesis, MIT, Cambridge, Mass.
Zurek, P.M. (1979). Measurements of binaural echo suppression. J. Acoust. Soc. Am. 66, 1750–1757.
Zurek, P.M. (1983). A predictive model for binaural advantages in speech intelligibility. J. Acoust. Soc. Am. 71, S87.
Zurek, P.M., Durlach, N.I., Colburn, H.S., Gabriel, K.J. (1983). Masker bandwidth and the MLD. J. Acoust. Soc. Am. 73, S77.
Zurek, P.M. (1986). Consequences of conductive auditory impairment for binaural hearing. J. Acoust. Soc. Am. 80, 466–472.

12

Some Modern Techniques and Devices Used to Preserve and Enhance the Spatial Qualities of Sound

CONSTANTINE TRAHIOTIS AND LESLIE R. BERNSTEIN

Our purpose in this chapter is to discuss various modern techniques that are used to enhance the spatial qualities of sound produced by typical stereophonic, high-fidelity, home-entertainment systems. One major shortcoming of the typical two-channel stereo array is that the "sound stage" is limited to the space between left and right loudspeakers. The listener experiences sounds whose apparent position cannot be left of the left loudspeaker or right of the right loudspeaker. In addition, the sounds lack depth in the sense that the apparent distance of the sounds is the distance of the listener from the loudspeaker. Also, listeners do not perceive dramatic differences in terms of how far away or how near sounds are to them or to other sounds.

This may be summarized by stating that the typical two-channel stereo sound stage is confined to a plane between and including the two loudspeakers that produce the sound. Of course, the size of the sound stage can be restricted by the size of the room and the radiation pattern of the loudspeakers themselves.

There have been several efforts to remedy the problem, and each of the attempts appears to have gained favor more than once in the last three or four decades. Hi-fi enthusiasts have been told for many years that a monaural (left and right channels mixed), or third channel placed between left and right loudspeakers, would add definition to the front-back relations of sound. This arrangement did not become popular possibly because the listener needed an additional loudspeaker, amplifier, a mixing network, and possibly, most importantly, the permission of a spouse who didn't relish seeing another ugly loudspeaker where the TV belongs.

Of course, quadraphonics was another effort to enhance the spatial qualities of sound. By placing four loudspeakers symmetrically around the listener and providing differing signals to each of the loudspeakers, it was hoped that a "more realistic environment" would be created. The attempt was short-lived for several reasons, including the fact that many people did not seem to want four loudspeakers, four amplifiers, etc. Those of us who did were frustrated by not knowing which of several decoding schemes was best and which would emerge as the standard, and by the lack of material recorded in a manner that would optimize the properties of four-channel sound. On the other hand, one could buy many records that would place one atop the net in a ping pong game, in the middle of an infinitely long, circular, railway train, or in the middle of their favorite orchestra. For many, the added expense, complexities, and sonic benefits were not desirable.

Perhaps the major potential benefit of four channels was the ability to provide the listener with the illusion of being in a large room. In such rooms the listener hears not only the direct sounds but also sounds that arrive later because they are reflected from nearby surfaces such as walls and ceilings. Other things being equal, the larger the room, the larger the delay between direct and reflected sounds. The illusion can be created by placing time-delayed information through loudspeakers placed behind or to the side of the listener. Presently, there are available several commercial devices that allow the listener virtually to be able to choose the apparent size of the listening room. We wish to make clear the distinction between the perceived locus of the sound (determined primarily by interaural differences) and the apparent size of the room (determined by variables such as reflections) within which the sounds are heard. We shall see that modern devices may enhance both of these aspects, and the reader is warned that the lay literature often confuses the two types of enhancement.

One inexpensive, simple, and still-utilized technique for adding ambience or the illusion of being in a large room is to use what is frequently called the "Hafler hookup." This involves using a second pair of loudspeakers such that the *differences* between the left and right channels are presented through loudspeakers placed beside or behind the listener. By extracting the differences between left and right channels, unique information recorded in each channel is recoverable. This information could be the reverberant sound in the room in which the recording was made. These signals can be recovered because the information common to the left and right channels (perhaps from recordings made using only two microphones) is suppressed owing to the subtraction of the two channels. The listener still has to purchase four loudspeakers, but it has been known for several years that the additional loudspeakers need not be of the highest quality, and, therefore, the cost for having the system may not be very great.

We shall soon see how the Hafler hookup has been essentially reborn in modern attempts to solve the problem of restricted *left-right* size of the sound stage.

Let us return to the statement that the position of sounds presented over normal, two-channel stereophonic systems cannot be left of the left loudspeaker or right of the right one. To see why this is so, the reader should note that the apparent locus of the sound is determined by the interaural time and intensity differences (see Chapters 1–4) that occur at the position of the listener. We begin by assuming an extreme case: information present in only the left channel. The listener will perceive the sounds as originating from the left loudspeaker. This should surprise no one. The listener will have as cues interaural time and intensity differences that are determined by the position of the loudspeakers and of the listener. We shall next consider adding information to the *right* channel. In order to make matters clear we consider three possibilities. First, suppose we add a replica of the information in the left channel to the right channel. As we increase the intensity in the right channel, the apparent locus of the sound moves from the left toward the right and is heard as being somewhere between the two, depending on the relative intensities of the sounds and on the position of the listener. A second case can be illustrated by assuming that *totally different* information from that occurring in the left channel is added to the right channel. By assumption, the information in the right channel is uncorrelated with that in the left and, equivalently, unique information is provided to each channel. Now the listener will hear *two* sources of sound, one originating at each loudspeaker. The third condition is one in which the information added to the right channel contains *some* information also present in the left channel. Each channel now contains both unique information and common information (present in the other channel). In this instance the listener can identify sources of sounds ranging from the left loudspeaker to the right loudspeaker, including all positions in between, depending on the actual amounts of sounds that are common and their relative intensities. Note that *all three conditions* are characterized by a sound stage bounded by the position of the loudspeakers. How can we produce a larger sound stage?

One way is simply to move the loudspeakers farther apart. By doing so, the interaural time and intensity differences will change and will be appropriate to sound sources located in a more lateral position. This may not be a practical solution because of the size of the room.

The question then becomes what else can one do to provide the illusion of a larger sound stage? The answer is to transform information in left and right channels in order to increase or to create the desired interaural differences at the position of the listener. Several modern devices have been designed to do just this. To understand how these devices accomplish their feats, we must return to our consideration of what occurs when a listener is positioned between two loudspeakers in a typical two-channel stereo array. Again, let us assume that only the left channel is active. Note that the information from the left loudspeaker will reach *both* left and right ears of the listener with appropriate interaural time and intensity differences (Fig. 12-1). Indeed, these differences enable the listener to locate the position of the loudspeaker. If the information reaching the

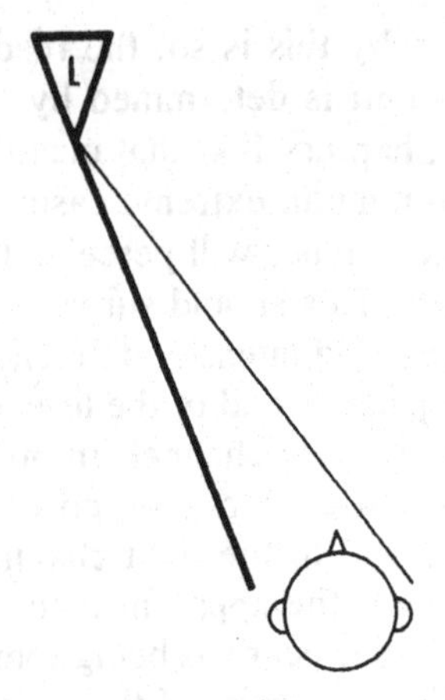

FIGURE 12.1. The sound sources arriving at the ears from a left loudspeaker. The difference in the two paths to the two ears (heavy and light lines) represents the interaural differences.

listener's right ear could be diminished or otherwise altered, the resulting interaural time and intensity differences may indicate that the sound originates not from the actual position of the loudspeaker, but from another position *more to the left* of the loudspeaker. That is, the alteration of information reaching the ear *opposite* the sound source, or what is referred to as "cross-talk" is the key to the operation of several modern devices. We now turn to a discussion of specific manners by which cross-talk is altered.

Figure 12-2 is a modification of Figure 12-1 and includes a second loudspeaker (right) that provides two kinds of information. First, it will provide information directly from the source (FM, tape, record, etc), and, in addition, it will provide a signal that is intended to *cancel* the information

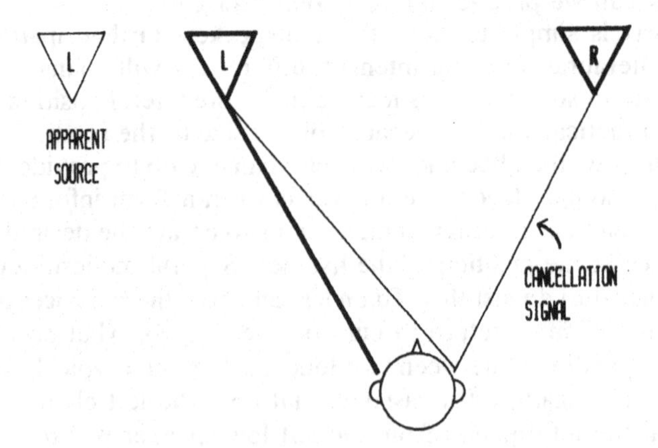

FIGURE 12.2. The perceived position of the left speaker is shifted more leftward if the cross-talk signal going to the right ear is cancelled (or partially cancelled) by the cancellation signal from the right speaker.

at the right ear which originates from the left loudspeaker, the cross-talk. The transformations that are used to derive the "cancellation signal" can be accomplished electronically in several ways. The general idea is to estimate the characteristics (e.g., amplitude and phase shifts) of the cross-talk and to present this signal through the right loudspeaker in a manner that will allow the cross-talk to be cancelled. Theoretically, the cross-talk would be completely cancelled if the right loudspeaker emitted a signal whose power spectrum and time of arrival at the right ear were identical to the cross-talk signal, but whose phase-spectrum was inverted with respect to the cross-talk signal. In practice, one does not and would not want to cancel totally the cross-talk. This is so because signals unique to one ear do not occur in sound-fields and are only heard when one listens through earphones that contain unique information. That is, although cancellation of some of the cross-talk is required, total cancellation is neither required nor desirable.

Incomplete cancellation (attenuation) of the cross-talk signal can be quite effective in that the apparent position of sources of sound is greatly extended in the left-right direction, and listeners are often amazed to hear sounds that appear to emanate from beyond the walls of the room.

Electronic devices designed to provide cancellation of cross-talk may, for convenience, be divided into three classes. As we attempt to explain how these devices work, we will continue to depict the functions of the devices for the left channel.

One class of devices derives the cancellation signal by splitting the normal left-channel signal into two parts, one destined for the left loudspeaker and the other destined to become the cancellation signal that will emanate from the right loudspeaker. The latter signal is scaled, inverted, and filtered according to the manufacturer's notions of what constitutes proper cross-talk cancellation. A second class of devices includes the former transformations and also includes a time-delay of the cancellation signal to account for the time it takes for the cross-talk to travel around the head. Both of these classes of devices are able to enhance greatly the width of the sound stage. In fact, it is not uncommon for listeners to experience sounds that appear to emanate from *behind them*. The specific transformations of the acoustic signals that each of these devices perform inevitably lead to two related problems. First, many listeners note that sounds that would normally appear to be directly in front of the appropriately placed listener can be altered greatly. For example, voices that normally would appear to be "center stage" may sound somewhat hollow, thin, and vaguely positioned. In addition, sounds may appear to have unnatural low-frequency or bass content. These effects can be understood by noting that *all* information common to left *and* right channels is simply subtracted when the original and inverted cancellation signals interact. Vocalists, other soloists, and much of the bass content of stereophonic recordings are often recorded at near-equal levels in each channel. Consequently, cancellation of these signals causes the listener to experience what is often described to be a "hole" in the middle of the sound stage.

The severity of this potential problem is quite difficult to quantify because manufacturers of these devices often include some amount of bass-boost and, in some instances, asymmetric scaling of the cancellation signals. Those devices that include a time-delay of the cancellation signal in addition to providing an inversion may sound "hollow" for a quite different reason. The constant time-delay will be equivalent to a phase-shift, which is a function of frequency. When the original and time-delayed information are added, frequencies for which the time-delay is equal to a 180 degree phase-shift will be severely attenuated, resulting in what is popularly known as a "comb-filtering" effect. Such spectra have an amplitude variation across frequency (see Chapter 10) and are perceived as "colored" or unnatural. Strong statements about how comb-filtering degrades sound quality are not warranted because the effect of the time-delay will interact with the specific type of bass-boost or spectral-shaping (equalization) used by the manufacturer.

A third class of electronic devices differs from the first two mainly because it uses the *difference between the right and left channels* to form the cancellation signal. As we will see, this modification precludes several of the problems associated with the degradations of signals that are common to both channels. In general terms, this class of devices passes the left (right) channel signal virtually unmodified to the left (right) loudspeaker. In addition, the right and left channels are split and subtracted from each other. The resulting difference signal (say, L-R) is delayed, equalized, (spectrally shaped), scaled, and, finally, is split once more. One of these modified difference signals (L-R) is added to the normal information in the left channel. The other difference signal is inverted (shifted 180 degrees), thereby becoming R-L [-(L-R)], and added to the normal information in the *right* channel. The result of these transformations is that the output of the left loudspeaker is the normal left channel information (L) plus the cancellation signal $(L\text{-}R)_\tau$ where τ represents the time-delay. The output of the right loudspeaker consists of the normal right information in the right channel (R) and the cancellation signal $(R\text{-}L)_\tau$. This situation is depicted in Figure 12-3. Note that the normal outputs of each loudspeaker are indicated in parentheses. For the sake of clarity, we again assume that only the left channel is active, resulting in the outputs of the loudspeakers being $L + L'_\tau$ $[L + (L - 0)_\tau]$ and $-L'_\tau$ $[0 + (0 - L)_\tau]$, respectively. The prime is used as a reminder that this signal has been shaped (filtered) spectrally. The reader is cautioned that the information that arrives at each ear is very different from that depicted in Figure 12-2. With this arrangement, the listener's *left* ear receives the normal information in the left channel plus a delayed, attenuated, spectrally-shaped version of exactly this information.

Next we examine what arrives at the *right* ear from this same *left* loud-

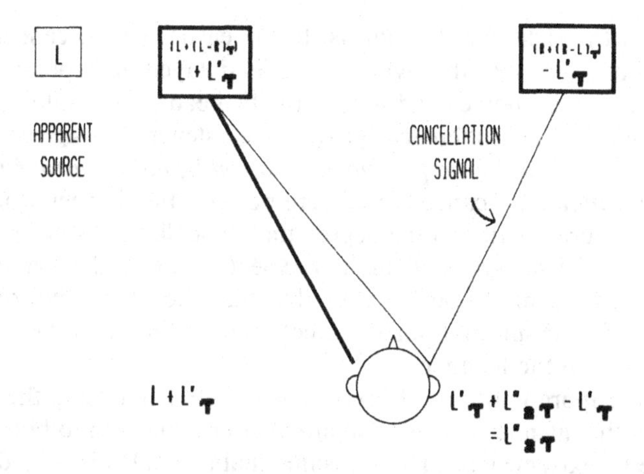

FIGURE 12.3. Another way to achieve cancellation by sending the original left signal plus a delayed version of the difference signal (left signal-right signal) form the left speaker *and* the original right signal plus the inverted difference signal (right signal-left signal) from the right speaker. The cross-talk will be partially cancelled by the right signal, resulting in an apparent shift of the left speaker more toward the left.

speaker. This is the cross-talk signal and consists of the components of $L + L'_\tau$, which travel around the head. Assuming that τ equals the normal travel-time around the head (a simple and, as shown in Chapter 1 by Kuhn, psychoacoustically inadequate assumption), this cross-talk signal will be an attenuated, spectrally-shaped (by the head) and delayed version of $L + L'_\tau$, which can, under our simplifying assumptions, be denoted by $L'_\tau + L''_\tau$. The right ear also receives $-L'_\tau$, the cancellation signal from the right loudspeaker. Assuming the cancellation signal $(-L'_\tau)$ is an exact, inverted replica of the *normal* cross-talk (which caused the problems in the first place), the result in the right ear is $L'_\tau + L''_\tau - L'_\tau$, which reduces to L''_τ.

Because the intensity of the delayed, spectrally-shaped component of the signal at the left ear (L'_τ) can be varied over a great range, its specific effect is impossible to determine a priori. Even if L'_τ is equal in level to L, probably the most severe effect to be expected would be comb-filtering at quite high frequencies. Assuming the manufacturer uses a typical delay of 125 μs, one would expect dynamically occurring spectral notches in the comb-filtered sound at 4 kHz, 12 kHz, 20 kHz, etc (see Chapter 10). On the other hand, L'_τ is usually attenuated relative to L, and the expected spectral notches may not be psychoacoustically salient.

Now let us concentrate on the difference between the signals that actually reach the two ears. Assuming L'_τ in the left channel is attenuated and not psychoacoustically significant, one would expect L in the left ear and L_T'' in the right ear. In contrast to the earlier cases, cancellation has resulted in a signal at the right ear that is delayed and spectrally shaped *twice:* first by the device and second by the head. The resulting large time delays and intensity differences provide the listener with signals that would normally or *naturally be produced by a loudspeaker farther to the left.* Even in principle, if cancellation were perfect, the listener would always have a binaural input. To the degree that cancellation is not perfect with any of these devices, one would not expect a monaural signal in practice. However, it seems to be the case that this use of the difference signal will less often result in dramatic effects such as instruments appearing to be adjacent to the listener.

A much more important benefit that results from using the difference signal in this manner concerns inputs that are common to both channels. To take an extreme case, let us assume that a vocalist is recorded in such a manner that left and right channels are perfectly correlated in time and are identical in level. Because in this case $R = L$, $R - L = 0$ and the device has absolutely no effect. That is, the bass cancellation, spectral shaping to ameliorate this effect, and the aforementioned "hole in the middle" simply do not occur. To the degree that information in the left and right channels differs (e.g., as a source of sound is moved farther away from the center), the effect of the device increases. This results in a fairly solid center stage, and apparent sources of sound are spaced rather widely. The discerning reader may be bothered by the possibility that the cancellation signal may, itself, produce cross-talk. Although this is the case, it should be mentioned that this aspect of cross-talk would be severely attenuated and delayed (first by the device and then by the head) and would be expected to be relatively unimportant by its time of arrival and decreased amplitude (relative to the main signal).

Shortly after electronic image-enhancers achieved popularity, it was noted that one could achieve similar transformations passively. This is accomplished by utilizing the normal two front loudspeakers and by adding a second pair of loudspeakers that produce the desired cancellation signals. In the interests of brevity and clarity, we focus on manners by which proper *difference* signals are used passively for cancellation. Following this discussion we discuss an alternative method of achieving cross-talk cancellation.

Reexamination of Figure 12-3 reveals that there are two sources of information emanating from each loudspeaker: L and $(L\text{-}R)_T$ from the left loudspeaker and R and $(R\text{-}L)_T$ from the right loudspeaker.

Let us begin with the normal two-channel stereo array, L and R, which are wired to the normal loudspeakers. To derive the appropriate difference signals (L-R and R-L) one begins by connecting a wire between the two "minus" or "ground" terminals of a second pair of loudspeakers. A second wire is then connected from the left "plus" or "hot" terminal of the receiver/amplifier to the "positive" terminal of the loudspeaker, which will provide L-R. The last wire is connected from the right "plus" or "hot" terminal of the receiver/amplifier to the "positive" terminal of the loudspeaker, which will produce R-L.

To achieve the desired delay of these difference signals, one merely places the two additional loudspeakers a few inches farther from the listening position than the two main loudspeakers. Sometimes it is convenient and not unattractive to place the additional loudspeakers atop the main loudspeakers. The level of the difference signal may be scaled by inserting a potentiometer in series with the two additional loudspeakers. In addition, the difference or cancellation signal may be spectrally-shaped via the addition of an appropriate passive filtering network. An alternative to spectral-shaping is to use rather inexpensive, small loudspeakers that restrict the frequency content to the mid-range.

Interestingly, this passive circuit for providing cancellation of cross-talk is identical to the one depicted earlier as the "Hafler hookup." The only change is that the additional loudspeakers have been moved from the rear or side of the room to atop, behind, or beside the main loudspeakers. Consumer-oriented articles often contain warnings to those who wish to employ this circuit that they check to make sure that their receiver/amplifier is "common ground." We heartily endorse this view and wish that we had done so before we destroyed completely the car stereo owned by one of the authors.

A second passive image-recovery system using two additional loudspeakers is described in the May 1983 edition of *Audio* magazine. This system is functionally similar in effect to the second class of electronic devices discussed earlier. The salient difference between this passive hookup and the "Hafler hookup" resides in the nature of the signal used to cancel the cross-talk. The setup described in *Audio* utilizes an inverted, scaled, delayed, and spectrally-shaped version of the information in each main channel to produce cancellation of the cross-talk. Therefore, one would expect the same disadvantages concerning bass-cancellation etc. discussed earlier.

One novel and well-received solution is to purchase commercially available loudspeakers that incorporate extra circuits and elements in order to provide enhanced imaging. The Polk SDA-1 and SDA-2 loudspeakers produce both the normal output in each channel and in addition produce the difference signals, $(R\text{-}L)_\tau$ and $(L\text{-}R)_\tau$. The difference signals are provided by additional, matched devices placed to the outside of each cabinet. The delay is provided by the difference in distance between the more medial drivers, which produce the main signal, and the more lateral drivers, which produce the difference signal. To our knowledge, the level of the difference signal appears not to be attenuated relative to the main signal. Still, our own experience agrees with that of several reviewers who have praised the Polk loudspeakers for their imaging qualities. In passing, we may note that one may, in principle, calibrate any of the home-made passive devices quickly and easily. By using a monaural input, say to the left channel, one needs only to adjust the level and the delay of the cancellation signal until an image far to the left is heard. In our experience, if the level of the cancellation signal is too great, the resulting sounds appear to be objectionably diffuse.

Evaluations of these types of devices will certainly depend upon the criteria one uses. For example, if one desires to achieve the best possible cancellation of cross-talk, then a number of psychoacoustic data would seem to be relevant. Certainly, one formidable difficulty would be to attempt to simulate the naturally-occurring, frequency-dependent interaural time-delays and intensity differences measured so painstakingly by Kuhn and covered in Chapter 1. Based on his findings, choosing one value of time-delay would not seem to be efficacious. In addition, as Damaske (1971) has demonstrated, the appropriate *level* and time-delay of the cancellation signal are highly frequency-dependent. It should be stressed that Damaske used more complex circuitry than we have discussed and than is available commercially and used accuracy of localization of sounds as his criterion.

If one chooses to evaluate these devices not by demanding accuracy but, instead, by noting what they contribute to the stereo illusion, one quickly becomes a fan. In our experience, even in such non-optimal environments as university classrooms, these devices produce dramatic and robust effects. Listeners never fail to experience enhanced left-right and front-back expansion of the sound stage and often state spontaneously that sounds appear to originate from beyond the walls of the room.

Perhaps the most stunning way in which to demonstrate the pleasant nature of the effects is to turn off the device abruptly. Virtually all listeners complain that the sounds appear to collapse back to a very small region bounded by the loudspeakers. This is an extremely robust effect and has been mentioned several times in consumer-oriented magazines.

As noted earlier, many attempts to enhance the spatial quality of stereophonic reproduction have occurred over the past few decades. Indeed, the readers of the modern consumer-oriented literature may be surprised to learn that the concept of cross-talk cancellation was discussed quite thoroughly by Bauer as early as 1961. Because there are so many facets to the problem, there exists a large and varied relevant literature both in scientifically oriented and in consumer-oriented publications.

Those who wish to learn more about enhancement of ambience may find it helpful to read scientifically oriented papers, including Schroeder and Logan (1961), Schroeder (1961), Schroeder (1979), Madsen (1970), Kotschy et al (1974), Ando (1977), and Chaudiere (1980). In addition, there are a number of relevant and interesting consumer-oriented articles including Cooper (1970), Cooper (1971), Hafler (1970), Hodges (1971), Hodges (1978), Stark (1971), Bauer (1973), and Mitchell (1978). Similarly, there are several articles regarding cross-talk and its cancellation, including

Bauer (1961), Damaske (1971), Schroeder et al (1974), Schroeder (1980), Sakamoto et al (1981), and Sakamoto et al (1982). In addition there are several articles in the consumer-oriented literature such as Carver (1982), Cohen (1982), and Kaufman (1983).

Finally, it should be pointed out that we have omitted discussion of the physical, physiological, and psychological correlates of auditory localization.

Besides finding a raft of relevant information in other chapters of this volume, readers may find it helpful to read three general consumer-oriented articles written by Toole (1973a,b, 1982). Toole discusses several pertinent issues in a manner that is clear and interesting.

Finally, readers who wish to investigate historically important, precisely presented, germinal papers that will require careful study should read Leakey (1959), Bauer (1960, 1961), Gardner (1968), and the English translation of Haas' classic article published by the Audio Engineering Society in 1972.

Although we have attempted to separate the notions of ambience and the width of the sound stage, we would be remiss if we failed to mention that both time-delay and cross-talk cancellation result in reduced interaural correlations at the listener's ears. The magnitude of the interaural correlation is known to be related (inversely) to subjective preference of concert halls (Schroeder, 1980).

Recommended Reading

Ando, Y. (1977). Subjective preference in relation to objective parameters of music sound fields with a single echo. J. Acoust. Soc. Am. 62, 1436–1441.

Bauer, B.B. (1960). Broadening the area of stereophonic perception. J. Aud. Eng. Soc. 8, (2), 91–94.

Bauer, B.B. (1961). Stereophonic earphones and binaural loudspeakers. J. Aud. Eng. Soc. 9, (2), 148–151.

Bauer, B.B. (1961). Phasor Analysis of some stereophonic phenomena. J. Acoust. Soc. Am. 33, (11), 1536–1539.

Bauer, B.B. (1973). Quadraphony needs directional loudspeakers. Audio 57, (3), 22–30.

Carver, R.W. (1982). Sonic holography. Audio, 66 (3), 26–35.

Cohen, J. (1982). Stereo image expander. Radio Electron. June, 45–48.

Cohen, J. (1982). Stereo image expander: Part II. Radio Electron. Sept, 63–91.

Chaudiere, H.T. (1980). Ambiophony: Has its time finally arrived? J. Aud. Eng. Soc. 28, (718), 500–509.

Cooper, D. (1970). How many channels? Audio 54, (11), 36–42.

Cooper, D. (1971). How many channels?: Part II. Audio 55 (2), 28.

Damaske, P. (1971). Head-related two-channel stereophony with loudspeaker reproduction. J. Acoust. Soc. Am. 50, (4), 1109–1115.

Gardner, M.B. (1968). Historical background of the Hass and/or precedence effect. J. Acoust. Soc. Am. 43, (6), 1243–1248.

Haas, H. (1972). The influence of a single echo on the audibiligy of speech. J. Aud. Eng. Soc. 20, (2), 146–159.
Hafler, D. (1970). A new quadraphonic system. Audio 54, (7), 24, 26, 56–57.
Hodges, R. (1971). The home experimenter's guide to multi-channel listening. Stereo Rev. 26, (4), 62–66.
Hodges, R. (1978). Audio Basics: Pseudo time delay. Stereo Rev. 41, (6), 37.
Kaufman, R.J. (1983). How to build a LOW-CO$T stereo enhancer. Audio, 67 (5), 58–61.
Kotschy, A., Tarnoczy, T., Vicsi, K. (1974). Subjective judgment of artificial reverberation processes. J. Acoust. Soc. Am. 56, (4), 1192–1194.
Leakey, D.M. (1959). Some measurements on the effects of interchannel intensity and time differences in two channel sound systems. J. Acoust. Soc. Am. 31, (7), 977–986.
Madsen, E.R. (1970). Extraction of ambiance information from ordinary recordings. J. Aud. Eng. Soc. 18, (5), 490–496.
Mitchell, P.W. (1978). In the continuing search for the perfect 'you are there' sound experience, audio designers' attentions now turn to time-delay systems. Stereo Rev. 41, (4), 87–92.
Sakamoto, N., Gotoh, T., Kogure, T., Shimbo, M., Cleeg, A.H. (1981). Controlling sound-image localization in stereophonic reproduction. J. Aud. Eng. Soc. 29, 794–799.
Sakamoto, N., Gotoh, T., Kogure, T., Shimbo, M., Cleeg, A.H. (1982). Controlling sound-image localization in stereophonic reproduction: Part II. J. Aud. Eng. Soc. 30, 719–722.
Schroeder, M.R. (1961). Improved quasi-stereophony and 'colorless' artifical reverberation. J. Acoust. Soc. Am. 33, (8) 1061–1064.
Schroeder, M.R. (1979). Binaural dissimilarity and optimum ceilings for concert halls: More lateral sound diffusion. J. Acoust. Soc. Am. 65, (4), 958–963.
Schroeder, M.R. (1980). Toward better acoustics for concert halls. Phys. Today 24–30.
Schroeder, M.R., Gottlob, D., Siebrasse, K.F. (1974). Comparative study of European concert halls: correlation of subjective preference with geometric and acoustic parameters. J. Acoust. Soc. Am. 56, (4), 1195–1201.
Schroeder, M.R., Logan, B.F. (1961). 'Colorless' artificial reverberation. J. Aud. Eng. Soc. 9, (3), 192–197.
Stark, C. (1971). One approach to four-channel sound. Audio 55, (10), 30–32.
Toole, F. (1973a). Stereo Hearing: A psycho-acoustician describes some of the physical and psycho-physiological factors that together govern our perception of the 'dimensions of auditory space.' Stereo Rev. 30, (1), 75–78.
Toole, F. (1973b). Stereo Hearing: (Part 2) The recording and reproduction of auditory space is still more art than science, despite out relatively complete understanding of the nature of auditory perception. Stereo Rev. 30, (2), 67–71.
Toole, F.E. (1982). Listening tests—turning opinion into Fact. J. Aud. Eng. Soc. 54 (11) 431–445.

Index of Names*

*Page numbers only refer to pages where the author's work is cited. In addition to this index, consult the reference sections at the end of each chapter.

Subject Index

U

V

W

Y

Printed in the United States
by Baker & Taylor Publisher Services

Printed in the United States
by Baker & Taylor Publisher Services